BARRON'S

HSPA

NEW JERSEY MATH

2ND EDITION

Eileen D. Arendt
Math Teacher
Deptford Township High School
Deptford, New Jersey

About the Author: Eileen D. Arendt is a high school mathematics teacher at Deptford Township High School, where she has been teaching since 2001. Previously, she taught for ten years at Long Branch High School. She earned her Bachelor of Arts from Trenton State College (now the College of New Jersey). She has also accumulated a number of graduate credits and has an avid interest in mathematics. In the summers, she teaches a course at Mercer County College.

Ms. Arendt is involved in numerous committees at school and is also a member of AMTNJ (Association of Mathematics Teachers of New Jersey) and NCTM (National Council of Teachers of Mathematics).

All inquiries should be addressed to:
Barron's Educational Series, Inc.
250 Wireless Boulevard
Hauppauge, NY 11788
www.barronseduc.com

ISBN-13: 978-0-7641-4027-3
ISBN-10: 0-7641-4027-2

Library of Congress Control Number: 2008927244

Printed in the United States of America

9 8 7 6 5 4 3 2 1

Contents

Introduction

New Jersey's 11th-grade mathematics test is a state-required exam that must be passed for graduation from any public high school in the state.

This exam assesses knowledge and skills in four content areas also sometimes referred to as "clusters." These are as follows.

I. Number Sense, Concepts, and Applications
II. Spatial Sense and Geometry
III. Data Analysis, Probability, Statistics, and Discrete Mathematics
IV. Patterns, Functions, and Algebra

Below is the suggested breakdown of test questions for the 11th-grade HSPA in the area of mathematics as given by various mathematics assessment committees.

Cluster Percent Distribution
11th Grade

I. 15%
II. 25%
III. 30%
IV. 30%

The test consists of three types of questions as described below.

MULTIPLE-CHOICE QUESTIONS

The multiple-choice questions are not the standard multiple-choice questions of the past but rather are questions requiring a little bit more thought and work. It is expected that each multiple-choice question will take a minute or two. Each question is worth one point and will be scored by a computer.

The assessment committees recommend between 26 and 31 multiple-choice questions.

SHORT CONSTRUCTED RESPONSE QUESTIONS

Since multiple-choice questions have their limitations, the test also includes questions that require the student to write his or her own response in some form. The short constructed response questions can require one to write numbers, equations, expressions, simple graphs, and so on. It is expected that each question in this section will take an average of 2 to 3 minutes. Each question is worth one point and will be hand-scored.

The assessment committees recommend about 4 of these types of questions.

OPEN-ENDED QUESTIONS

These questions require a student-written numerical and/or graphical response to a question and often an explanation as well. Most of the time, an open-ended question consists of several parts. It is expected that each question of this type will take about 10 minutes. Each question is worth three points and will be hand-scored.

The assessment committees recommend a total of seven open-ended questions on the 11th-grade test with one from Cluster I and two from each of the remaining three clusters. A student can score between 0 and 3 based on how well he or she responds. The guidelines for scoring provided by the New Jersey State Department of Education appear at the end of this introduction.

Typically an open-ended question will ask the student to do one of the following.

1. Explain the steps used to solve a problem.
2. Determine if a certain process or result is valid.
3. List items that meet a certain criteria.
4. Draw a diagram that meets given conditions.
5. Describe or extend a pattern.
6. Indicate what might happen if a certain aspect of a problem is changed.
7. Measure.

In order to ensure success on the problems, be sure to do the following.

1. Write in complete sentences but do not be excessively wordy.
2. Explore different possibilities for a problem as appropriate.
3. Answer all parts of the question.
4. Answer the questions being asked.
5. Draw diagrams to assist you with explanations as necessary.
6. Provide an example of what you are describing if appropriate.
7. Double-check your computations.

Remember, open-ended questions often have more than one correct answer and/or approach!

CALCULATORS

The New Jersey State Department of Education has indicated that the use of any calculator other than one with a QWERTY keyboard (similar to a typewriter) is permissible and encouraged on the exam, if not essential for some questions.

Minimally, your calculator should have

1. Built-in algebraic logic (meaning the calculator follows the order of operations) and a parentheses key
2. Exponent and root keys
3. Change-in-sign key
4. Percents
5. Fraction keys
6. Reset button (a way to clear all memory).

Two important things to realize about calculator usage for the exam:

1. You should use a calculator you are comfortable with and familiar with.
2. The exam will not tell you when or when not to use the calculator; you must decide if it will be helpful or not.

It's important that prior to the exam you familiarize yourself with how to use YOUR calculator. Make sure you know the syntax for the following calculations (please consult the manual that came with your calculator):

- Fractions
- Converting decimals and fractions
- Percents
- Exponents
- Roots
- Mode (be sure your calculator is in degree mode)
- Scientific Notation (if your calculator does this)

MEMORIZATION OF FORMULAS

It is not necessary to memorize formulas for the HSPA Exam. When you take the exam, you will be provided with a mathematics reference sheet. This sheet contains formulas you might need while taking the test, as well as any specific "tools" (such as cutout rulers and protractors) you will need to use on the test. You are not permitted to have any materials or tools for use during the exam other than a calculator and the provided reference sheet.

A sample reference sheet is included at the end of this introduction.

USING THIS BOOK EFFECTIVELY

This book has been designed as a comprehensive review of all topics you should be familiar with before taking the New Jersey HSPA Exam in Math. Throughout, you will notice many sample questions that will help you prepare for the exam. Be sure to use these to your advantage! They include a mixture of multiple-choice, short response, and open-ended questions—a good preparation for the variety of types of questions on the actual HSPA Exam.

Your first step in using this book to prepare should be to *take and score the diagnostic test* beginning on page 3. It's a good idea to take it alone and stay focused until all 25 questions are completed. Work as quickly as you can without sacrificing accuracy. The result of this diagnostic test should give you a fairly good idea of your basis of knowledge before starting work on the individual math topics relevant to the HSPA test.

Again, we recommend that as you progress through this study guide, you answer and check all of the "Mixed Practice" questions (the Answer Key begins on page 209). Then, when you reach the end, you will find *two* Practice HSPA Exams that are modeled directly on the actual HSPA Math Exam. These should be a fair indication of how much you've learned.

Keep in mind that on the HSPA Exam, you will be allowed to use a ruler and a calculator, and you will be provided with a Mathematics Reference Sheet (a sample is provided on page xi of this book). These will all be useful tools in solving problems and helping you achieve your highest possible score.

Good luck!

Holistic Scoring Guide for Mathematics Open-Ended Items (Generic Rubric)

3-Point Response

The response shows complete understanding of the problem's essential mathematical concepts. The student executes procedures completely and gives relevant responses to all parts of the task. The response contains few minor errors, if any. The response contains a clear, effective explanation detailing how the problem was solved so that the reader does not need to infer how and why decisions were made.

2-Point Response

The response shows nearly complete understanding of the problem's essential mathematical concepts. The student executes nearly all procedures and gives relevant responses to most parts of the task. The response may have minor errors. The explanation detailing how the problem was solved may not be clear, causing the reader to make some inferences.

1-Point Response

The response shows limited understanding of the problem's essential mathematical concepts. The response and procedures may be incomplete and/or may contain major errors. An incomplete explanation of how the problem was solved may contribute to questions as to how and why decisions were made.

0-Point Response

The response shows insufficient understanding of the problem's essential mathematical concepts. The procedures, if any, contain major errors. There may be no explanation of the solution or the reader may not be able to understand the explanation. The reader may not be able to understand how and why decisions were made.

HIGH SCHOOL PROFICIENCY ASSESSMENT MATHEMATICS REFERENCE SHEET

Use the information below, as needed, to answer questions on the Mathematics Section of the High School Proficiency Assessment.

$\pi = 3.14$ or $\frac{22}{7}$

Circle

Area = πr^2

Circumference = $2\pi r$

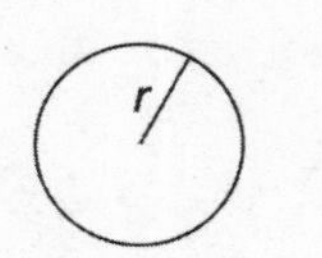

Trapezoid

Area = $\frac{1}{2}(b_1 + b_2)h$

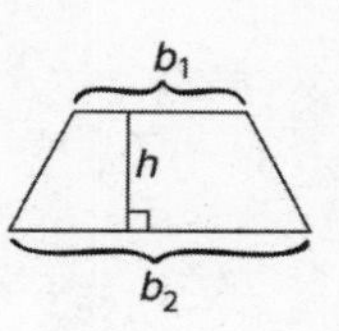

Rectangle

Area = lw

Perimeter = $2(l + w)$

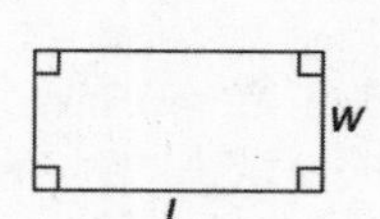

Triangle

Area = $\frac{1}{2}bh$

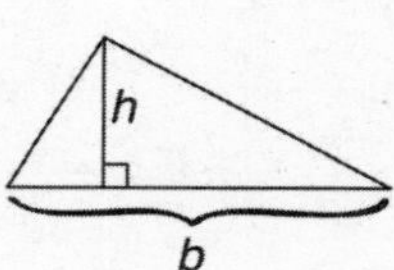

Parallelogram

Area = bh

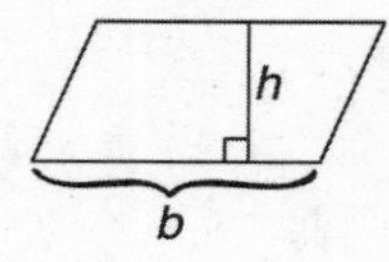

Pythagorean Formula

$c^2 = a^2 + b^2$

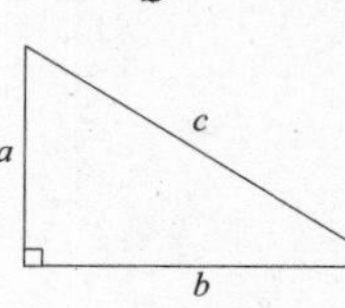

Rectangular Prism

Volume = lwh

Surface Area = $2lw + 2wh + 2lh$

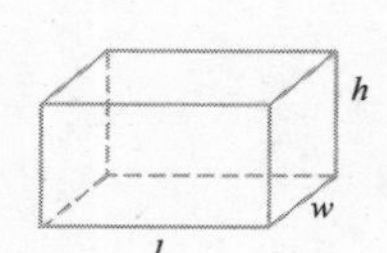

Cone

Volume = $\frac{1}{3}\pi r^2 h$

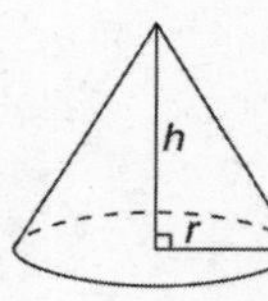

Sphere

Volume = $\frac{4}{3}\pi r^3$

Surface Area = $4\pi r^2$

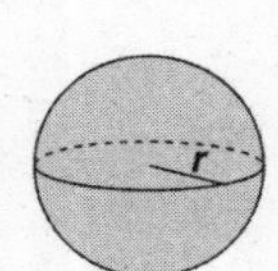

Cylinder

Volume = $\pi r^2 h$

Surface Area = $2\pi rh + 2\pi r^2$

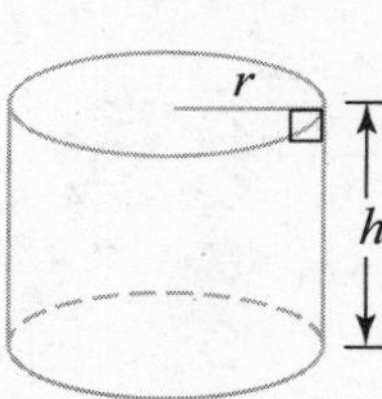

The sum of the measures of the interior angles of a triangle = 180°
The measure of a circle is 360° or 2π radians
Distance = rate × time   Interest = principal × rate × time

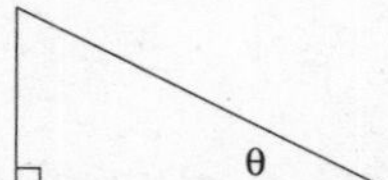

Given a right triangle:

$\sin\theta = \frac{\text{opposite side}}{\text{hypotenuse}}$ $\cos\theta = \frac{\text{adjacent side}}{\text{hypotenuse}}$ $\tan\theta = \frac{\text{opposite side}}{\text{adjacent side}}$

Simple Interest Formula: $A = p + prt$ **Compound Interest Formula:** $A = p(1+\frac{r}{n})^{nt}$

A = amount after t years; p = principal; r = annual interest rate; t = number of years;
n = number of times compounded per year

The number of combinations of n elements taken r at a time is given by $\frac{n!}{(n-r)!r!}$

The number of permutations of n elements taken r at a time is given by $\frac{n!}{(n-r)!}$

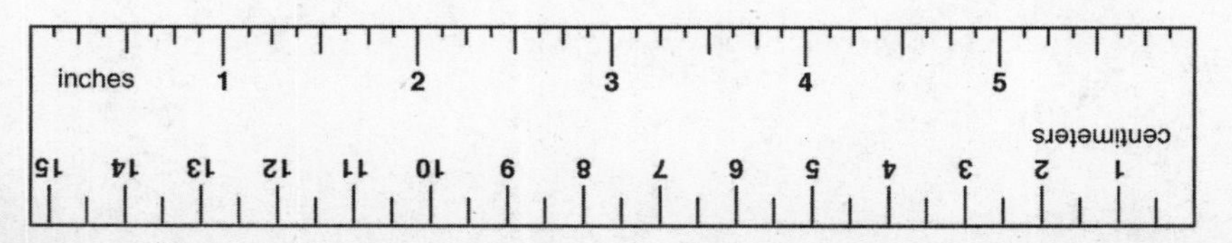

Use the following equivalents for your calculations

12 inches = 1 foot
3 feet = 1 yard
36 inches = 1 yard
5,280 feet = 1 mile
1,760 yards = 1 mile
100 centimeters = 1 meter
1000 meters = 1 kilometer

1000 milliliters (mL) = 1 liter (L)

60 seconds = 1 minute
60 minutes = 1 hour
24 hours = 1 day
7 days = 1 week
52 weeks = 1 year

16 ounces = 1 pound

1000 milligrams = 1 gram
100 centigrams = 1 gram
10 grams = 1 dekagram
1000 grams = 1 kilogram

8 fluid ounces = 1 cup
2 cups = 1 pint
2 pints = 1 quart
4 quarts = 1 gallon

Given the points (x_1, y_1), (x_2, y_2),

Distance between two points:

$d = \sqrt{(x_2 - x_1)^2 + (y_2 - y_1)^2}$

Slope Formula:

$m = \frac{\text{rise}}{\text{run}} = \frac{y_2 - y_1}{x_2 - x_1}$

Slope-intercept form of a line:

$y = mx + b$

DIAGNOSTIC TEST AND ANSWERS

ANSWER SHEET: DIAGNOSTIC TEST

1. Ⓐ Ⓑ Ⓒ Ⓓ
2. Ⓐ Ⓑ Ⓒ Ⓓ
3. Ⓐ Ⓑ Ⓒ Ⓓ
4. Ⓐ Ⓑ Ⓒ Ⓓ
5. Ⓐ Ⓑ Ⓒ Ⓓ
6. Ⓐ Ⓑ Ⓒ Ⓓ
7. Ⓐ Ⓑ Ⓒ Ⓓ
8. Ⓐ Ⓑ Ⓒ Ⓓ
9. Ⓐ Ⓑ Ⓒ Ⓓ
10. Ⓐ Ⓑ Ⓒ Ⓓ
11. Ⓐ Ⓑ Ⓒ Ⓓ
12. Ⓐ Ⓑ Ⓒ Ⓓ
13. Ⓐ Ⓑ Ⓒ Ⓓ
14. Ⓐ Ⓑ Ⓒ Ⓓ
15. Ⓐ Ⓑ Ⓒ Ⓓ
16. Ⓐ Ⓑ Ⓒ Ⓓ
17. Ⓐ Ⓑ Ⓒ Ⓓ
18. Ⓐ Ⓑ Ⓒ Ⓓ
19. Ⓐ Ⓑ Ⓒ Ⓓ
20. Ⓐ Ⓑ Ⓒ Ⓓ
21. Ⓐ Ⓑ Ⓒ Ⓓ
22. Ⓐ Ⓑ Ⓒ Ⓓ
23. Ⓐ Ⓑ Ⓒ Ⓓ
24. Ⓐ Ⓑ Ⓒ Ⓓ
25. Ⓐ Ⓑ Ⓒ Ⓓ

Cut along dotted line.

DIAGNOSTIC TEST QUESTIONS

1. If the numbers 1 through 12 are each written on a slip of paper, what is the probability that a slip chosen at random will contain a number that is a factor of 12?

A. $\frac{1}{12}$ **B.** $\frac{1}{3}$ **C.** $\frac{1}{2}$ **D.** $\frac{2}{3}$

2. If a computer network is to be wired so that each employee can communicate with the boss and *exactly* two co-workers and so that the boss can communicate with each employee, draw a possible network.

3. Tanya works 35 hours per week and earns $9.50 an hour. Which of the following best approximates her yearly salary?

A. $335 **B.** $12,000 **C.** $17,000 **D.** $25,000

4. The weather forecast calls for a 25% chance of snow on Thursday and a 40% chance of snow on Friday. What is the probability that it will snow both days?

A. 1% **B.** 10% **C.** 65% **D.** 100%

5. A turkey needs 24 minutes per pound to cook thoroughly. If the Smiths' turkey weighs 12 pounds and the Jones' turkey weighs $15\frac{1}{2}$ pounds, how much longer does the Jones' turkey have to be cooked?

A. 3.5 hours **B.** 2.8 hours **C.** 1.9 hours **D.** 1.4 hours

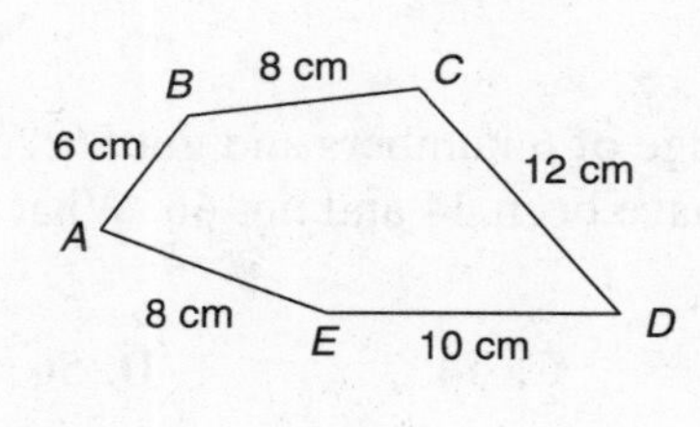

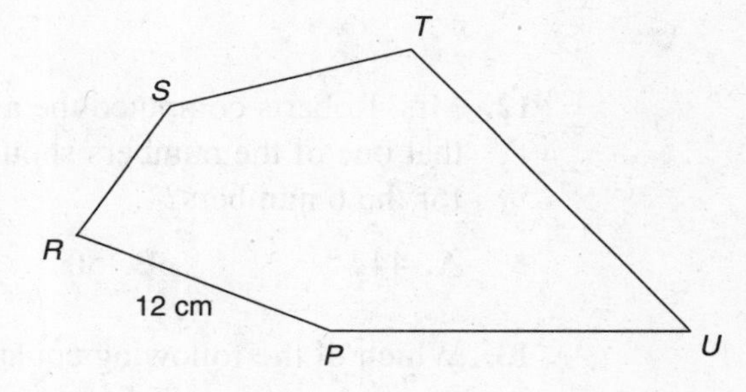

6. Figure *ABCDE* is similar to figure *RSTUV*. Find the measure of $\overline{TU}$.

A. 8 centimeters
B. 12 centimeters
C. 15 centimeters
D. 18 centimeters

7. Homes recently sold in a certain area cost $150,000; $135,000; $170,000; $164,000; $170,000; $140,000; and $175,000.

(a) Find the mean, median, mode, and range for the set of data.
(b) What would happen to the mean if the house costing $175,000 ended up selling for $140,000?
(c) If an eighth house is sold, how much would it have to sell for to keep the median price the same?

8. The label on a box of crackers states that the box contains 16 servings, each containing 2.5 grams of sugar. If Connie eats half the box of crackers, how many grams of sugar has she eaten?

A. 8 grams **B.** 10 grams **C.** 20 grams **D.** 40 grams

9. A box 10 inches high, 6 inches wide, and $1\frac{1}{2}$ inches tall has the same volume as a box 9 inches high, 2 inches wide, and ________ inches tall.

A. 2.5 inches **B.** 5 inches **C.** $2\frac{3}{4}$ inches **D.** 4 inches

10. A wooden crate holding 15 CDs weighs 4.3 pounds. The box weighs .7 pounds when it is empty. Which of the following is a correct expression if c is the weight of one CD?

A. $15c + .7 = 4.3$
B. $15c - .6 = 4.3$
C. $15\ (c + .7) = 4.3$
D. $15c + 7 = 43$

11. If Mark is going to take a bath, sketch a graph showing what will happen to the water level in the tub. Explain your graph.

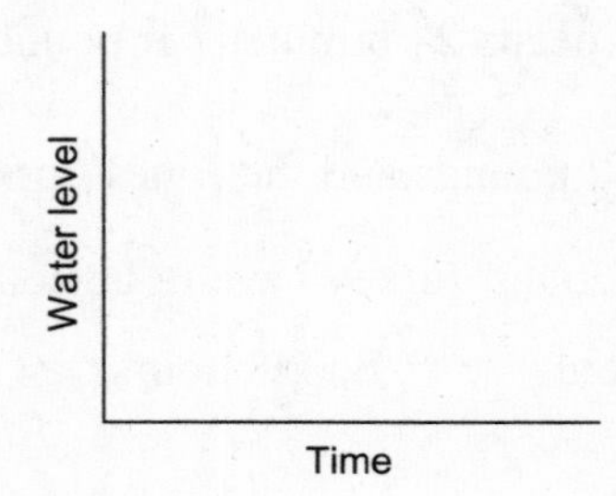

12. Mrs. Roberts computed the average of 6 numbers and got 56. Afterward, she realized that one of the numbers should have been 34 and not 46. What is the correct average for the 6 numbers?

A. 44 **B.** 50 **C.** 54 **D.** 56

13. Which of the following equations describes the data in the table below?

A. $y = x + 3$ **B.** $y = 2x + 1$ **C.** $y = 3x - 2$ **D.** $y = 3x - 1$

x	2	3	4
y	5	7	9

14. A human hand contains 27 different bones. The human body contains 206 bones altogether. About what percent of the bones in the human body are *not* in the hands?

A. about 98.7% **B.** about 97.4% **C.** about 87% **D.** about 74%

15. An insurance company randomly selects 300 statements at the end of the year to check for errors. Out of the 300 statements, 4 contained errors. Based on this information, about how many of the 13,000 statements sent out during the year had errors?

A. 4 **B.** 175 **C.** 1750 **D.** 4000

16. Which list shows temperatures in order from highest to lowest?

A. –35°, –10°, 32°, 50°
B. –10°, –35°, 32°, 50°
C. 50°, 32°, –10°, –35°
D. 50°, 32°, –35°, –10°

17. ABC Company wants to lower costs for the laminate covering on their boxes; that is, they need to reduce the surface area of the box while keeping the volume the same.

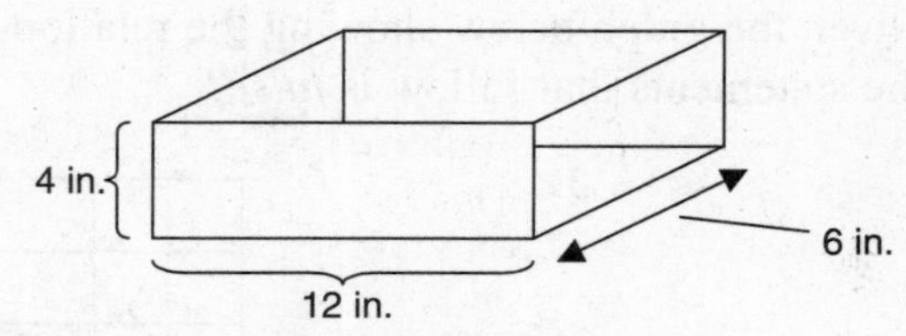

(a) What is the volume of the box? Show your work.
(b) What is the surface area of the box? Again, show your work.
(c) Provide ABC Company with dimensions for a new box that would have the same volume but less surface area. Describe how you found your answer.

18. Stereo World is having a special sale, and every hour beginning at 11:00 A.M., all merchandise will be reduced 15%. This will continue throughout the day until all the merchandise is sold. If Tom wants to buy a stereo system costing $180 but has only $145, what is the earliest time of the day that he could purchase the stereo assuming they are not sold out by then? Show your work.

19. In the cube below, give an edge that is parallel to face *AEHD*.

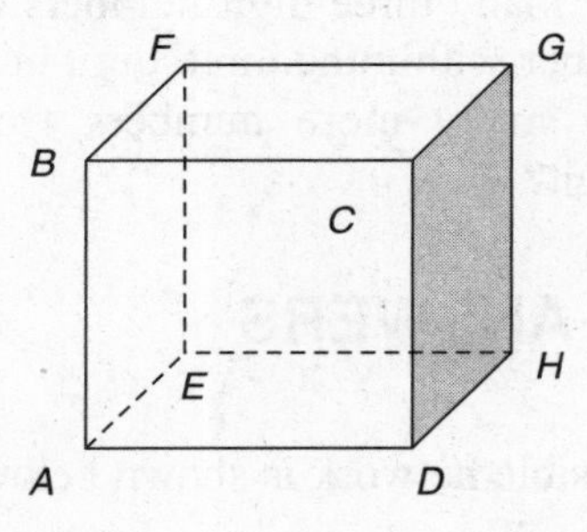

20. For a holiday party, Susan makes a trail mix containing 2 pounds of almonds, 3 pounds of raisins, and 4 pounds of M&M's. In order to make 45 pounds of trail mix, how many pounds of almonds does she need?

A. 5 pounds **B.** 10 pounds **C.** 9 pounds **D.** 18 pounds

21. Suppose you begin with the set of numbers from 10 to 50, inclusive, and

remove all prime numbers
remove all factors of 40
remove all multiples of 6
remove all perfect squares, and
remove all numbers in the sequence 1, 1, 2, 3, 5, 8, 13, 21, ...

How many numbers are left?

22. If the angles of a triangle are in a ratio of 1 : 3 : 5, what kind of triangle is it?

A. acute **B.** right **C.** obtuse **D.** isosceles

23. At what point does the graph of $4x - 2y = 8$ cross the x-axis?

A. (0, –4) **B.** (–4, 0) **C.** (0, 2) **D.** (2, 0)

24. Given the graph below showing the relationship between distance and time, which of the statements that follow is *false*?

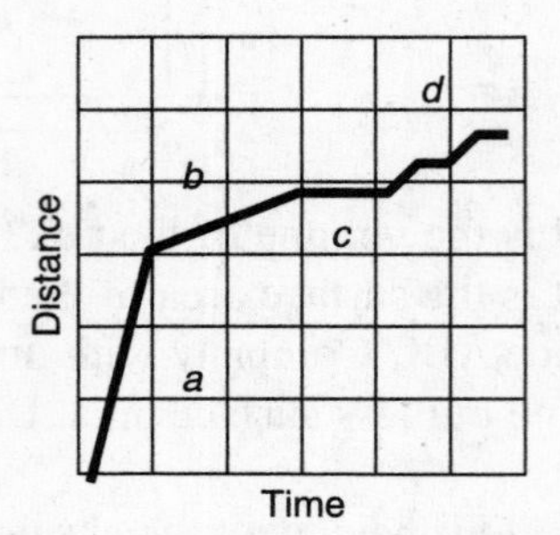

A. The driver is going really fast during the part on the graph labeled *b*.
B. The driver is going faster during the part labeled *a* than during the part labeled *b*.
C. The driver is stopped on the part of the graph labeled *c*.
D. The driver is in stop-and-go traffic during the part labeled *d*.

25. The digits 1, 2, 3, 4, and 5 are being used to form three-digit numbers.

(a) How many three-digit numbers can be formed if you are not allowed to repeat a number within the three-digit number?
(b) How many more numbers would there be if you were allowed to repeat a digit?

DIAGNOSTIC TEST ANSWERS

1. **C**
2. One possible network is shown below:

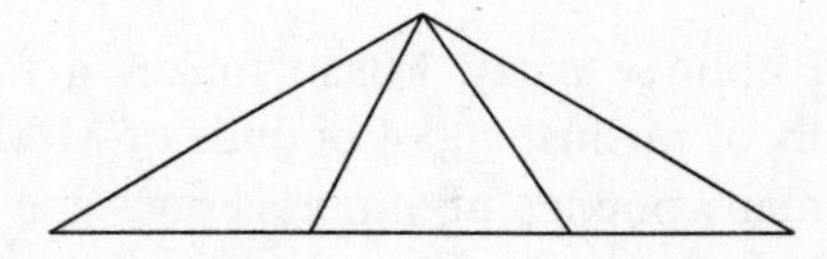

3. **C**
4. **B**
5. **D**
6. **D**
7. **(a)** mean ≈ \$157,714; median = \$164,000; mode: \$170,000; range = \$40,000
   (b) The mean would decrease by \$5,000.
   (c) \$164,000
8. **C**
9. **B**
10. **A**

11. One possible graph is shown below

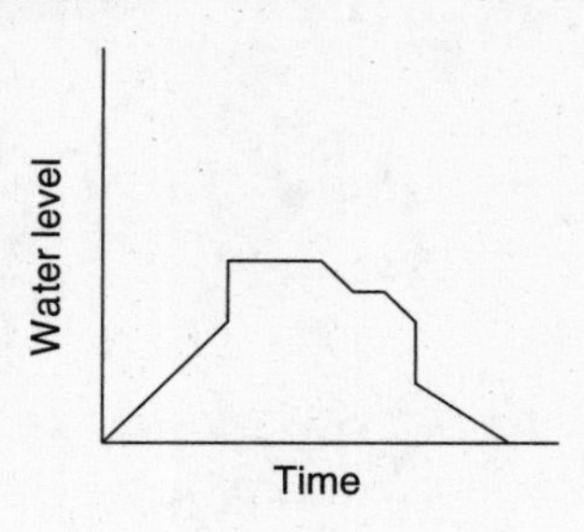

12. **C**
13. **B**
14. **D**
15. **B**
16. **C**
17. **(a)** $V = l \times w \times h = 12'' \times 6'' \times 4'' = 288\text{ in}^3$

    (b) Surface Area $= 12\text{ in} \times 4\text{ in} \times 2 = 96\text{ in}^2$ (front/back)
    $= 6\text{ in} \times 4\text{ in} \times 2 = 48\text{ in}^2$ (sides)
    $= 12\text{ in} \times 6\text{ in} \times 2 = 144\text{ in}^2$ (top/bottom)

    Thus, the total surface area is $96\text{ in}^2 + 48\text{ in}^2 + 144\text{ in}^2 =$ **288 in^2**.

    (c) One possible box would be 8″ by 6″ by 6″, which still has a volume of 288 in^3 but has a surface area of only 264 in^2.
18. He will be able to get the stereo beginning at 12:00 P.M.
    At 11:00 A.M., the first discount is $180 \times .15 = \$27$ bringing the price to \$153.
    At 12:00 P.M., the discount is $153 \times .15 = \$22.95$, bringing the price to \$130.05.
19. $\overline{BF}$, $\overline{CG}$, $\overline{BC}$, or $\overline{FG}$
20. **B**
21. **15** numbers are left
22. **C**
23. **D**
24. **A**
25. **(a)** **60**

    (b) $125 - 60 =$ **65** more numbers

CLUSTER I

NUMBER SENSE

MACRO A

TYPES OF NUMBERS

Of course, many of us associate mathematics with numbers. There are several types and formats of numbers you will need to be familiar with for the New Jersey HSPA Exam in Mathematics.

Until one studies Algebra II or Precalculus and beyond, all numbers are **real numbers**. Real numbers consist of the following special types of numbers:

Integers: $\{\ldots, -3, -2, -1, 0, 1, 2, 3, \ldots\}$
Whole Numbers: $\{0, 1, 2, 3, \ldots\}$
Natural Numbers (counting numbers): $\{1, 2, 3, 4, \ldots\}$
Rational Numbers: Numbers that can be expressed as a fraction or, by formal definition, numbers that can be written as $\frac{a}{b}$, where a and b are each integers with b not equal to 0

Note: All integers, whole numbers, and natural numbers are rational since, for example, we can write 3 as $\frac{3}{1}$.

Irrational Numbers: Numbers that are not rational or reasonable. Irrational numbers include non-terminating, non-repeating decimals such as pi (not 3.14 but the real value of pi), and square roots of many non-perfect squares such as 30.

Many rational numbers are **decimals**. Decimals sometimes *terminate* or *repeat*. The following are some examples of the types of decimals.

Examples

A. The rational number (fraction) $\frac{5}{8}$ can be converted to a decimal by dividing the numerator (top number) by the denominator (bottom number) as follows.

$$\begin{array}{r} .625 \\ 8\overline{)5.000} \\ 4\,8 \\ 0\,20 \\ 0\,16 \\ 0\,040 \\ 0\,040 \\ 0\,000 \end{array}$$

Since the resulting decimal ends (or terminates), we call it a **terminating decimal**.

B. The rational number (fraction) $\frac{2}{3}$ can be changed to a decimal by dividing 2 by 3 as follows.

```
     .666
  3)2.000
    1 8
    0 20
    0 18
    0 020
```

Since the resulting decimal consists of a number that repeats over and over without end, we call it a **repeating decimal**.

C. The rational number (fraction) $\frac{6}{11}$ can be converted to a decimal by dividing 11 into 6 as follows.

```
      .5454
  11)6.000
     5 5
     0 50
     0 44
     0 060
     0 055
     0 0050
```

Since the result is a set of two numbers that repeat, it is also considered a repeating decimal.

PRACTICE SET

1. Give a number that is a whole number but not a natural number.

2. Show that −4 is rational.

3. Is $\frac{7}{9}$ a terminating or a repeating decimal?

4. Is $\frac{3}{8}$ a terminating or a repeating decimal?

5. Is $\frac{14}{33}$ a terminating or a repeating decimal?

6. *Open-Ended Question*: An operation is *closed* over a set or group of numbers if you can perform the operation using *any* two numbers in the set and the answer is a number also in the set. Is the set of natural numbers closed for subtraction? Explain.

ESTIMATING DECIMALS

In the result for Example A, .625, recall that the 6 is in the **tenths place**, the 2 is in the **hundredths place**, and the 5 is in the **thousandths place**. Often you will be required to estimate numbers involving decimals. To round to a specified place, follow these steps:

1. Identify the place to be rounded to.
2. Look at the digit to the right of this place.
3. If the digit to the right is 5 or higher, round up.
4. If the digit to the right is 4 or lower, keep the number the same.

Examples

A. Round 5.83 to the nearest whole number.

In any decimal, the number to the left of the decimal point is the *whole* and the number to the right of the decimal point represents the *part*. So, to round to the nearest whole, we look at the first digit to the right of the decimal point (in this case 8). Since 8 is 5 or more, we round 5 up to 6.

B. Round 3.912 to the nearest hundredth.

Hundredth place is 1 so we look to the right of 1 and since it is 4 or less (it's 2), we keep it and thus round the number to 3.91.

C. Round 26.4738 to the nearest tenth.

Tenth place is 4, and when we look to the right of 4, the 7 means we round up; so the result is 26.5.

Practice Set

1. Round 11.34 to the nearest whole number.
2. Round 28.657 to the nearest tenth.
3. Round 34.2341 to the nearest thousandth.
4. Round 56.876 to the nearest hundredth.

OPERATIONS WITH SIGNED NUMBERS

One of the most common types of computations you will encounter involves signed numbers. Signed numbers "crop up" in equations in algebra, in statistical computations, and in other areas too. So, it is important that you have a good grasp of how to add, subtract, multiply, and divide positive and negative numbers.

On a number line, negative numbers fall to the left of zero and positive numbers fall to the right of zero. **Absolute value** measures how far a number is from zero. Absolute value is symbolized with two vertical lines like this:

$$|5| \quad \text{or} \quad |-3|$$

When you see an absolute value, you have to determine how many units it is from zero to the number between the absolute value bars. Since it's measuring distance, it has to be a positive value. Regardless of the direction you go (left or right), the distance is still a positive number. So, in Example A, since 5 is 5 units to the right of zero, $|5| = 5$. In Example B, since -3 is 3 units to the left of zero, $|-3| = 3$.

We will use absolute value when working with the addition and subtraction of signed numbers. In the **addition** of positive and negative numbers, there are two basic rules to follow. If the numbers have the same sign, add the absolute values and keep the sign. If the numbers are opposite in sign, subtract the absolute values and take the sign of the larger absolute value.

Examples

A. $3 + 5 = 8$ — In this problem, you simply add and the answer is positive.

B. $-5 + (-2) = -7$ — In this problem, you still add the absolute values, but the answer is negative.

C. $-6 + 8 = 2$ — In this problem, you subtract $|8| - |-6| = 8 - 6 = 2$ and it is positive since 8 has a larger absolute value than -6.

D. $-10 + 4 = -6$ — In this problem, you subtract $|-10| - |4| = 10 - 4 = 6$ and it is negative since -10 has a larger absolute value than 4.

It is sometimes helpful to think of money when working with the addition of positive and negative numbers. For instance, in Example C above, think of it as "owing 6" since it's negative and "having 8" since it's positive. If you owe someone \$6 and have \$8, you are able to pay them what you owe and still have \$2 left (positive). In Example D, on the other hand, you owe 10 and only have 4; therefore you can only pay \$4 and you still owe \$6 (negative).

Numbers that are the same distance from zero have the same absolute value and are called *opposites*. For example, -3 and 3 are opposites; so are -11 and 11. Opposites always add up to zero.

The **subtraction** of signed numbers is really defined as addition of the opposite. So, eventually the rules of addition will apply to many subtraction problems once you rewrite them as I am about to describe. There are still, of course, standard subtraction problems like $12 - 5 = 7$. This problem does not need to be handled any differently than it was in the past.

However, other problems have to be changed as follows.

$10 - 13 = 10 + (-13) = -3$ — Change subtraction to addition, and the second number to its opposite.

$3 - (-4) = 3 + 4 = 7$ — Change subtraction to addition, and the second number to its opposite.

$-2 - (-5) = -2 + 5 = 3$ — Change subtraction to addition, and the second number to its opposite.

$-8 - (-2) = -8 + 2 = -6$ — Change subtraction to addition, and the second number to its opposite.

$-3 - 6 = -3 + (-6) = -9$ — Change subtraction to addition, and the second number to its opposite.

Practice Set

Add or subtract as indicated:

1. $-2 + (-6)$ **2.** $-3 + 10$ **3.** $12 - 15$ **4.** $-1 - 6$
5. $-8 + 8$ **6.** $-2 - (-9)$ **7.** $4 - (-3)$ **8.** $-5 - (-3)$
9. $-17 + 9$ **10.** $-8 - (-13)$

Signed numbers go "hand in hand" with another topic: vectors. You will also see vectors if you study Physics and/or Calculus. A **vector** is a directional ray. Basically, what that means is that it's an arrow showing direction and length (usually referred to as magnitude or strength). For example, we can draw a vector diagram to model $-3 + 5$ as follows:

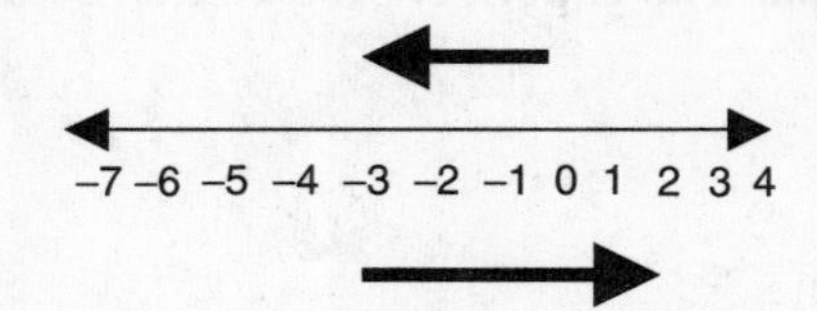

This can also be used as a technique for adding signed numbers using a number line. We will revisit the topic of vectors later on to show you other directions in which vectors can move and how they can be applied to real-world problems.

Multiplication and **division** of signed numbers is relatively easy since there's one "blanket" or general rule to remember: If the two numbers have the same sign, the answer (product or quotient) is positive. If the two numbers have different signs, the answer (product or quotient) is negative.

Examples

A. $3 \times 5 = 15$ — Since 3 and 5 both have the same sign, the answer is positive.

B. $-2 \times 8 = -16$ — Since -2 and 8 have different signs, the answer is negative.

C. $6 \times -10 = -60$ — Since the numbers have different signs, the answer is negative.

D. $-20 \div (-4) = 5$ — Since the numbers are both negative (the same), the answer is positive.

E. $30 \div (-6) = -5$ — Since the numbers have different signs, the quotient is negative.

PRACTICE SET

1. -4×-6 **2.** -3×8 **3.** -10×4
4. $24 \div (-8)$ **5.** $-40 \div (-4)$ **6.** $-60 \div 5$

7. *Short Constructed Response*: If $a > 0$ and $b < 0$, determine if each of the following is never, sometimes, or always true.

(a) $a \div b > 0$
(b) $a + b < 0$
(c) $a - b > 0$

Be sure to explain and justify your response.

EXPONENTS AND ROOTS

Another common operation or manipulation is that indicated by an exponent. In a problem like the one on page 18, the 2 is called the **base** and the 3 is called the **exponent**. Exponents give a way of showing repeated multiplication in a shortened form. The *exponent tells you how many times to multiply the base by itself.*

$$2^4 = 2 \times 2 \times 2 \times 2 = 16$$

This expression is read, by the way, as "2 to the fourth power." Exponents can also be done on a calculator by using a special key that usually says x^y or y^x. To use this special key to calculate an exponent problem, let's do a second example.

Examples

A. Calculate 3^5.

On a calculator, this problem can be done by first entering the base (3).
Next, hit the x^y key and then enter the exponent (5).
Hit the equal sign, and you should get the correct result, which is 81.

B. Find 5^2.

$5 \times 5 = 25$

Note: 5^2 is often read "5 squared," which is a special way of saying "5 to the second power."

C. Evaluate 4^3.

$4 \times 4 \times 4 = 64$

Note: 4^3 is often read "4 cubed," which is a special way of saying "4 to the third power."

D. Calculate $(-2)^4$.

$-2 \times -2 \times -2 \times -2 = 4 \times -2 \times -2 = -8 \times -2 = 16$

E. Find $(-3)^3$.

$-3 \times -3 \times -3 = 9 \times -3 = -27$

PRACTICE SET

1. 2^5 **2.** $(-4)^2$ **3.** $(-2)^3$ **4.** 3^8

5. *Short Constructed Response*: Based on the clues given below, determine the number. Then explain how you figured out the number.

(a) I am a positive integer that equals between 100 and 250 when squared.
(b) I am greater than 2000 when cubed.
(c) I am even.

6. *Open-Ended Response*: A mathematician spills some coffee on his papers. He needs to determine what the value of a base is and all he remembers is that the final answer is more than 4000 but less than 5000. Can you help him figure out what the missing base value is? Explain how you figured out the number.

$$?^4 + 125$$

PROPERTIES OF EXPONENTS

In some problems, you will need to be able to apply various rules for exponents when involved in computations. The most important thing is that *you must have the same base* in order to work with two or more exponential quantities in a computation.

Examples

A. What does $2^3 \times 2^5$ equal?

To find the rule for this problem, simply rely on what you already know about exponents.

That is, write the problem as $2 \times 2 \times 2 \times 2 \times 2 \times 2 \times 2 \times 2$, which gives us 2 times itself a total of 8 times, or 2^8.

A shortcut would be to *add the exponents* if you are multiplying two numbers with the same base.

B. What does $5^8 \div 5^2$ equal?

Again, to find the rule for this situation, we can expand each part to a product of 5's. So, the problem becomes

$$\frac{5\times5\times5\times5\times5\times5\times5\times5}{5\times5}$$

Well, at this point, we can use the fact that $5 \div 5 = 1$ and we can "cancel" two pairs of 5's. This leaves us with six 5's or 5^6.

A shortcut would be to *subtract the exponents* if you are dividing two numbers with the same base.

C. What does $(4^5)^3$ equal?

Once more, we can come up with a rule by following what we already know. For this problem, we can rely on what we know about an exponent of 3. It means that we multiply the base by itself 3 times. So, in this problem, we have 4^5 raised to the power of 3. So, we have $4^5 \times 4^5 \times 4^5$. At this point, we can use the rule we established in Example A; we are multiplying and we have the same base, so we can *add* the exponents and get 4^{15}. We could have also found this result in one step by the shortcut: *multiply the exponents* if you are taking a power to another power.

PRACTICE SET

1. $6^2 \times 6^5$ **2.** $(7^3)^4$ **3.** $\frac{3^{10}}{3^4}$ **4.** $5^3 \times 5^7$ **5.** $\frac{4^{12}}{4^6}$

As stated earlier, these rules for exponents apply only if numbers have the same base. Sometimes, however, you will be faced with a problem in which the exponents are not the same, but can be rewritten to have the same base.

Examples

A. Simplify: $2^5 \times 4^3$

Since $4 = 2^2$, we write this problem as $2^5 \times (2^2)^3$, which then becomes $2^5 \times 2^6$ (power to power means we multiply exponents, and so $2^{2\times 3}$), and then since we are multiplying, we add the exponents to arrive at a final answer of 2^{5+6}, or 2^{11}.

B. Simplify: $\dfrac{27^5}{9^4}$

Again, we can work with a problem like this only if the bases match. So, we write both 27 and 9 in base 3. This gives us

$$\frac{(3^3)^5}{(3^2)^4}$$

Then, we apply our rule for power to power (multiply the exponents). This gives us

$$\frac{3^{15}}{3^8}$$

Finally, we apply the rule for division (subtract the exponents), and we get 3^7.

PRACTICE SET

1. Simplify: $9^3 \times 27^4$

2. Simplify: $\dfrac{25^6}{125^3}$

One other interesting manipulation of exponents occurs when you are asked to look at the sum of two numbers with the same base and consecutive exponents. For example, consider $3^8 + 3^9$. This can actually be written as $3^8(1 + 3)$ since $3^8 \times 1 = 3^8$ and $3^8 \times 3^1 = 3^9$. Thus, $3^8 + 3^9 = 4 \times 3^8$.

PATTERNS IN EXPONENTS

Since an exponent indicates repeated multiplication by the same number (the base), it is natural to encounter a pattern in the end number of the resulting product. That is, the *units digit* or *ones place* ends in a certain pattern of numbers.

For example, consider the powers of 2:

$$2^1 = 2 \quad 2^2 = 4 \quad 2^3 = 8 \quad 2^4 = 16 \quad 2^5 = 32 \quad 2^6 = 64 \quad 2^7 = 128$$

Notice the units place in the results: 2, 4, 8, 6, 2, 4, 8, 6, ...

The units place occurs in a "cycle" of four. If we make use of this idea coupled with the concept of a remainder in long division, we can figure out what the units digit in a power of 2 is without using a calculator or multiplying the entire product out by hand.

What if you were asked for the units digit in 2^{19}? First, it should be noted that even a calculator is not going to give you this answer because the result for 2^{19} is too large to fit on the calculator's screen. However, if we keep in mind that the units digit occurs in cycles of four (2, 4, 8, 6, 2, 4, 8, 6, . . .), we can perform long division to get the correct units digit as follows.

$$4\overset{4}{\overline{)19}}$$

Since 4 times 4 is 16, the remainder (19 – 16) is 3. This means that there are four complete cycles of 2, 4, 8, 6 and then 3 remaining digits, which, means there is an 8 in the units place in 2^{19}.

SCIENTIFIC NOTATION

Scientific notation is a special notation or format for numbers in mathematics and science that is very closely connected to exponents. Typically, it is used to write numbers that are either very large or very small using a shortened form. Scientific notation is what we sometimes refer to as **base 10** because everything is written times 10 to a power.

Examples

A. Write 320,000 in scientific notation.

Your first job is to look at just the number without the zeros (32) and write this as a number between 1 and 10. So, we'll use 3.2.

Now, 3.2×10^5 since we would have to move the decimal 5 places to the right from 3.2 to get 320,000.

B. Write 4,890,000 in scientific notation.

Again, look at the numbers without the zeros (489) and insert a decimal point so that you create a number between 1 and 10 (4.89). Then, from 4.89 we would have to move the decimal point 6 places to the right, so we say $4{,}890{,}000 = 4.89 \times 10^6$.

C. Write .00076 in scientific notation.

Again, we look at just the 76 and ignore the zeros at first. We insert a decimal point so that we get a number between 1 and 10 (7.6). Now, since we are moving the decimal left, we use a negative exponent. So, $.00076 = 7.6 \times 10^{-4}$.

PRACTICE SET

1. Write 780,000,000 in scientific notation.
2. Write .000031 in scientific notation.
3. Write 9,400,000 in scientific notation.
4. Write .0044 in scientific notation.

Frequently, on the HSPA exam, rather than simply converting a number to scientific notation, you will be asked to determine a range of values that are possible for a product written in scientific notation. For example, you may be asked, What is the range of values for the following product?

$$(3.824 \times 10^3) \times (5.\_\_\_ \times 10^6)$$

To respond to this question, you must first realize that you need to take the extreme values for the second number. That is, you must take 5.000 and 5.999, which are the smallest and largest values. Then, multiply your two products:

$$3.824 \times 10^3 \times 5.000 \times 10^6 \quad \text{and} \quad 3.824 \times 10^3 \times 5.999 \times 10^6$$

Multiply the numbers and then add the exponents to get

$$19.12 \times 10^9 \quad \text{and} \quad 22.940176 \times 10^9$$

The second value can be rounded to give 22.9402×10^9. Thus, the range of values is between 19.12×10^9 and 22.9402×10^9. This can also be expressed using inequality notation as follows.

$$19.12 \times 10^9 \leq p \leq 22.9402 \times 10^9$$

PRACTICE SET

1. Give the range of values for $(6.738 \times 10^3) \times (8.\_\_\_ \times 10^4)$.

2. What are the possible values for $(5.261 \times 10^2) \times (3.\_\_\_ \times 10^6)$?

ROOTS

The inverse of evaluating an exponent is taking a root, most commonly a square root (the inverse of an exponent of 2). A square root "undoes" squaring or raising to a power of 2. The symbol for a root is $\sqrt{\ }$. If we raise 4 to the second power, of course, $4^2 = 16$. The square root of 16 is then 4; that is, $\sqrt{16} = 4$ since $4 \times 4 = 16$.

Every positive number has a square root; however, not every square root is an integer. In fact, only a certain set of numbers have integral square roots. These are called **perfect squares**.

Examples

A. $\sqrt{25} = 5$ since 5×5 or 5^2 equals 25.

B. $\sqrt{64} = 8$ since 8×8 or 8^2 equals 64.

C. $\sqrt{121} = 11$ since 11×11 or 11^2 equals 121.

Now, if you don't recognize the square root off the top of your head, you can use a calculator. On most calculators, there is a $\sqrt{\ }$ key and you simply hit this key and then the number you wish to take the square root of. You will need to check the syntax of your calculator by consulting the manual that came with it.

If the number you are taking the square root of is not a perfect square, you will definitely need a calculator to get the value.

Examples

A. Find $\sqrt{45}$.

Since 45 is close to 49 (a perfect square), we expect the result to be close to 7 (the square root of 49). By using a calculator, we get $\sqrt{45} \approx 6.708$. *Note*: We have to approximate the result; the actual square root of 45 is a nonterminating, nonrepeating decimal (an irrational number).

> **Take Note:** Many times on the HSPA exam, you will need to classify certain square roots as "irrational." Both of these examples are irrational since the result on the calculator is a lengthy decimal filling the screen.

B. Find $\sqrt{103}$.

Since 103 is a little more than 100 (a perfect square), we expect the result to be a little more than 10 (the square root of 100). By using a calculator, we get $\sqrt{103} \approx 10.149$.

PRACTICE SET

Find each square root shown below. If necessary, round the answer to the nearest thousandth.

1. $\sqrt{81}$ **2.** $\sqrt{30}$ **3.** $\sqrt{119}$ **4.** $\sqrt{36}$ **5.** $\sqrt{200}$

ALGEBRAIC EXPRESSIONS INVOLVING EXPONENTS AND ROOTS

For some problems on the HSPA exam, you will need to know how to evaluate algebraic expressions that may involve exponents. Remember, to evaluate an expression, you simply "plug in" the value for the variable(s).

Examples

[Note: • is an alternative sign for multiplication, now that we are using "x" as a variable.]

A. Evaluate $5x^2$ if $x = 3$.

$5x^2$ means $5 \cdot x \cdot x$, so $5 \cdot 3 \cdot 3 = 5 \cdot 9 = 45$

B. Evaluate $(3x)^2$ if $x = 4$.

Note that this problem differs from Example A because of the parentheses. We first multiply $3 \cdot x$ because it's in parentheses and of course we get 12. This result is then raised to the second power to get $12^2 = 144$.

C. Evaluate $2x^3$ for $x = -3$.

$2x^3$ means $2 \cdot x \cdot x \cdot x = 2 \cdot (-3) \cdot (-3) \cdot (-3) = 2 \cdot (-27) = -54$

PRACTICE SET

Given $x = 3$, $y = 4$, and $z = -5$, evaluate each of the following.

1. $2x^2$ **2.** $(2y)^2$ **3.** $2z^3$ **4.** $3y^2$ **5.** $(2x)^2$

ORDER OF OPERATIONS

In order to establish correctness and consistency, there is what is called an **order of operations** in mathematics. Calculators, for the most part, already have this built in. (It's what we refer to as algebraic logic.) If you are faced with a lengthy computation-based question on the HSPA exam, it is important to enter it into your calculator in the exact order and in the exact way that it is given. If you are asked to explain the order of operations or determine if a problem has been done correctly, you may need to know the order of operations yourself.

The correct order of operations is as follows.

1. Parentheses, brackets, or other grouping symbols (which actually alter or change the standard order of operations)
2. Exponents
3. Multiplication or division from left to right
4. Addition or subtraction from left to right

Examples

A. $30 \div 2 \times 5 - 20 + 7$

Since there are no parentheses or exponents, we begin this problem with multiplication and division from left to right. Thus, we get

$$15 \times 5 - 20 + 7$$
$$75 - 20 + 7$$

We continue with addition and subtraction from left to right. This yields

$$55 + 7$$
$$62$$

B. $18 \div (2 \times 3) + 4^2$

We begin this problem by completing the smaller problem within the parentheses. This gives

$$18 \div 6 + 4^2$$

Next we evaluate the exponent, which makes the problem

$$18 \div 6 + 16$$

Finally, we divide and then add:

$$3 + 16$$
$$19$$

C. $56 - [(12 - 7)^2 + 2 \times 3]$

In this problem, we first complete the problem within the brackets (which are just another form of parentheses). Within the brackets, we must still follow the order of operations. Thus, we evaluate $12 - 7$ first and then the exponent; then we perform multiplication and finally addition:

$$56-[5^2+2\times 3]$$
$$56-[25+6]$$
$$56-31$$
$$25$$

PRACTICE SET

1. $24 \div 3 \times 2 - 9 + 4$

2. $108 \div (4 + 5) - 2^3$

3. $92 - [37 + (8 - 3)^2 \times 2]$

PROPERTIES OF NUMBERS

The **commutative property** states that an operation is true when performed in either order. That is, addition is commutative since $5 + 8$ is the same as $8 + 5$. Not all operations are commutative, however. For instance, subtraction is *not* commutative since $3 - 7 \neq 7 - 3$. Division is not commutative either since $15 \div 3 \neq 3 \div 15$. Multiplication is commutative since $3 \cdot 6 = 6 \cdot 3$.

The **associative property** states that an operation is true when three or more numbers are grouped in different ways. For example, addition is associative since $8 + (12 + 7) = (8 + 12) + 7$. Multiplication is also associative. For instance, $(2 \times 8) \times 5 = 2 \times (8 \times 5)$. Subtraction and division are *not* associative.

These two properties are useful at times in allowing us to easily perform mental arithmetic by carefully grouping numbers to create easy sums or products. For example, suppose you have to calculate the following.

$$6\times 4\times 5$$

It would be a lot easier if you grouped the numbers 4 and 5 to get $6 \times (4 \times 5) = 6 \times 20 = 120$ rather than having to evaluate 6×4 and then multiply by 5, or 24×5.

As a second example, consider

$$18+36+12$$

To make this computation easier, we can use the commutative property to make it $18 + 12 + 36$ and then group it as $(18 + 12) + 36$, which equals $30 + 36 = 66$.

An **equivalence relationship** is a relationship of equality or sameness. There are three essential properties that must be satisfied in order to determine that a given operation or statement is an equivalence relationship:

Reflexive Property ($a = a$): A number is equal to itself or the property applies between an object and itself.

Symmetric Property (if $a = b$, then $b = a$): This is very similar to commutativity; it says that the property applies in both directions.
Transitive Property (if $a = b$ and $b = c$, then $a = c$): This property is like a transference.

Examples

A. Is addition an equivalence relation?

Reflexive: $6 + 2 = 6 + 2$

Symmetric: $2 + 6 = 6 + 2$

Transitive: $8 = 2 + 6$ and $2 + 6 = 5 + 3$, so $8 = 5 + 3$

Since all three properties are satisfied, addition is considered an equivalence relation.

B. Is > an equivalence relation?

Reflexive: $5 > 5$

This is *not* true.

Symmetric: If $8 > 3$, then $3 > 8$.

This is *not* true.

Transitive: If $3 > 1$ and $1 > -4$, then $3 > -4$

This is true.

However, since not all three properties are satisfied > is *not* an equivalence relation.

C. Is the relationship "is parallel to" transitive? Show why or why not.

First, when asked a question like this, you should translate the property in question into words and/or diagrams. In this case, you should ask, If $\overleftrightarrow{AB} \parallel \overleftrightarrow{CD}$ and $\overleftrightarrow{CD} \parallel \overleftrightarrow{EF}$, then is it true that $\overleftrightarrow{AB} \parallel \overleftrightarrow{EF}$? In preparing your diagrams and argument, be sure you are considering all the possibilities. A drawing might look like this:

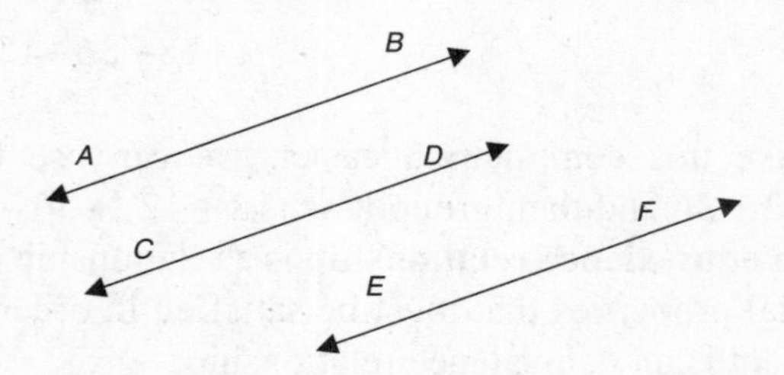

In the diagram, one can clearly see that the first (AB) and third (EF) lines are indeed parallel and thus "is parallel to" is transitive.

PRACTICE SET

1. If a new operation is defined using the symbol # and $a \# b$ means $a^2 + b^2$, is this new operation # commutative? Is it associative? Use examples to justify your answer.

2. Is the relationship "is perpendicular to" transitive?

3. Explain how to make the problem 12 + 45 + 28 + 25 easiest to do in your head without a calculator. What properties of arithmetic are helpful?

4. John argues that the relationship "is sitting next to" is transitive. Is he correct? Justify your decision.

PRIMES, FACTORS, AND MULTIPLES

Numbers are said to be either **prime** or **composite**. Whether a number is considered prime or composite is determined by the number of factors it has. A **factor** of a number is a number that divides evenly into the given number. Or, in other words, a factor is a number that can be multiplied by another number to give the specified number. For example, the factors of 8 are 1 and 8 and 2 and 4 since $1 \times 8 = 8$ and $2 \times 4 = 8$.

Examples

A. Give all the factors of 20.

Begin with the obvious pair (1 and 20). Then continue with 2 and 10. Skip 3 since 20 is not evenly divided by 3. The next pair of factors is 4 and 5.

Thus, all the factors of 20 are 1, 2, 4, 5, 10, 20.

B. State all the factors of 36.

Again begin with 1 and 36.

Continue testing 2, 3, 4, 5, ... The factors continue with 2 and 18, 3 and 12, 4 and 9, and 6 and 6. Thus, the entire list of factors for 36 is 1, 2, 3, 4, 6, 9, 12, 18, 36.

PRACTICE SET

1. Give all the factors of 40.

2. State every factor of 64.

3. Give all the factors of 45.

Consider a number like 13. If you try the list of numbers 2, 3, 4, and so on, you will find that no number divides evenly into 13. Since the only factors of 13 are 1 and 13, we call 13 a **prime number**. All numbers that are not prime are considered **composite numbers**.

Several other prime numbers are 2, 3, 5, 7, 11, 13, 17, 19, 23, 29, 31.

Take Note: Remember that 1 is NOT prime!!! It is an exception. And, 2 is the only even prime!

Composite numbers can be factored into what is referred to as a **prime factorization**. Prime factorization refers to the writing of a number as a product of primes. Typically, to get prime factorization, we make what is called a *factor tree*.

Examples

A. State the prime factorization for 30.

30

3 10

∧

2 5

∧

Thus, the prime factorization of 30 is $2 \times 3 \times 5$.

B. Give the prime factorization of 162.

162

2 81

∧

3 27

∧

3 9

∧

3 3

∧

Thus, the prime factorization of 162 is 2×3^4.

PRACTICE SET

1. What is the sum of the first five prime numbers?

2. Is 57 prime or composite?

3. Give the prime factorization for each of the following numbers: 40, 95, and 120.

If, instead of "breaking a number apart" you multiply it by another number, you get what is called a **multiple** of that number. For example, the multiples of 4 are $4 \times 2 = 8$, $4 \times 3 = 12$, $4 \times 4 = 16$, $4 \times 5 = 20$, $4 \times 6 = 24$, and so on. Unlike factors, the list of multiples never ends.

Examples

A. List the first five multiples of 6.

$6 \times 1 = 6$ $6 \times 2 = 12$ $6 \times 3 = 18$ $6 \times 4 = 24$ $6 \times 5 = 30$

B. Give a multiple of 4 that is also a multiple of 6.

The solutions may vary, but the key is that the number chosen must be divisible by both 4 and 6, such as 12, 24, or 36.

PRACTICE SET

1. Give the first five multiples of 8.

2. State a multiple of 6 that is also a multiple of 8.

APPLYING FACTORS AND MULTIPLES

One of the most important reasons to know about factors and multiples is to be able to use these concepts to find common factors and multiples.

If you list the factors of two or more given numbers, the largest shared factor is what is called the **greatest common factor (GCF)**. The GCF is useful especially when trying to reduce a ratio or fraction to lowest terms.

To find the GCF for a set of numbers, follow these steps:

1. List the factors of each number in order.
2. Identify the largest number that is common to both lists.

Examples

A. Find the GCF for 12 and 30.

Factors of 12: 1, 2, 3, 4, 6, 12
Factors of 30: 1, 2, 3, 5, 6, 10, 15, 30

The largest or greatest factor that is common to both is 6. Therefore, the GCF for 12 and 30 is 6.

B. Find the GCF for 24, 32, and 56.

Factors of 24: 1, 2, 3, 4, 6, 8, 12, 24
Factors of 32: 1, 2, 4, 8, 16, 32
Factors of 56: 1, 2, 4, 7, 8, 14, 28, 56

The largest factor that is shared by all three numbers is 8, which is the GCF.

PRACTICE SET

1. What is the GCF of 18 and 45?

2. Find the GCF for 35, 56, and 84.

With multiples, the **least common multiple (LCM)** is the most useful. The LCM refers to the first or smallest multiple that is common to two or more numbers. To find the LCM for two or more numbers, follow these steps:

1. List several multiples for each number.
2. Check to see if there is a common multiple in all the lists.
3. If not, list several more multiples of each number until you find a common multiple.

Examples

A. Find the LCM for 8 and 12.

Multiples of 8: 8, 16, 24, 32, 40, 48, . . .
Multiples of 12: 12, 24, 36, . . .

The first multiple (the lowest) is 24, which is the LCM.

B. Find the LCM for 6, 12, and 20.

Multiples of 6: 6, 12, 18, 24, 30, 36, 42, 48, 54, 60, 66, 72, . . .
Multiples of 12: 12, 24, 36, 48, 60, 72, 84, . . .
Multiples of 20: 20, 40, 60, 80, 100, . . .

The LCM is 60.

PRACTICE SET

1. What is the LCM for 8 and 10?

2. Find the LCM for 6, 15, and 18.

It is common to be expected to apply the idea of an LCM to a situation similar to the following. If a store is offering a special promotion and gives every 12th customer a free pen and every 15th customer a free T-shirt, how many people will get both items free if there are 300 customers all day? The first thing to realize here is that this problem deals with multiples. Every 12th customer means the 12th, the 24th, the 36th, and so on (the multiples of 12). Every 15th customer means the 15th, the 30th, the 45th, and so on (the multiples of 15). So, if we find the LCM of 12 and 15, this will indicate which person will be the first to get both items free.

Multiples of 12: 12, 24, 36, 48, 60, 72, 84, and so on
Multiples of 15: 15, 30, 45, 60, 75, and so on

The first (or lowest) common multiple is 60; so this is the LCM. Thus, every 60th person will get both items free. So, the people who will get both free are the 60th, 120th, 180th,

240th, and 300th (we stop at 300 since there were 300 customers). Five people will get both free.

PRACTICE SET

1. If it takes Mark 8 minutes to run a lap and it takes Steve 10 minutes to run a lap, how many times will Mark and Steve cross the start line at the same time if they run for 3 hours?

2. On an assembly line, one worker checks every 12th item for painting defects and another worker inspects every 18th item for assembly defects. If 250 items pass by these employees, how many will be checked for both painting and assembly defects?

MIXED PRACTICE, CLUSTER I, MACRO A

1. Approximate: $3.8 \times .42$.

   A. 8 **B.** 1.6 **C.** 16 **D.** 4.3

2. Which of the following statements are false?

   A. $\frac{8^6}{8^2} = 8^4$ **B.** $8^3 \cdot 8^2 = 8^6$ **C.** $(8^2)^5 = 8^{10}$ **D.** $\sqrt{8^{12}} = 8^6$

3. If $x^2 = 27$, what type of number is x?

   A. repeating decimal
   B. rational number
   C. fraction
   D. irrational number

4. Given that $h = 30t - 6t^2$ gives the height, h (in meters), of an object thrown up in the air at an initial speed of 30 meters per second after t seconds, find the height after 3 seconds.

5. Select the phrase that could be written in the blank in the statement below to get the best true sentence:

   4^3 ________ 3^4

   A. is about half of
   B. is approximately $\frac{2}{3}$ as much as
   C. is about $1\frac{1}{2}$ times
   D. is about double

6. If the correct answer to a question for which the solution is to be rounded to the nearest hundred is \$1400, what is the smallest value of the actual answer? What is the largest value of the actual answer?

7. Which of these expressions is *not* equivalent to the others?

   A. $3^6 \cdot 9^6$ **B.** $(3^2)^9$ **C.** 3^{3^3} **D.** $(27^2)^3$

8. Which of these numbers is closest to $.032 \times .0607$?

A. .2 **B.** .02 **C.** .002 **D.** .0002

9. Which number is in the units (or ones) place in the number equivalent to 7^{83}?

A. 1 **B.** 3 **C.** 7 **D.** 9

10. Is $4^2 + 4^5$ equal to 4^7? Explain.

11. Which of these numbers is irrational?

A. $\sqrt{50}$ **B.** $\sqrt{100}$ **C.** $5\frac{17}{32}$ **D.** $4.\overline{371}$

12. Which of the following is an irrational number?

A. 6.08 **C.** $6.0\overline{2}$

B. $6\frac{9}{100}$ **D.** 6.02002000200002 . . .

13. Which of the following is a multiple of *both* 4 and 9?

A. 2028 **B.** 3582 **C.** 3744 **D.** 8325

14. Assuming that the origin is the starting point, what addition problem is modeled by the vector diagram below?

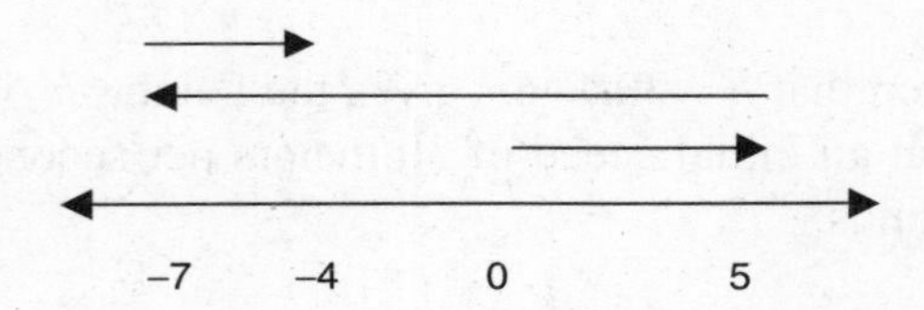

15. In a prehistoric cave, the volume of water that some cavemen predicted would cover the earth once the Ice Age ended was chiseled into the wall of the cave. However, part of the rock has eroded so that part of one number is illegible. The volume of the water was written as $(3.821 \times 10^4) \times (1.3\#\# \times 10^6)$. Find the range of possible values.

16. Find all the three-digit numbers that satisfy all four conditions listed below. Explain how you figured out the answer.

(a) The number is a multiple of 9.
(b) The number is odd.
(c) The number is more than 480.
(d) The number is less than 550.

17. Students are working together on a classroom project in groups of 3 or 7. When all the students are present, there is always one student left over. Which of the following could *not* be the number of students in the class?

A. 25 **B.** 24 **C.** 32 **D.** 26

18. A new operation is defined using & and $x \& y = x^2 + 3xy$. Find the value of $2 \& -5$; show all your work.

19. Which of the following numbers is the largest prime number less than 120?

A. 119 **B.** 117 **C.** 115 **D.** 113

20. Find all numbers that meet all the following conditions.

(a) The number is 4 more than a perfect square.
(b) The number is less than 100.
(c) The number is odd.
(d) The sum of the digits of the number is a prime number.

21. Two whole numbers are referred to as *relatively prime* if their only common factor is 1. For instance, 3 and 10 are relatively prime. How many of the following pairs are relatively prime?

7 and 9 21 and 24 54 and 93 43 and 100

A. One **B.** Two **C.** Three **D.** Four

22. Is it possible for a multiple of 13, greater than 13, to be a prime number? Why or why not?

23. If you are making up packages and each package must contain the same combination of pens as pencils, what is the greatest number of packages you can make using 63 pens and 72 pencils. How many of each will each package contain? Explain how you got your answer.

24. Every Tuesday night the local ice cream parlor gives every 10th customer a free ice cream cone and every 12th customer a free sundae. Last Tuesday there were 195 customers.

(a) How many free ice cream cones did the store give away? How many sundaes did it give away?
(b) When Tom went to the store, the cashier told him he was the first person to get both a free ice cream cone and a sundae. How many customers were ahead of Tom? How many other people got both items free?

[illegible]

Which of the following numbers has the largest prime [illegible] less than 100?

[illegible]

Which [illegible] of the following [illegible]

(a) [illegible] perfect square [illegible]

(b) [illegible]

(c) The [illegible]

(d) [illegible] the digits of the [illegible]

[illegible]

[illegible]

[illegible]

[illegible]

[illegible]

[illegible]

[illegible]

MACRO B

FRACTIONS

A **fraction** is used to express any part of a whole. The **numerator** is the top number of the fraction and represents the part we are taking. The **denominator** is the bottom number of the fraction and represents the number of parts into which the whole is divided. A fraction is written using a horizontal bar:

$$\frac{3}{8} \quad \frac{9}{4} \quad 3\frac{5}{12}$$

The first fraction $\left(\frac{3}{8}\right)$ is a standard fraction in reduced form. It means that we have a whole divided into 8 parts and we are using 3 out of these 8 parts.

The second fraction $\left(\frac{9}{4}\right)$ is called an **improper fraction** because the numerator is larger than the denominator. Any improper fraction can be turned into a mixed number by dividing the denominator (the bottom number) into the numerator (the top number). In this case, we divide $9 \div 4$ to get 2 with a remainder of 1. The remainder is out of the divisor, the number we are *dividing by*. So, the mixed number is $2\frac{1}{4}$.

It is important to realize that many fractions are equivalent to any given fraction. For example, consider the fraction $\frac{1}{2}$. Depending on the whole (the denominator) and the part (the numerator) that one uses, many fractions are equivalent to $\frac{1}{2}$. For instance, $\frac{5}{10}$ is equivalent to $\frac{1}{2}$. The fraction $\frac{7}{14}$ is also equal to $\frac{1}{2}$. To get a fraction equivalent to a given fraction, simply multiply *Both* the numerator and the denominator by the same number.

Examples

A. Change $\frac{12}{5}$ to a mixed number.

To change an improper fraction to a mixed number, recall that you divide the numerator by the denominator. In this case $12 \div 5 = 2$ with a remainder of 2. The remainder is out of 5 since that is what we were dividing by. Thus, the mixed number is $2\frac{2}{5}$.

Calculator Tip: Most scientific calculators can convert an improper fraction to a mixed number. Be sure you're familiar with working with fractions on your calculator before taking the HSPA.

B. Write $3\frac{7}{12}$ as an improper fraction.

To go from a mixed number to an improper fraction, you multiply the whole number and the denominator and then add the numerator. So, in this case $3 \times 12 + 7 = 36 + 7 = 43$. The result is put over the denominator to get $\frac{43}{12}$.

C. Give three fractions equivalent to $\frac{3}{5}$.

To make a list of equivalent fractions, you simply multiply the numerator and denominator by the same number. For example, we can multiply by 2 to get

$$\frac{3}{5} \times \frac{2}{2} = \frac{6}{10}$$

Or, we can multiply by 4 to get

$$\frac{3}{5} \times \frac{4}{4} = \frac{12}{20}$$

Finally, we can multiply by 5 to get:

$$\frac{3}{5} \times \frac{5}{5} = \frac{15}{25}$$

Thus, the fractions $\frac{6}{10}$, $\frac{12}{20}$, and $\frac{15}{25}$ are all equivalent to $\frac{3}{5}$.

PRACTICE SET

1. Write the mixed number $4\frac{2}{5}$ as an improper fraction.

2. Write $\frac{14}{3}$ as a mixed number.

3. State three fractions that are equal to $\frac{5}{8}$.

You will also be expected to be able to reduce fractions to their lowest terms. To do this, look for the GCF between the numerator and the denominator and then divide both the numerator and the denominator by this number.

Examples

A. Reduce $\frac{12}{20}$ to lowest terms.

When asked to reduce a fraction to its lowest terms, look for the GCF between the numerator and the denominator. So, in this case, we want the GCF for 12 and 20.

Factors of 12: 1, 2, 3, 4, 6, 12
Factors of 20: 1, 2, 4, 5, 10, 20

Thus, the GCF for 12 and 20 is 4. The fraction is then "reduced by 4" as follows.

$$\frac{12 \div 4}{20 \div 4} = \frac{3}{5}$$

So, the reduced form of $\frac{12}{20}$ is $\frac{3}{5}$.

B. What is the fraction $\frac{28}{42}$ in lowest terms?

This example will show you that it is not absolutely essential to use the GCF to reduce a fraction. You can reduce the fraction in steps.

Suppose that you notice the numerator and denominator both can be divided by 2. It is fine to start with that, and little by little, the fraction's numerator and denominator will "shrink."

$$\frac{28 \div 2}{42 \div 2} = \frac{14}{21}$$

Now, we continue to reduce the resulting fraction. This fraction's numerator and denominator can now each be divided by 7 with the following results:

$$\frac{14 \div 7}{21 \div 7} = \frac{2}{3}$$

Thus, the final reduced form of $\frac{28}{42}$ is $\frac{2}{3}$.

Calculator Tip: If you hit equal or SIMP once you enter a fraction on your calculator, the fraction will automatically be reduced. Be sure to review this process in your calculator's manual if necessary before the exam.

PRACTICE SET

Reduce each fraction to its lowest terms.

1. $\frac{18}{24}$ **2.** $\frac{12}{30}$ **3.** $\frac{32}{96}$ **4.** $\frac{36}{60}$

COMPARING FRACTIONS

A very useful skill associated with fractions is being able to compare fractions to one another and having an intuitive sense of the size of a given fraction.

It is helpful to realize that the further apart a numerator and denominator are, the smaller the fraction. The closer together the numerator and denominator are, the closer to 1 the fraction is. For example, a fraction such as $\frac{3}{20}$ is relatively small since 3 and 20 are further apart. The fraction $\frac{23}{25}$, on the other hand, is very close to 1 (or a whole) since 23 and 25 are close together.

Sometimes, you might need to be able to determine how close a fraction is to $\frac{1}{2}$. For instance, consider the fraction $\frac{11}{20}$. Ask yourself what half of the denominator is; in this case, What is half of 20? Since half of 20 is 10 and the numerator is 11, which is more than 10, this fraction is slightly more than $\frac{1}{2}$.

Examples

If you are asked to compare two fractions, you can use various strategies as described below.

A. *Same Denominator*: This is the easiest case of comparing fractions.

$$\frac{3}{8} \text{ and } \frac{5}{8}$$

Since the denominators are the same, we have two wholes both cut into the same number of pieces (8). Clearly, the more pieces we take, the bigger the fraction. So,

$$\frac{3}{8} < \frac{5}{8}$$

B. *Same Numerator*: Consider

$$\frac{5}{12} \text{ and } \frac{5}{18}$$

In this case, we have two wholes cut into different numbers of pieces (12 and 18) and we are taking 5 pieces from each. So, the question becomes, which pieces are bigger?

Clearly, if you divide a whole into 12 parts compared to 18 parts, the one divided into 12 parts contains bigger pieces, and so

$$\frac{5}{12} > \frac{5}{18}$$

C. *Comparing to* $\frac{1}{2}$: Sometimes it is helpful to look at fractions as they compare to $\frac{1}{2}$.

For instance, consider

$$\frac{5}{9} \text{ and } \frac{7}{15}$$

Since half of 9 is 4.5 and the numerator is 5, the fraction $\frac{5}{9}$ is slightly more than $\frac{1}{2}$.

Since half of 15 is 7.5 and the numerator is 7, the fraction $\frac{7}{15}$ is slightly less than $\frac{1}{2}$. So, since one fraction is more than $\frac{1}{2}$ and one fraction is less than $\frac{1}{2}$, the one that is less than $\frac{1}{2}$ is the smaller fraction. Thus

$$\frac{5}{9} > \frac{7}{15}$$

D. *General Case*: Now, since it is not always going to be the case that we have a special circumstance like those described above in Examples A through C, we have to have a general rule to follow to compare fractions.

To compare fractions that do not fit a special mold, we need a common denominator, which is the LCM of the two denominators. For example, compare

$$\frac{7}{10} \text{ and } \frac{5}{8}$$

Since both of these fractions are above $\frac{1}{2}$ and have neither the same numerator nor the same denominator, we must have a different method of comparing than that described in Examples A through C.

Multiples of 10: 10, 20, 30, 40, 50, 60, . . .
Multiples of 8: 8, 16, 24, 32, 40, 48, . . .

The LCM or least common denominator (LCD) of 10 and 8 is 40. So, in terms of a denominator of 40:

$$\frac{7\times 4}{10\times 4} = \frac{28}{40} \quad \frac{5\times 5}{8\times 5} = \frac{25}{40}$$

Now we can compare just the numerators since the denominators are both 40. Since 28 is larger than 25, the fraction $\frac{7}{10}$ is larger.

Practice Set

1. Which number is the fraction $\frac{15}{19}$ closest to? 0? $\frac{1}{2}$? 1?

2. Which number is the fraction $\frac{7}{13}$ closest to? 0? $\frac{1}{2}$? 1?

3. Which number is the fraction $\frac{5}{34}$ closest to? 0? $\frac{1}{2}$? 1?

4. Is $\frac{16}{25}$ more or less than $\frac{1}{2}$? How do you know?

5. Insert < or > between each pair of fractions to make a true statement.

(a) $\frac{6}{11}$ $\quad$ $\frac{8}{17}$ $\quad$ **(b)** $\frac{7}{17}$ $\quad$ $\frac{7}{10}$ $\quad$ **(c)** $\frac{9}{16}$ $\quad$ $\frac{5}{16}$ $\quad$ **(d)** $\frac{7}{12}$ $\quad$ $\frac{8}{15}$

RATIOS

A fraction can be considered a **ratio** if it is viewed in the sense of comparing two groups rather than as a part out of a whole. For example, if we count the number of males and the number of females in a group, the ratio of boys to girls can be expressed as a fraction or ratio:

$\frac{12}{17}$ means there are 12 boys and 17 girls in the group

A ratio can be written in two main other ways other than as a fraction. The above ratio can also be written as

12:17 or 12 to 17

Just like fractions, a ratio can sometimes be reduced. For example, had this ratio been 12 to 18, we could reduce it by dividing by 6 to get 2 to 3. This would mean that for every 2 males there are 3 females.

Examples

A. Write the ratio of the number of days in September to the number of days in October

30 to 31 or 30:31 or $\frac{30}{31}$

B. Write the ratio of the number of stars to the number of circles in the diagram below.

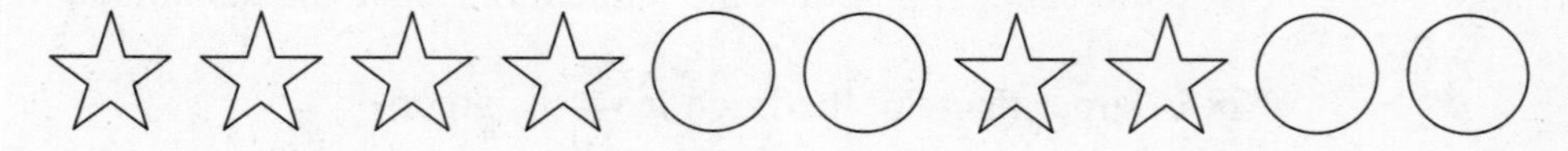

First, count the number of stars and the number of circles. Just be sure to write the ratio in the correct order. Since we want the ratio of stars to circles, we must put stars in the numerator and circles in the denominator so the ratio is $\frac{6}{4}$, which can be reduced to $\frac{3}{2}$ by dividing by 2. This means that for every 3 stars, there are 2 circles.

Take Note: A ratio should **never** be written as a mixed number.

Practice Set

1. Write the ratio of the number of odd numbers to the number of even numbers in the set below.

$$\{1,5,10,12,8,7,3,4,6,2,12,6\}$$

2. Write the ratio of squares to hearts in the figure below.

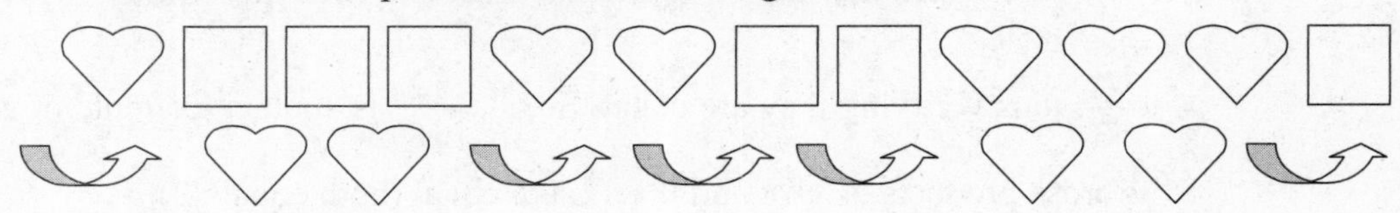

3. Using the above diagram, give the ratio of arrows to squares.

UNIT RATE

A special ratio in which the denominator is 1 is called a **unit rate**. Basically, a unit rate is the amount per *one* of something, whether it be the price per pound, the pay rate per hour, the number of words a person types per minute, and so on.

Examples

A. If apples cost $3.50 for a 5-pound bag, what is the unit rate per pound?

To find a unit rate, we divide: 3.50 ÷ 5 = .70/1 or .70 per pound.

B. If Donna can type a document containing 3250 words in 50 minutes, what is her typing speed per minute?

Again, to get a unit rate, we divide: 3250 ÷ 50 = 65/1 or 65 words per minute.

C. If a car takes 8 hours to make a trip that is 496 miles, what is its average speed per hour?

Divide: 496 ÷ 8 = 52/1 or 52 miles per hour (mph)

Practice Set

1. Find the unit rate per hour: Janet earns $630 working 35 hours per week. What is her hourly rate of pay?

2. If a bag of oranges cost $6.25 and weighs 5 pounds, what is the rate per pound?

3. Tommy bicycles 45 miles in 5 hours. What is his unit rate of speed per hour?

PROPORTIONS

Earlier in this unit, we talked about equivalent fractions. If there is an equality between two fractions or ratios, we have what is called a **proportion**. There are special properties and applications of proportions that are important for the HSPA exam.

In a proportion, the cross-products are equal. By *cross-products* we mean that we multiply diagonally (one numerator times the other denominator, and vice versa). For instance, we know the fractions $\frac{1}{2}$ and $\frac{5}{10}$ are equal or equivalent. So, the statement $\frac{1}{2}=\frac{5}{10}$ is a proportion. The cross-products are 1×10 and 2×5 and, as we said, the cross-products are equal: $1 \times 10 = 2 \times 5 = 10$.

Consider another set of equivalent fractions such as $\frac{4}{6}$ and $\frac{12}{18}$, which both reduce to $\frac{2}{3}$, thus verifying they are equal. So, $\frac{4}{6}=\frac{12}{18}$ is another example of a proportion, and its cross-products, 4×18 and 6×12, are equal (both equal 72).

Solving Proportions

The way in which you will be expected to use proportions is by solving them. By solving a proportion, we mean finding a missing numerator or denominator so that the proportion is true. To do this, we cross-multiply and solve the resulting basic equation.

Examples

A. Solve for x: $\frac{3}{4}=\frac{x}{12}$

Remember, in a proportion, cross-products are equal. So, $3 \cdot 12 = 4 \cdot x$. Since $3 \cdot 12 = 36$, we divide 36 by 4 to get the missing value of 9 for x.

B. Solve for x: $\frac{5}{3}=\frac{20}{x}$

Again, the cross-products are equal: $5 \cdot x = 3 \cdot 20$, which means $5 \cdot x = 60$. Dividing by 5, we find the missing value of x to be 12.

C. Solve for x: $\frac{3}{7}=\frac{x}{28}$

For this problem, we are going to look at an alternative solution to cross-multiplying. Recall the idea of equivalent fractions and ask yourself, What did we multiply 7 by to get 28? We multiplied by 4. Thus, we multiply the numerator (3) by the same number and $x = 12$.

To verify that $\frac{3}{7}=\frac{12}{28}$ is a proportion, we can cross-multiply: $3 \cdot 28 = 7 \cdot 12 = 84$.

PRACTICE SET

Find x in each proportion below.

1. $\frac{2}{3}=\frac{x}{12}$ **2.** $\frac{5}{8}=\frac{10}{x}$ **3.** $\frac{7}{2}=\frac{x}{6}$ **4.** $\frac{3}{4}=\frac{x}{20}$ **5.** $\frac{4}{5}=\frac{24}{x}$

Applying Proportions

A proportion can be set up to find a missing quantity if we know the ratio of one group to another and we know the number in one group.

The main area of application for proportions is in creating *scale drawings*. A common scale drawing is a map. Obviously, a map is not drawn its exact size but rather on what is called a scale. This means that each inch or each centimeter represents a certain actual distance.

When setting up a proportion, the main area of importance is keeping the ratios in the same order. That is, if the map length is in one of the numerators, it must be in the other numerator.

Examples

A. If the ratio of rabbits to guinea pigs in a pet store is 3:5 and there are 40 guinea pigs, how many rabbits are there?

When setting up a proportion, simply pay attention to the meaning of each number in the ratio and set up the second ratio in the same manner. In this problem, there are 3 rabbits for every 5 guinea pigs. So,

$$\frac{3}{5}=\frac{x}{40}$$

Notice that in the second ratio, the 40 is in the denominator because it represents the number of guinea pigs, as does 5.

Now, cross-multiplication gives $3 \cdot 40 = 5 \cdot x = 120$; so $120 \div 5$ means that $x = 24$.

B. On a map of New Jersey, 1 inch represents 16 miles. If the distance between Delran and Toms River is 4.5 inches, what is the actual distance between these two towns?

To solve a problem about a scale drawing, set up a proportion:

$$\frac{1}{16}=\frac{4.5}{x}$$

Again, notice that since the 1 represents inches, we put the 4.5 in the same "spot."

Now, cross-multiplying yields

$$x=4.5 \cdot 16=72$$

Therefore, the actual distance between Delran and Toms River is 72 miles.

C. If 2 out of 3 dentists surveyed prefer Bright Teeth toothpaste, how many dentists would prefer this brand if we surveyed 342 dentists?

Again, set up a proportion, being careful to place numbers in the corresponding location according to what they mean. In this problem, the correct proportion is

$$\frac{2}{3} = \frac{x}{342}$$

Notice that 2 is how many dentists like the toothpaste and 3 is how many dentists were asked. So, since we will survey 342 dentists, this number has to correspond with 3 and is therefore also in the denominator.

Now, we cross-multiply to get

$$2 \cdot 342 = 3 \cdot x = 684$$
$$684 \div 3 = 228$$

So, out of 342 dentists surveyed, 228 would prefer Bright Teeth toothpaste.

> **Take Note:** Each time you are given a ratio of one quantity to another, you can always re-write it. For example, if you are told a ratio of wins to losses is 5:8, you can say the team has won 5 out of 13 games.

D. The ratio of blue to red in a mixture for violet paint is 3 ounces to 7 ounces. If an artist is going to make 50 ounces of this paint, how many ounces of blue paint does he or she need?

This problem is different because it focuses on the *total* amount in one "group" of the mixture. If there are 3 ounces of blue paint for every 7 ounces of red paint, we have a group of 10 ounces of paint. Thus, in 50 ounces of paint, there are 5 groups of 10.

So, the blue and red are each multiplied by 5. Since the problem asks for the number of ounces of blue paint, we multiply 3 ounces by 5 to get 15 ounces of blue paint.

The diagram below might help.

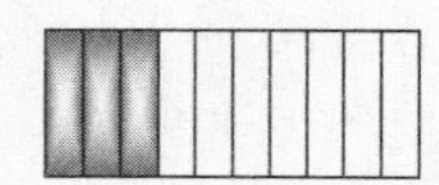

One "group": 3 ounces blue and 7 ounces red, making a total of 10 ounces.

To get 50 ounces total, we would need 5 of these groups. Thus, each part would be multiplied by 5.

PRACTICE SET

1. On a map, 2 centimeters represent 25 kilometers. If the distance between two cities is 650 kilometers, how many centimeters does this correspond to on the map?

2. If 3 out of 8 doctors recommend No More Aches aspirin, how many doctors were asked if 75 in the group recommended this brand?

3. The ratio of cashews to pecans in a trail mix is 3 pounds to 4 pounds. If we have 68 pounds of pecans, how many pounds of cashews would we need to use? What would be the weight of the entire mixture?

4. On a blueprint, 1 inch represents 5 feet. If the length of a room is 2.25 inches, how long is the room?

5. The ratio of rainy days to sunny days over a month on a tropical island is 4 to 11. If a family is planning a vacation there for 60 days, how many rainy days should they expect?

PERCENTS

Very closely related to fractions and proportions is percents. **Percent** means "out of 100" (per cent). We symbolize percent with the sign %. Thus, 40% means 40 out of 100.

Percents can also be written as fractions or decimals as well. To write a percent as a fraction, simply put the percent over 100. Then, if possible, reduce the fraction.

Examples

A. Write 35% as a fraction.

Since percent means "out of 100," we write 35% as $\frac{35}{100}$, which can be reduced by dividing by 5:

$$\frac{35 \div 5}{100 \div 5} = \frac{7}{20}$$

B. Write 84% as a fraction.

Again, percent means "out of 100," so we write 84% as $\frac{84}{100}$, which can be reduced to $\frac{21}{25}$.

Percents can also be written as decimals. To write a percent as a decimal, simply move the decimal point back (to the left) two places. This is the same as dividing by 100.

Examples

A. Write 28% as a decimal.

28.0% (remember, the decimal point is to the right of the whole number) = .28.

B. Write 62.5% as a decimal.

Move the decimal two places to the left to get .625.

C. Write .03% as a decimal.

Again, move the decimal two places to the left: .0003.

You will also be asked on occasion to go in the opposite direction; that is, you will be expected to be able to change a fraction or a decimal to a percent.

Let's look at changing a decimal to a percent first since it's the easier case. To change a decimal to a percent, simply move the decimal point two places to the right.

Examples

A. Change .83 to a percent.

Move the decimal point two places to the right to get 83%.

B. Change .914 to a percent.

Again, move the decimal point two places to the right: 91.4%.

C. Change .6 to a percent.

Move two places to the right, adding a zero: $.6 = 60\%$.

If you are asked to change a fraction to a percent, two basic cases emerge as described below.

Examples

Case I—"100-Friendly" Denominator: Since percent means "out of 100," the fraction corresponding to a percent has a denominator of 100. So, if you are fortunate enough to have a denominator that can easily be changed into 100, your task is a little bit easier.

A. Change $\frac{9}{20}$ to a percent.

Ask yourself: $\frac{9}{20} = \frac{?}{100}$

To get from 20 to 100, we multiply by 5, so we will do the same thing to the numerator (9) to get 45. Since it's $\frac{45}{100}$ or 45 out of 100, it is 45%.

B. Change $\frac{13}{25}$ to a percent.

Again, attempt to solve: $\frac{13}{25} = \frac{?}{100}$

To get from 25 to 100, we multiply by 4; so $13 \cdot 4 = 52$ and $\frac{52}{100} = 52\%$.

Case II—"Not-100-Friendly" Denominator: Not all fractions have a denominator that can easily be changed to 100. For example, consider the following.

A. $\frac{3}{8}$ represents what percent?

When you attempt to set up equivalent fractions here, you get

$$\frac{3}{8}=\frac{?}{100}$$

It is, unfortunately, not as easy a task to find what the missing number is because 8 has to be multiplied by a decimal answer to equal 100. Thus, we solve this problem by cross-multiplying:

$$8 \cdot x = 3 \cdot 100 = 300$$

So, to find x we divide: $300 \div 8$. We get 37.5, and since it's over 100, this is the percent (37.5%).

B. Change $\frac{2}{3}$ to a percent.

Again, the proportion is not easy to solve; so we need a calculator.

$$\frac{2}{3}=\frac{x}{100}$$

When we cross-multiply, we get $3x = 200$; so dividing, $200 \div 3$, gives us $x = 66.\overline{6}\%$, which is also $66\frac{2}{3}\%$.

C. Change $\frac{6}{7}$ to a percent.

For this problem, we present an alternative solution to change a fraction to a percent. Instead of setting up a proportion, you can simply divide the numerator by the denominator and then change the decimal to a percent by moving the decimal point two places to the right.

In this problem, we can use a calculator to divide: $6 \div 7 = .857142\ldots$ Moving the decimal point two places to the right and rounding to the nearest tenth, we find that this fraction is equal to 85.7%.

Sometimes you will be asked to find *all* equivalent forms (decimal, fraction, percent) for a given quantity.

Example

Write .326 as a fraction and as a percent.

When asked to write a decimal as a fraction, one of the easiest ways to do so is to concentrate on how to properly read the given decimal using the place value. For instance, here we do not say "point 3, 2, 6." Rather, we say "326 thousandths."

Then, as a fraction, .326, 326 thousandths, is just $\frac{326}{1000}$, which can be reduced by 2 to give $\frac{163}{500}$.

To find the percent, simply move the decimal two places to the right to get 32.6%.

PRACTICE SET

Fill in all missing boxes in the chart below.

Fraction	Decimal	Percent
$\frac{13}{20}$		
	.86	
		24%
$\frac{5}{8}$		
	.325	
		78%

Applying Percents

It is important to be able to calculate problems involving percents. There are two main methods of solving a percent sentence: proportion and equation.

When setting up a proportion, use the following structure:

$$\frac{Part}{Whole} = \frac{Percent}{100}$$

When setting up an equation, use the decimal equivalence for the percent and write an algebraic sentence.

Examples

A. What number is 40% of 75?

Setup 1: Proportion

$$\frac{x}{75} = \frac{40}{100}$$

Reducing the fraction gives

$$\frac{x}{75} = \frac{2}{5}$$

Cross-multiplication gives

$$5x = 150$$

Divide:

$$150 \div 5 = 30$$

Therefore, 40% of 75 is 30.

Setup 2: Equation

What number is 40% of 75? That is, what number (x) is (=) .40 (the decimal equivalent of 40%) of (times) 75?

$$x = .40 \cdot 75$$
$$= 30$$

B. 52 is what percent of 80?

Setup 1: Proportion

$$\frac{52}{80} = \frac{x}{100}$$

Cross-multiplying gives

$$80x = 5200$$

Divide:

$$5200 \div 80 = 65$$

Therefore, 52 is 65% of 80.

Setup 2: Equation

52 is what percent of 80? That is, 52 is (=) what percent (x%) of (times) 80?

$$52 = x\% \cdot 80$$

Solve by dividing:

$$52 \div 80 = .65$$
$$.65 = 65\%$$

C. 154.7 is 91% of what number?

Setup 1: Proportion

$$\frac{154.7}{x} = \frac{91}{100}$$

Cross-multiplying gives

$$91x = 15{,}470$$

Divide:

$$15{,}470 \div 91$$
$$x = 170$$

So, 154.7 is 91% of 170.

Setup 2: Equation

154.7 is 91% of what number? That is, 154.7 is (=) 91% (.91 as a decimal) of (times) what number (x).

$$154.7 = .91x$$

Divide:

$$154.7 \div .91$$
$$x = 170$$

PRACTICE SET

1. What number is 75% of 92?
2. 10.5 is 15% of what number?
3. Find 45% of 60.
4. 36 is what percent of 180?
5. 112 is 64% of what number?
6. 98 is what percent of 140?

PERCENT OF INCREASE OR DECREASE, TAXES, DISCOUNTS

More important than just finding percents or numbers associated with percents is applying percentage to real-life situations such as taxes and discounts, which represent percents of increase and decrease, respectively.

If a quantity changes, to measure the percent of the change, follow the formula

$$\frac{\text{Change}}{\text{Original}} = \frac{\%}{100}$$

Examples

A. The price of an item goes from \$85 to \$68. What is the percent of decrease?

To find the change, subtract: 85 – 68 = 17. Thus, the price change is \$17 and the price was originally \$85. So

$$\frac{17}{85} = \frac{\%}{100}$$

Cross-multiply:

$$85x = 1700$$

Divide to solve:

$$1700 \div 85 = 20$$

Thus, the decrease is 20%.

B. A person's hourly wage goes from \$10.50 to \$13.65. By what percent did their wage increase?

The change is $13.65 - 10.50 = 3.15$. Following the formula,

$$\frac{3.15}{10.50} = \frac{\%}{100}$$

Cross-multiply:

$$10.50x = 315$$

Divide:

$$315 \div 10.50 = .3$$

$$.3 = 30\%$$

So, the wage went up 30%.

Sometimes, instead of being given the two quantities and being asked to measure the percent of change, you are given information about the percent of change (tax or discount) and asked to find the final quantity.

A tax is added to the price of various items that are subject to a tax rate. The rate of tax varies from area to area.
Tips are added to the total bill.
Discounts are subtracted from a price.

Examples

A. Mark purchases a stereo system for \$280. If he has to pay a 6% sales tax, what is the final price of the stereo?

We must find 6% of \$280. So, we multiply $.06 \times 280$ to get 16.8. Thus, the tax is \$16.80, which is added to the price. The total price is $280 + 16.80 = \$296.80$.

Alternative Solution: Sometimes on the HSPA exam, you may be asked to demonstrate knowledge of more than one way to solve a problem. In this case, since Mark has to pay 6% tax above and beyond the price of the stereo system, he is really paying 100% (the whole price) plus 6% of the price, for a total of 106%. So, in one step, we could simply find 106% of \$280 by multiplying 1.06 (106% as a decimal) by 280 to get \$296.80.

Take Note: Pay close attention to the alternative solution here. Remember, a discount always leaves a percent to be paid. For example, a 15% discount leaves 85% to pay. Tax gets added to the price so a 5% tax would mean the price you pay is 105% of the original price.

B. A customary tip is 15%. If Tom's dinner bill is \$45, approximately how much should he leave for a tip?

To find 15% of 45 we multiply $.15 \times 45$ to get 6.75. So, Tom should leave a tip of about \$6.75.

C. Shopper's World is having a 30%-off sale. If Donna buys a patio set originally priced at \$325, what is the sale price?

We find 30% of 325 by multiplying $.30 \times 325$, and the result is 97.5. Thus, this amount is subtracted from the price, and the sale price is $325 - 97.50 = \$227.50$.

Alternative Solution: Once again, we can solve this problem with a different approach if need be or if we find it easier. Since Donna is getting the patio set for 30% off, that means the store is taking the whole price (100%) and deducting 30%, which leaves Donna with 70% to pay. This is very much like grades that students earn—the percent they get incorrect is subtracted from 100% to leave a percentage. So, we can multiply $\$325 \times .70$ (70% as a decimal) to get \$227.50.

D. A house was reduced by \$4500, which represented a 3% drop in the asking price. What was the original selling price?

Always remember to concentrate on what represents a part and what represents the whole. In this case, \$4500 is the part (the amount by which the price decreased). So, the setup is as follows.

$$\frac{4500}{x} = \frac{3}{100}$$

Cross-multiplication yields

$$3x = 450{,}000$$

Divide:

$$450{,}000 \div 3 = 150{,}000$$

So, the whole (or the original price) was \$150,000.

E. In a survey, 240 people preferred raspberry to orange. This number represented 40%. How many people liked orange?

Again, concentrate on the part and the whole. 240 is the part that liked raspberry, which is 40% of the whole. The setup is

$$\frac{240}{x} = \frac{40}{100}$$

Reducing the fraction gives

$$\frac{240}{x} = \frac{2}{5}$$

Cross-multiplying yields

$$2x = 1200$$

Divide to solve:

$$1200 \div 2 = 600$$

So, 600 represents the whole, and the other part is 600 – 240 = 360. Thus, 360 people prefer orange.

PRACTICE SET

1. If Pete buys a vacuum cleaner for $149 and has to pay 7% tax, what is the final price?

2. Furniture Unlimited is having a weekend sale; everything is 35% off. If a dining room set is normally priced at $3400, how much is it during the sale?

3. The results of a magazine survey show that 432 people prefer autumn to summer. If 64% prefer autumn, how many prefer summer?

4. Mr. Jones' sales went up by 1400, which represents a 40% increase. How much did Mr. Jones sell previously?

5. If an item costing $80 is on sale for 20% off for 2 weeks and then the price is decreased an extra 10% during a bonus sale, how much does it cost?

MIXED PRACTICE, CLUSTER I, MACRO B

1. On a map, $\frac{1}{4}$ inch represents 50 miles. How far apart are two towns that are $1\frac{1}{2}$ inches apart on this map?

2. In a survey, it was noted that 75% of the participants prefer country music to jazz. If 450 like country music, how many like jazz?

3. A group of waiters and waitresses divide their tips in a ratio of 2:4:6. If their tips are $216, how much is the largest share?

4. If the annual inflation rate is 4% and a car currently costs $18,000, approximately how much will the same car cost in 5 years?

   A. $19,480 **B.** $21,600 **C.** $21,900 **D.** $25,000

5. If 7 out of 12 pieces manufactured by a factory are satisfactory, *about* how many out of 450 are satisfactory?

   A. 27 **B.** 60 **C.** 265 **D.** 840

6. Which symbol inserted in the blank will make a true statement:

$$12\frac{3}{8} \_\_\_\_\_ \ 12\frac{13}{34}$$

   A. > **B.** < **C.** ≥ **D.** =

7. A piece of cord is cut into two pieces in a ratio of 2 : 3. If the string is 40 inches long, how long is the shorter piece?

A. 8 inches **B.** 16 inches **C.** 24 inches **D.** 5 inches

8. A Spin-the-Wheel game has 16 sections. Ten of these sections are gray or black, and the rest are white. What percent are white?

A. 60% **B.** 62.5% **C.** 37.5% **D.** 20%

9. If an electrician has to drill a hole just slightly larger than $\frac{9}{16}$ inch in diameter, which of the following is the smallest but still larger than $\frac{9}{16}$ inches?

A. $\frac{1}{2}''$ **B.** $\frac{5}{8}''$ **C.** $\frac{7}{16}''$ **D.** $\frac{3}{4}''$

10. At the end of the summer season, a grill is discounted $33\frac{1}{3}\%$ and costs $108.66 with the markdown. What was the original price of the grill?

A. $36.22 **B.** $162.99 **C.** $72.44 **D.** $144.88

11. A supersonic transporter goes 342 miles in 36 minutes. At this speed, how many miles will it travel in 2 hours and 12 minutes?

12. Mr. Dennis sells computer systems. He earns a 6.5% commission on the base prices of the systems and an extra 10% on upgrades. Find his total commission for November based on the chart of information below.

Date	System Price ($)	Upgrade ($)
11/10	1199.99	399.99
11/16	999.99	199.99
11/22	1899.99	599.99
11/30	1499.99	299.99

13. An advertisement in the paper shows a leftover car model for $17,500. If the manufacturer's suggested retail price is $20,795, what is the approximate percent of discount?

A. 11.4% **B.** 12.7% **C.** 15.8% **D.** 18.8%

Take Note: When reading a question, watch for words like "not." Many times the HSPA will not draw attention to this wording and questions can be tricky.

14. Which of these improper fractions does *not* meet *all* three characteristics listed below.

I. It is less than 3.
II. It represents a repeating decimal.
III. It has a 6 in the hundredths place.

A. $\frac{254}{99}$ **B.** $\frac{25}{11}$ **C.** $\frac{284}{99}$ **D.** $\frac{125}{99}$

15. The Environmental Club at Whoville High School planted 252 tree saplings this year for Earth Day, which was 210% of the number they planted last year. How many did they plant last year?

A. 102 **B.** 108 **C.** 114 **D.** 120

16. Pedro can type 6 pages in 35 minutes. If he starts typing a 30-page manuscript at 9:00 A.M., when will he be finished? (Assume he takes no breaks.)

17. Sneakers-R-Us has the newest New Balance sneakers on sale for 30% off the original price of $120. Sneaks Unlimited has the same shoe for 35% off the original price of $135. Which is the better buy? Or is the price at the two stores the same? How much, if anything, do you save?

18. A home originally cost $95,000. If the owner wants to make a 55% profit, what price should he ask for the house?

A. $133,250 **B.** $139,250 **C.** $143,250 **D.** $147,250

19. Which is the best buy for liquid soap? *Note*: 1 quart = 32 ounces.

12 ounces	16 ounces	24 ounces	1 quart
$2.25	$2.95	$4.25	$5.75

A. 12 oz **B.** 16 oz **C.** 24 oz **D.** 1 quart

20. The number of juniors at Anytown High School who took the PSATs this year is 25% higher than the number who took them last year. If 268 took the exam this year, how many took it last year?

21. If a fraction's numerator is multiplied by 4 and its denominator is divided by 3, how many times larger does the whole fraction become?

22. A child's record player plays special records called "40 rpm," meaning the record rotates 40 times per minute. How long does it take one of these records to rotate once?

23. Which of the following changes reflects the greatest percent of increase?

A. 18 to 20 **B.** 16 to 18 **C.** 95 to 100 **D.** 600 to 650

Take Note: Successive discounts are often posed as trick questions. In #24, a 20% discount followed by a 10% discount is **not** the same as a single 30% discount.

24. A store has a $240 television on sale. The store offers two successive discounts of 20% and then 10%. How much does the television end up costing?

A. $210 **B.** $168 **C.** $172.80 **D.** $174.20

25. If a house is appraised at $160,000 and this price represents a 25% increase in value, what was it originally worth?

A. $120,000 **B.** $124,000 **C.** $128,000 **D.** $132,000

26. A group of co-workers go out for dinner. Their dinners cost $16.60, $10.75, $20.25, and $15.45. If there is a 6% tax and they leave a 15% tip (before tax), what is the total bill?

A. $73.92 **B.** $75.29 **C.** $76.29 **D.** $76.86

27. A factory produces 12,000 widgets per day. The inspectors find that 3% are defective. If there's a .27 profit on every good widget, how much is earned on an average day?

A. $314.28 **B.** $3142.80 **C.** $3240 **D.** $3540

28. If Linda buys a car for $3400 plus 5% tax in an urban enterprise zone, how much does she save if the standard tax rate is 6%?

A. $34 **B.** $68 **C.** $204 **D.** $408

29. Tyshawn works part-time after school. His gross salary per week is $135.40. Federal tax is 15%, state tax is 3%, and Social Security (FICA) is 7.7%. What is Tyshawn's net pay per week (after these deductions) to the nearest penny?

A. $104.10 **B.** $102.74 **C.** $101.56 **D.** $100.60

30. During a business trip, Jimmy paid a .50 sales tax on a $14.25 dinner in one state and a .45 sales tax on an $8.95 lunch in another state. What is the percent difference in tax rates between these two states?

A. 1% **B.** 1.2% **C.** 1.5% **D.** 1.8%

CLUSTER II

SPATIAL SENSE AND GEOMETRY

MACRO A

BASIC GEOMETRIC TERMS

The basic geometric component is a **point**. A point is a precise location in space. A point is very small—so small that it cannot be seen with a naked eye. However, we exaggerate the size of a point so that we can see it. A point is labeled using a "dot" marked by a capital letter and we call it point A or point C, depending on the letter chosen.

If we connect two points with a straightedge, we get a line, a line segment, or a ray. There are differences among these that are important to note and understand.

A **line** goes through the two points that name it but does not stop. We indicate this idea of not stopping by putting arrows on each end of a line.

A line is named by using two points on it. In this case, the line above is named $\overleftrightarrow{BC}$ or $\overleftrightarrow{CB}$.

Note: There are infinitely many points on any given straight line. We typically label only two or perhaps three points and then use any combination of two of these points on the line in any order to name the line.

A **line segment** is similar to a line except that it stops at both ends. It is completely contained between two points. A line segment is labeled using its two end points.

Again, like a line, a line segment contains many, many points in between, but we need to use only two to label it; use the end points determining the line segment. Thus, the above line segment is named $\overline{AB}$ or $\overline{BA}$.

A **ray** is somewhat of a combination of a line and a line segment. A ray has one endpoint and continues infinitely in the other direction through a second point. A ray must be named using its endpoint first and then using any other point that it goes through.

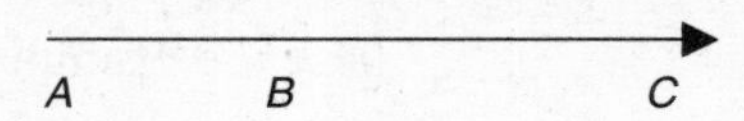

This ray can be named either $\overrightarrow{AB}$ or $\overrightarrow{AC}$. Note that ray BC would be different, as it would only start from B and then continue through C rather than starting at A.

If we join two rays at a common end point called a **vertex**, we get an **angle**.

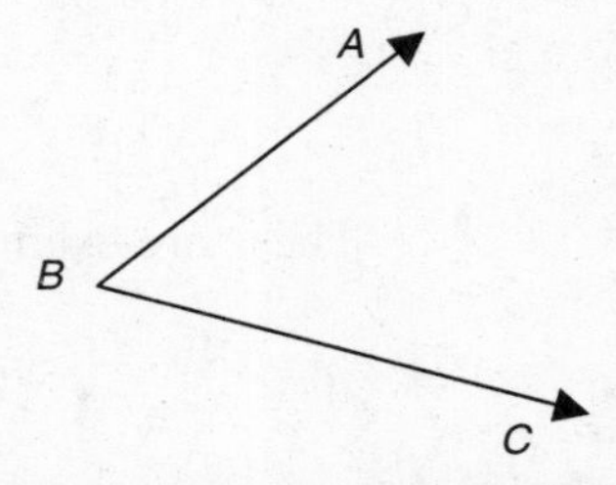

An angle is named by using three points. We use a symbol that resembles a tiny angle followed by three letters. The letters must be in an order that follows how they would be traced out to make the angle. The vertex (or corner), the common end point for the rays, must be in the center when naming the angle. The angle on page 59 could be named either $\angle ABC$ or $\angle CBA$. If it's very clear, it is possible to simply name an angle by its vertex. Thus, the angle on page 59 could just be named $\angle B$.

However, in a diagram such as the one below, you cannot get away with naming the angle using only the vertex point since there is more than one angle with B as a vertex.

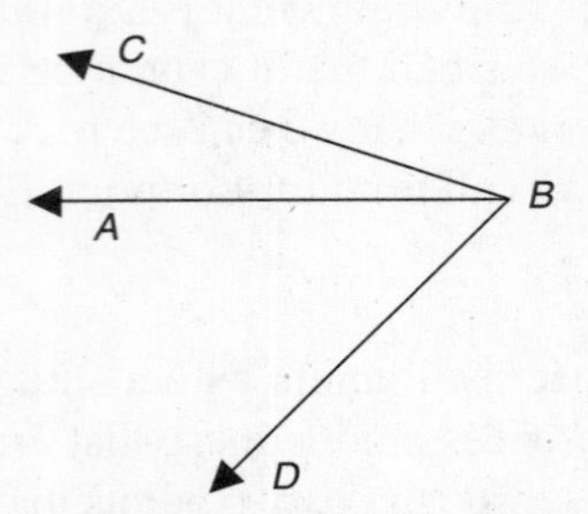

Angles are classified by size. A **right angle** is an angle measuring exactly 90°, making a perfect corner. Small angles (less than 90°) are called **acute angles**. Larger angles (more than 90°) are called **obtuse angles**. An angle that measures exactly 180° is called a **straight angle**, and makes a straight line.

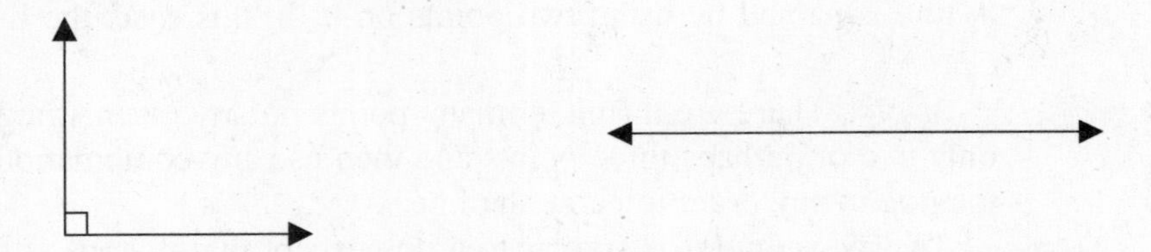

This is an example of a right angle. This is an example of a straight angle.

Note: The tiny square symbol in the diagram of the right angle is a standard way to indicate a right angle in a drawing.

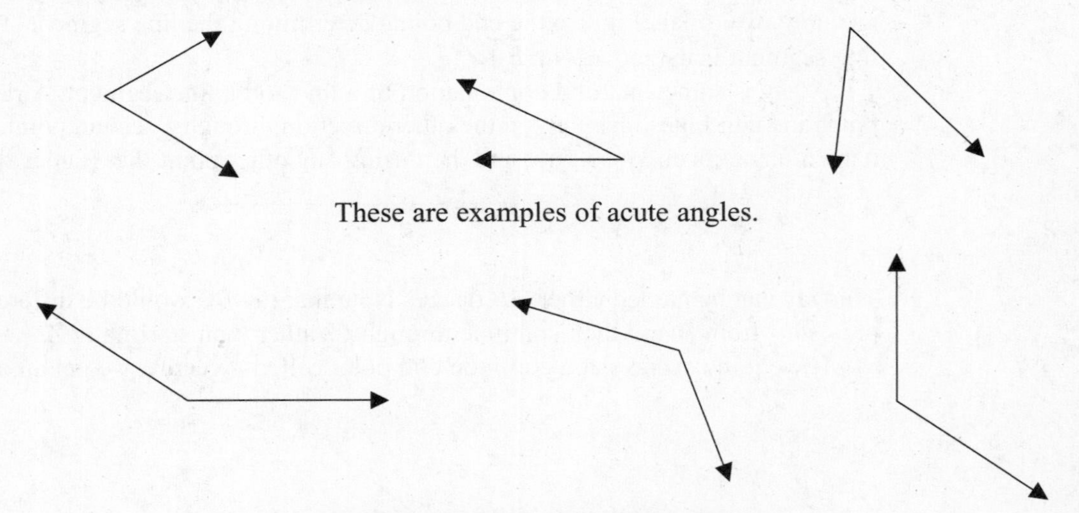

These are examples of acute angles.

These are examples of obtuse angles.

PRACTICE SET

Identify each angle as acute, right, obtuse, or straight.

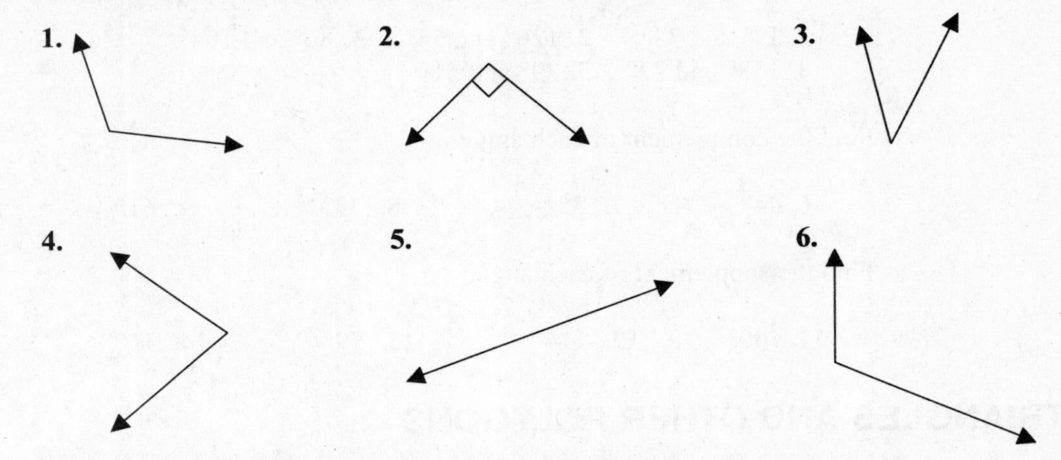

Two angles are **complements** or **complementary** if they add up to 90°.
Two angles are **supplements** or **supplementary** if they add up to 180°.

> **Take Note:** To remember the difference between complements and supplements, you may find it helpful to associate the word "complement" with "corner" and "supplement" with "straight."

Examples

A. 30° and 60° are complements because 30° + 60° = 90°.

B. 110° and 70° are supplements because 110° + 70° = 180°.

C. 46° and 44° are complements since 46° + 44° = 90°.

D. 137° and 43° are supplements since 137° + 43° = 180°.

E. What is the complement of 68°?

Complements add to 90°, so we have to subtract 68° from 90° to get the complement, which is 22°.

F. What is the supplement of 125°?

Supplements add to 180°, so we have to subtract 125° from 180° to get the supplement of 125°, which is 55°.

Practice Set

State whether each pair of angles is complementary, supplementary, or neither.

1. 17° and 73° **2.** 126° and 54° **3.** 86° and 104°
4. 158° and 22° **5.** 39° and 51°

Find the complement of each angle:

6. 44° **7.** 85° **8.** 38° **9.** 61° **10.** 27°

Find the supplement of each angle:

11. 100° **12.** 114° **13.** 79° **14.** 157° **15.** 32°

TRIANGLES AND OTHER POLYGONS

If you connect three or more points with a straightedge, you get a polygon (closed shape). Polygons are classified or named by the number of sides or vertices (corners) they have. A polygon with three sides is called a **triangle**. If a polygon has four sides, it is called a **quadrilateral**. A **pentagon** is a polygon with five sides. A **hexagon** has six sides. An **octagon** has eight sides.

> **Take Note:** This page contains a lot of important vocabulary to remember for the HSPA.

Triangles can be classified by the lengths of their sides and/or the measures of their angles. *Every* triangle has an angle sum of 180; this means that the three angles of any triangle total up to 180°.

If one angle of a triangle is 90°, or a right angle, the triangle is called a **right triangle**.

If one angle of a triangle is greater than 90°, or an obtuse angle, the triangle is called an **obtuse triangle**.

If all the angles of a triangle are less than 90°, or acute angles, the triangle is called an **acute triangle**.

To find the missing angle(s) of a triangle,

1. Add all the given angles.
2. Subtract this number from 180°.
3. Divide by 2 or 3 (if necessary—that is, if two or three angles are all equal to one another).

Examples

Classify these triangles as right, acute, or obtuse:

A.

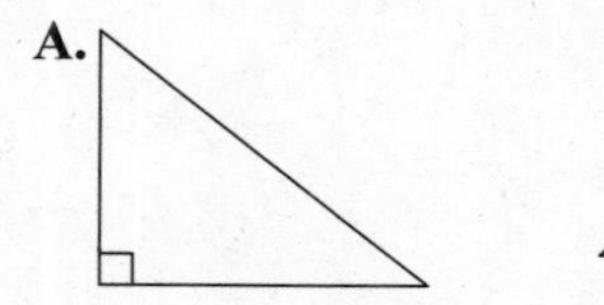

B.

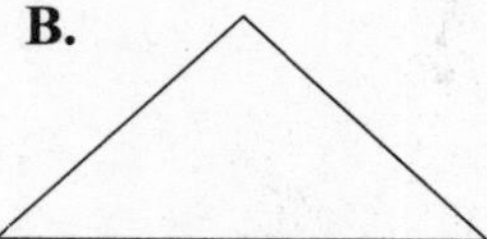

C.

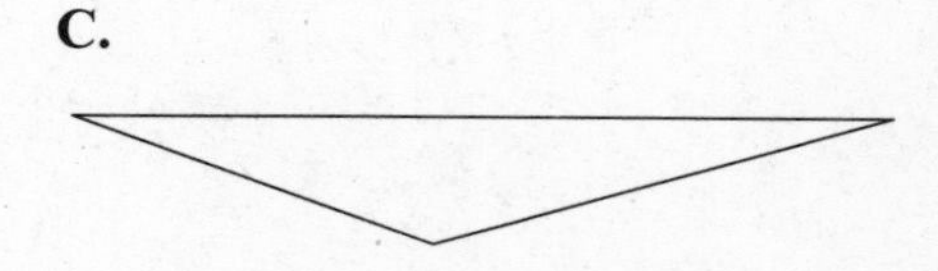

A. Since there is a little square symbol in the bottom left corner, we have a right angle and therefore a right triangle.

B. All the angles are acute; so this is an acute triangle.

C. The bottom angle is obtuse; so this is an obtuse triangle.

Find the missing angle(s) in the triangles below:

D.

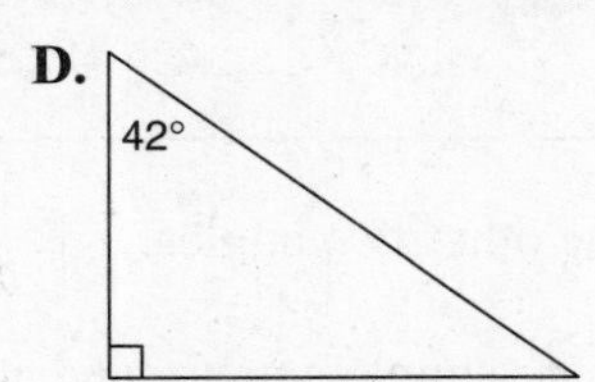

E.

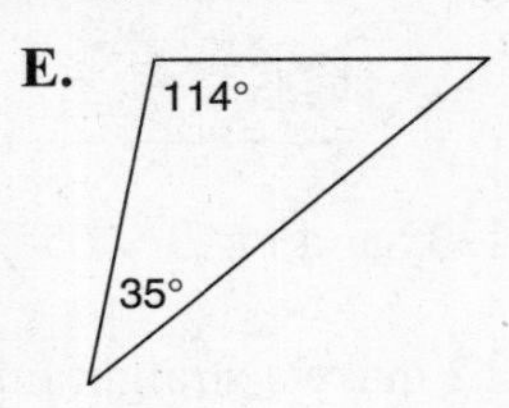

F.

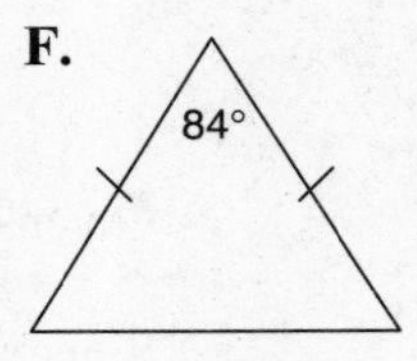

D. Add: $90° + 42° = 132°$. Subtract: $180° - 132° = 48°$, which is the missing angle.

E. Add: $35° + 114° = 149°$. Subtract: $180° - 149° = 31°$, which is the missing angle.

F. The marks on the two sides (legs) of this isosceles triangle indicate that the two sides are equal. This means the two base angles are also equal. So, $180° - 84° = 96°$. Divide: $96° \div 2 = 48°$, which are the missing angles.

PRACTICE SET

Tell whether each triangle below is acute, right, or obtuse:

1.

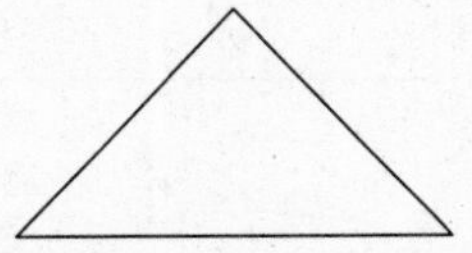

2.

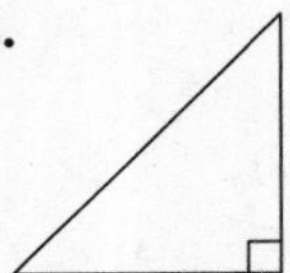

3.

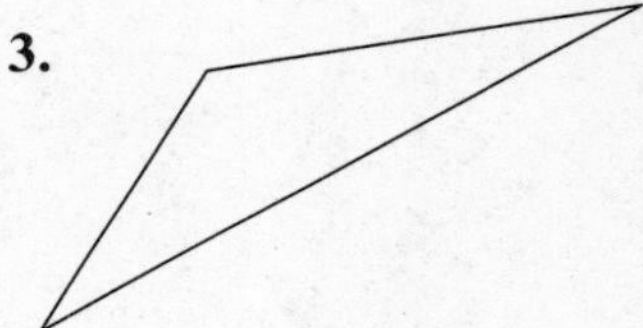

Find the missing angle(s) in each triangle below:

4.

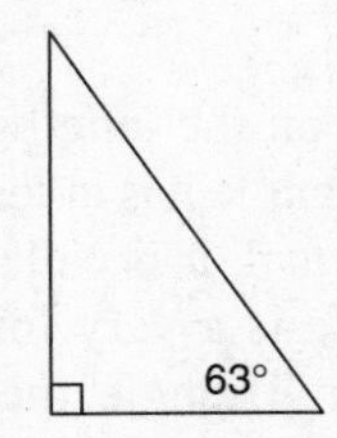

5.

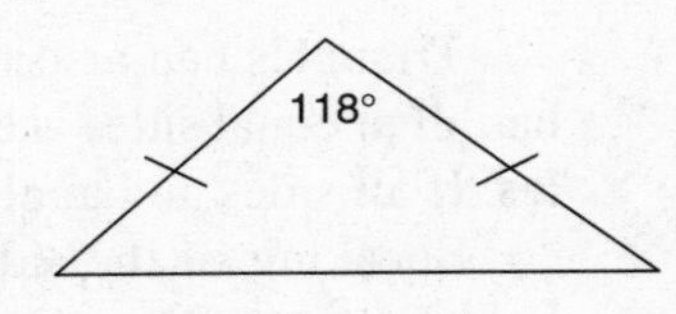

Example

Sometimes you need to combine the calculation of a supplementary angle with that of finding missing angles in a triangle. Consider the following diagram.

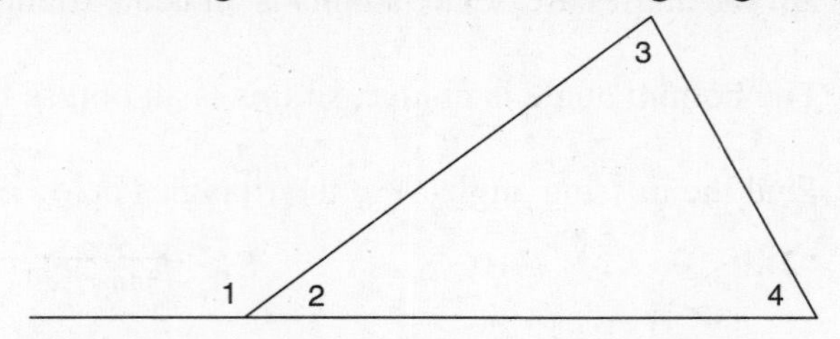

Given m∠1 = 133° and m∠3 = 65°, find the other two angles.

Since ∠1 and ∠2 make a straight line, they are supplementary and add to 180°. So, to get ∠2, we subtract: 180° – 133° = 47°. Then, ∠2, ∠3, and ∠4 should all add to 180° since they form the triangle. So, add: 47° + 65° = 112° and subtract: 180° – 112° = 68°, which is m∠4.

PRACTICE SET

Find the remaining angles in each diagram below.

1.

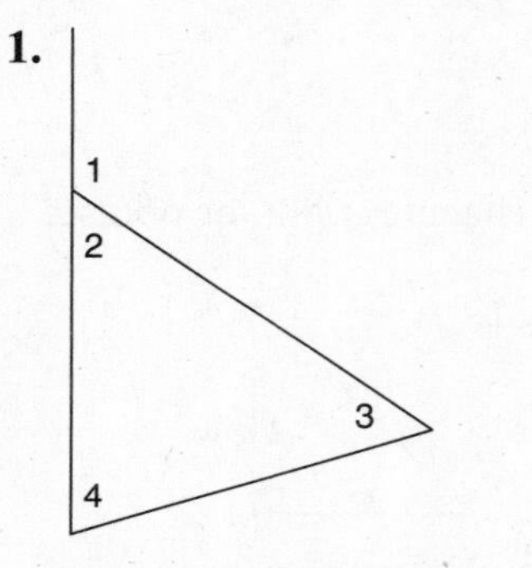

Given: m∠1 = 108°
m∠4 = 66°

2.

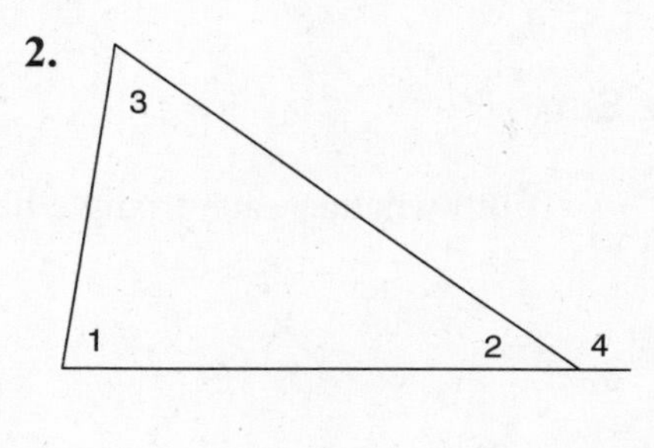

Given: m∠4 = 141°
m∠3 = 43°

Triangles can also be categorized based on the lengths of their sides. A triangle that has all unequal sides is called **scalene**. If a triangle has at least two equal sides, it is **isosceles**. If all sides and angles of a triangle are equal, it is called **equilateral** or **equiangular**.

Speaking of the sides of a triangle, there is a very important relationship that must hold between the sides of a triangle, as the Triangle Inequality. In every triangle, the lengths of the two shortest sides must add up to more than the length of the longest side. For the lengths 5, 10, and 12: 5 + 10 > 12.

PRACTICE SET

Verify that each set of lengths can form a triangle. Then, state if the triangle is scalene, isosceles, or equilateral.

1. 3″, 7″, 7″
2. 5 cm, 8 cm, 10 cm
3. 8″, 8″, 8″
4. 3′, 4′, 5′
5. 6 cm, 11 cm, 6 cm

ANGLE SUM OF OTHER POLYGONS

Since we know that the angle sum of any triangle is 180°, we can use this to find the angle sum of any polygon by drawing diagonals in the polygon *from one vertex* and creating triangles, each of which adds to 180°.

Examples

A. Find the angle sum of a quadrilateral by dividing it into triangles using diagonals from one vertex.

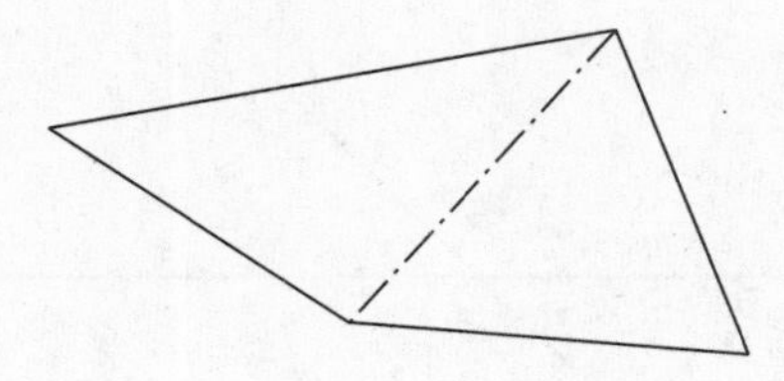

From any vertex of the quadrilateral, we can draw only one diagonal. Thus, we create two triangles, each 180°; so $2 \cdot 180° = 360°$ (the angle sum of any quadrilateral).

B. Find the angle sum of any pentagon using diagonals.

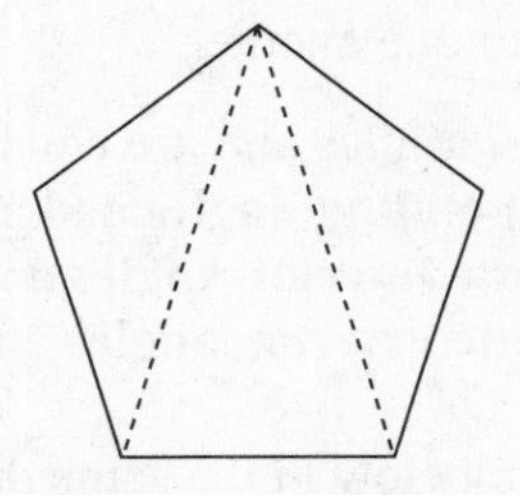

From any vertex of the pentagon, we can draw two diagonals. Thus, we create three triangles, each 180°; so $3 \cdot 180° = 540°$ (the angle sum of any pentagon).

If you continue with the strategy of dividing a figure into triangles using diagonals, you will realize that there is a formula to follow to find the angle sum of any "n-sided" figure:

Angle sum = $180° \ (n - 2)$, where n is the number of sides the polygon has.

PRACTICE SET

1. Find the angle sum of any hexagon.

2. Find the angle sum of any polygon with ten sides.

3. Find the missing angle of a quadrilateral with angles 60°, 140°, and 150°.

4. Find the missing angle of a pentagon with angles 80°, 95°, 95°, and 165°.

5. Find each angle measure of a regular octagon.

PARALLEL LINES

Parallel lines are two lines that never intersect. The symbol for parallel is ||. Or, we sometimes put arrows on lines to indicate that they are parallel:

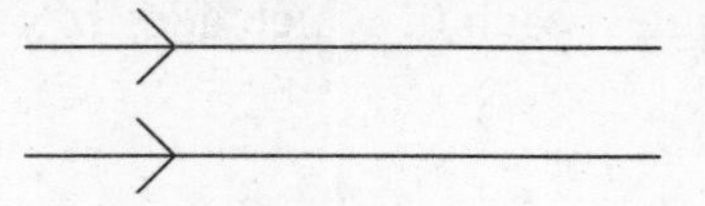

Parallel lines are often cut by what is called a *transversal*, which is simply a line going through the two parallel lines to create some angles. For example, this drawing shows two parallel lines cut by a transversal:

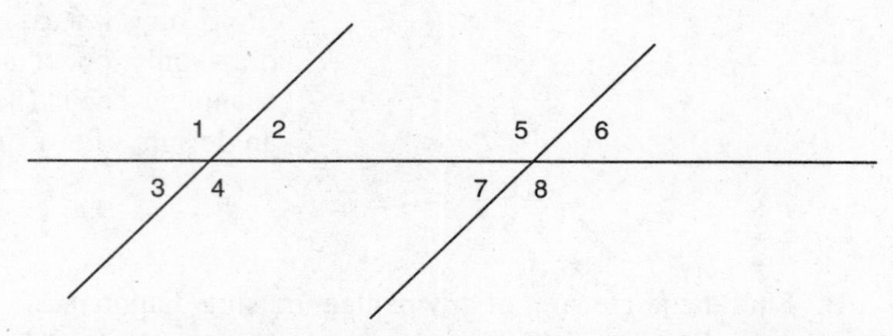

Eight angles are formed when a transversal crosses two parallel lines. Many of these angles have special names and are equal to one another.

$\angle 1$ and $\angle 4$ are an example of **vertical angles** and are equal in measure.
$\angle 2$ and $\angle 6$ are an example of **corresponding angles** and are equal in measure.
$\angle 2$ and $\angle 7$ are an example of **alternate interior angles** and are equal in measure.
$\angle 1$ and $\angle 8$ are an example of **alternate exterior angles** and are equal in measure.

If it is given that $m\angle 1 = 135$ as shown below in the same diagram, we can find every other angle.

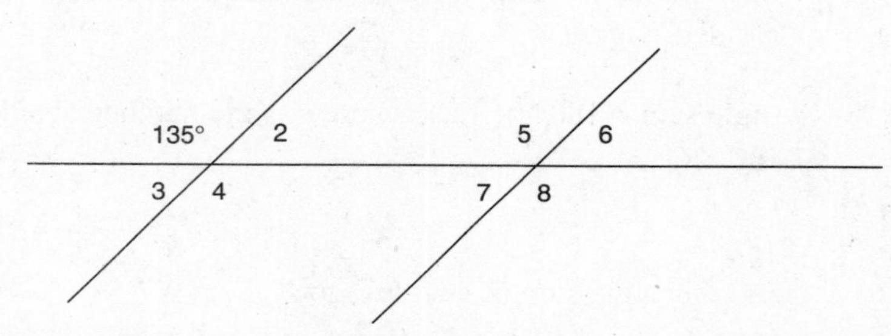

$\angle 4$ is also 135° (since it is a vertical angle with $\angle 1$). Then, $\angle 5$ is also 135° (since it is an alternate interior angle with $\angle 4$). Then, $\angle 8 = 135°$ (either because it is a vertical angle with $\angle 5$ or because it corresponds with $\angle 4$).

Using supplements, $\angle 2 = 180° - 135° = 45°$. Since $\angle 3$ is its vertical angle, it is also 45°. Then, $\angle 7$ is an alternate interior angle with $\angle 2$; so it is also 45°. And, finally $\angle 6 = 45°$ because it is a vertical angle with $\angle 7$.

Practice Set

Given m∠6 = 73°, find the remaining angles.

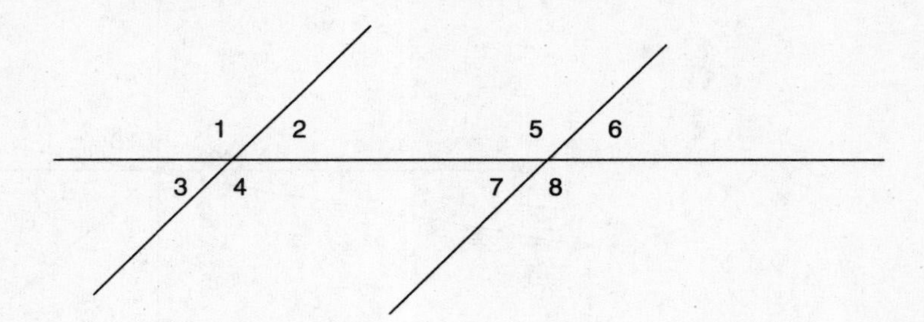

PERPENDICULAR LINES

Perpendicular lines are lines that meet at a right angle (90°). The symbol for perpendicular is ⊥.

Examples

A. Given $\overleftrightarrow{AB} \perp \overleftrightarrow{CD}$ and m∠1 = 62°, find the missing angles.

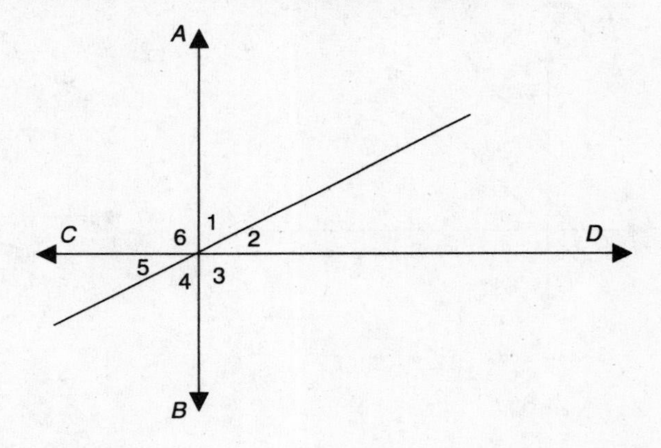

∠4 is vertical to ∠1, so it is also equal to 62°. Since the lines are perpendicular, the four main angles are each 90° and thus ∠3 and ∠6 are each 90°. The sum of ∠1 and ∠2 is 90°, and so is the sum of ∠4 and ∠5. So, 90° – 62° = 28°, which means that ∠2 and ∠5 are each 28°.

B. Given $m \perp n$ and the indicated angles, find all the other angle measures.

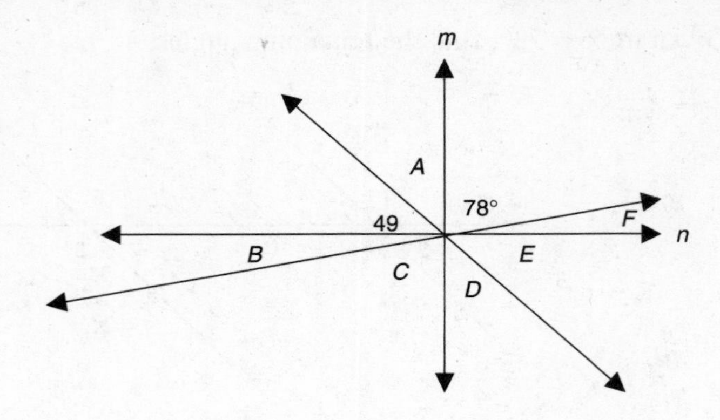

By vertical angles, $m\angle C = 78°$ and $m\angle E = 49°$. By complementary angles, $m\angle A = 41°$ and $m\angle F = 12°$. Then, by vertical angles, $m\angle D = 41°$ and $m\angle B = 12°$.

Practice Set

Given that $x \perp y$ and the angle measures indicated in each of the diagrams below, find the missing angles.

1. **2.**

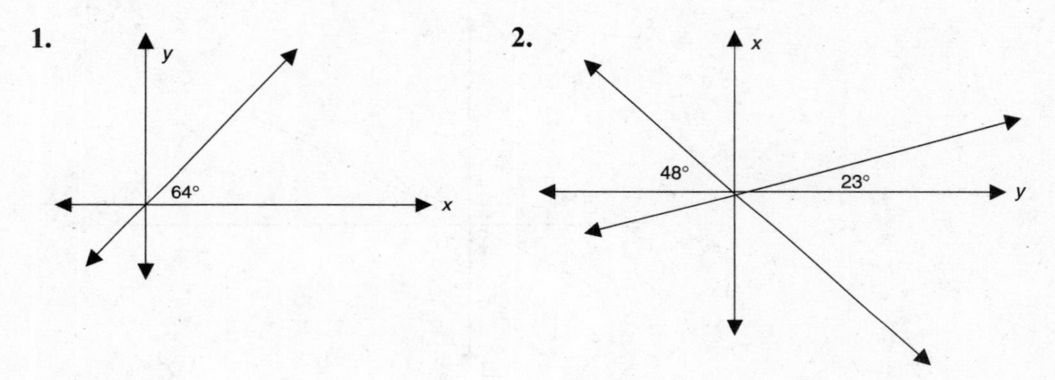

QUADRILATERALS

A polygon with four sides is known as a **quadrilateral**. Quadrilaterals can be grouped into various special categories based on certain properties. A quadrilateral with one pair of parallel sides is called a **trapezoid**. If a quadrilateral has two pairs of parallel sides, it is a **parallelogram**.

Parallelograms can be further categorized by the relationships between sides and angles. If a parallelogram has four equal sides, it is known as a **rhombus**. If a parallelogram has four right angles, it is a **rectangle**. And, the most special type of parallelogram is a **square**, which has both four right angles and four equal sides.

The types of quadrilaterals can be summarized as follows.

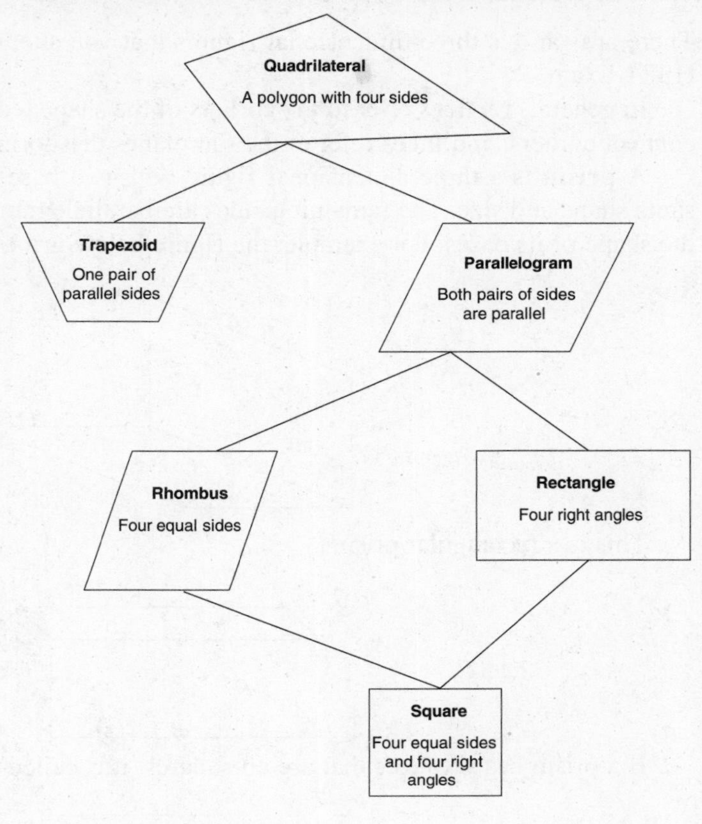

Using this visual organizer, you can clearly see the relationship among the various quadrilaterals. For example, every square is a rectangle, a parallelogram, a rhombus, and a quadrilateral.

Take Note: A favorite question to ask about this topic has to do with squares and rectangles. Note that all squares are rectangles but not all rectangles are squares.

Practice Set

Choose as many names as apply: quadrilateral, parallelogram, trapezoid, rectangle, rhombus, or square.

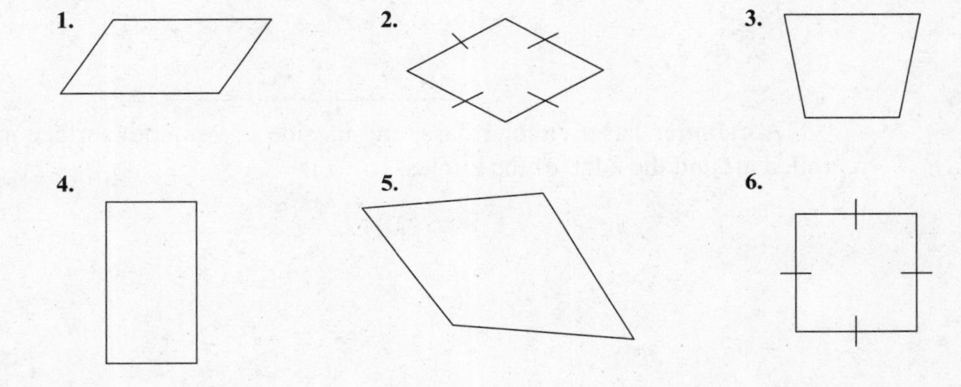

THREE-DIMENSIONAL FIGURES

There are various three-dimensional figures that you should be able to recognize for the HSPA Exam.

In general, **vertices** refer to the corners of the shape, **edges** refer to line segments that connect corners, and **faces** refer to the flat planes that form the sides of the figure.

A **prism** is a three-dimensional figure with two bases that are parallel faces of the same shape and size. The remaining sides are parallelograms. A prism is named based on the shape of its bases. For example, the figure below is a **triangular prism**.

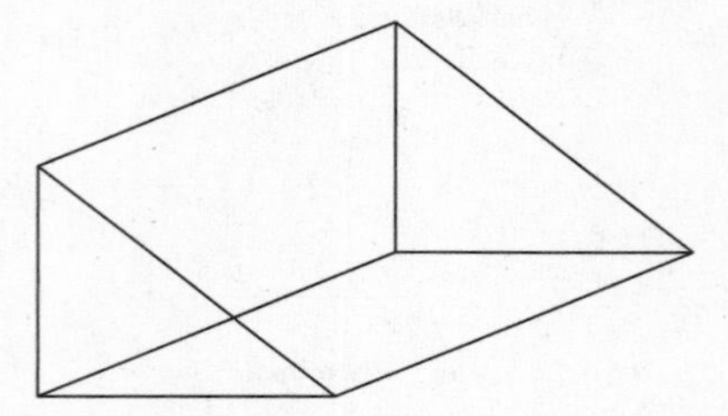

This is a rectangular prism:

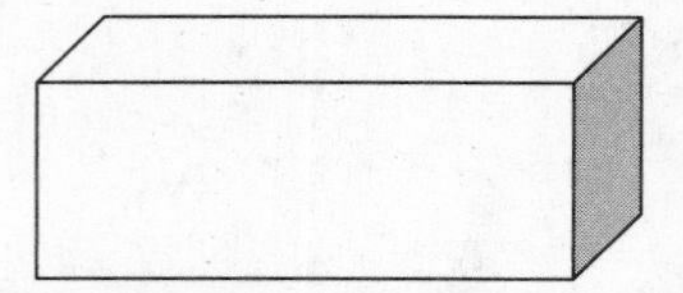

If a prism has six faces that are all squares, it is called a **cube**.

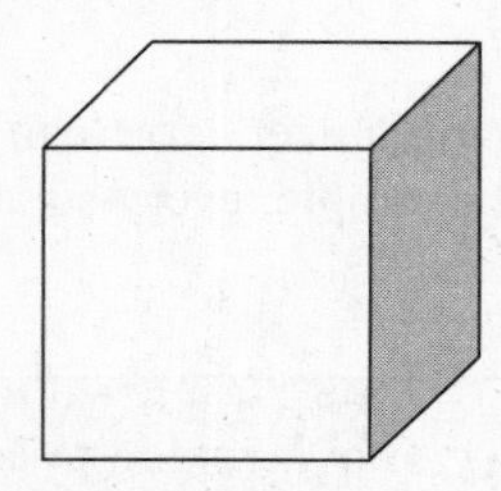

A **pyramid** is a three-dimensional figure that, unlike a prism, has only one base; all its other faces are triangles. A pyramid is also named based on the shape of its one base. For example, the following is a triangular pyramid (sometimes called a **tetrahedron**).

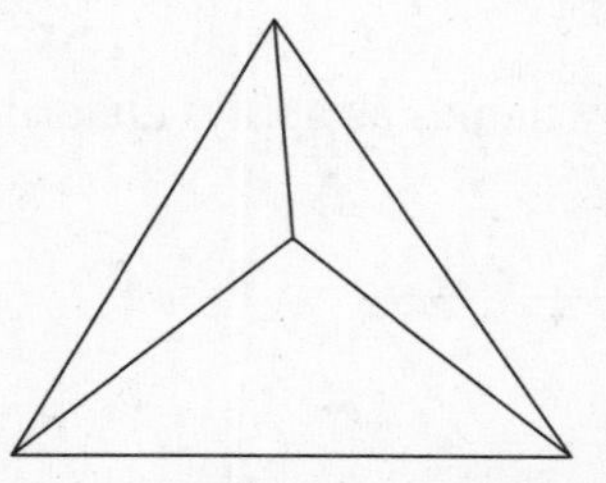

A **cylinder** has a circular base, and its side is a smooth surface made of a rectangle rolled around the edge of the circles.

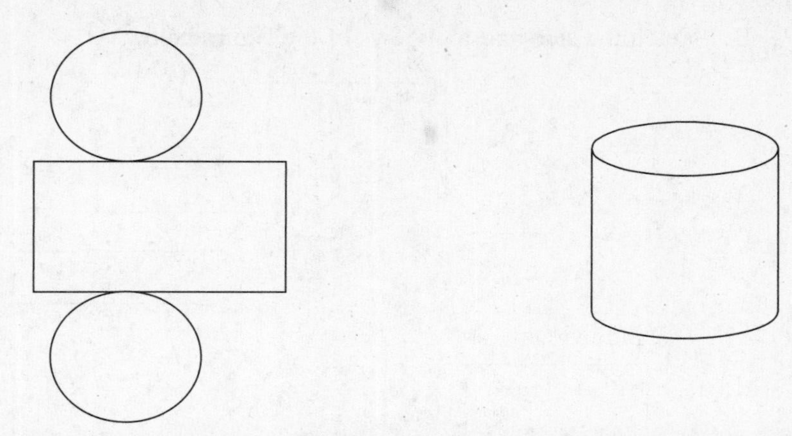

The shape that forms a three-dimensional shape when assembled is sometimes referred to as its **net**.

A **cone** is formed using a circular base and a triangle rolled around the circle to form a smooth side that slants.

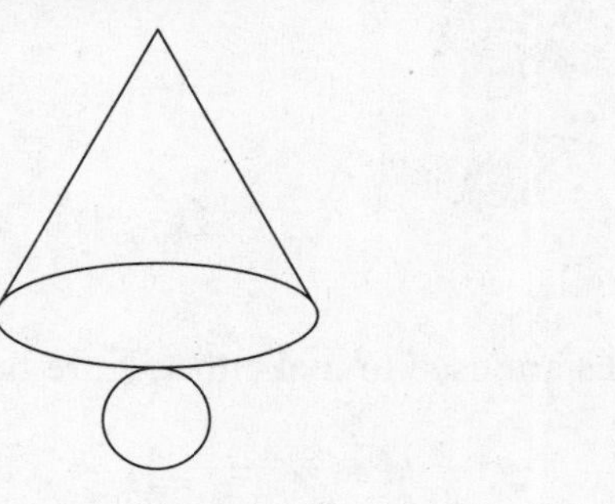

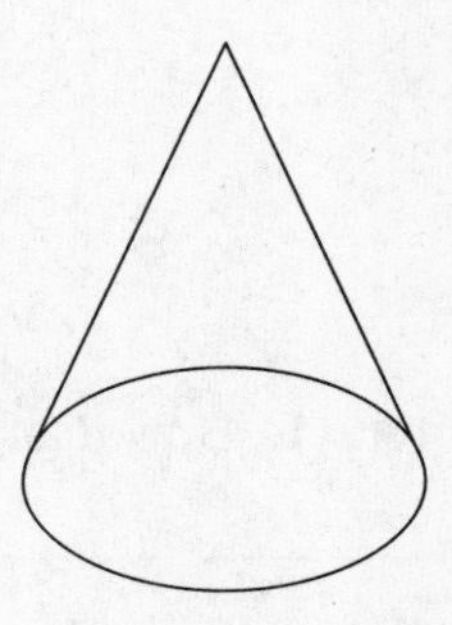

A **sphere** is like a ball.

Many times on the HSPA, you will be asked to count the number of smaller cubes that make up a larger figure or to sketch various views of a three-dimensional construction sometimes made up of small cubes.

Examples

A. How many small cubes are used to make the following figure?

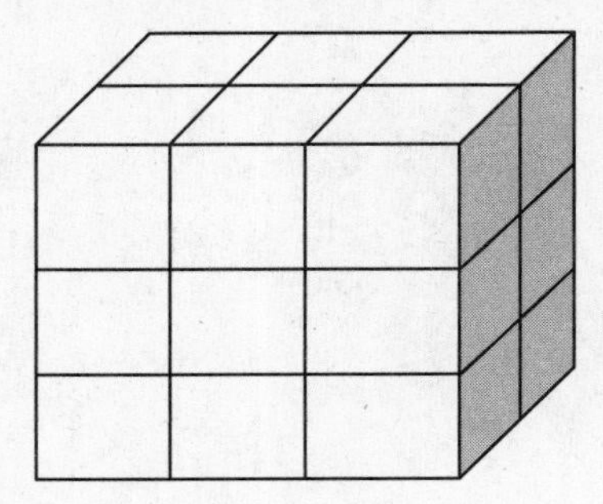

The easiest way to handle this type of problem is to count in "layers." For example, the top layer here is 3 across and 2 deep, for a total of 6 cubes. The entire structure is 3 layers tall, for a total of 18 cubes.

B. Sketch the side and top views of the figure below.

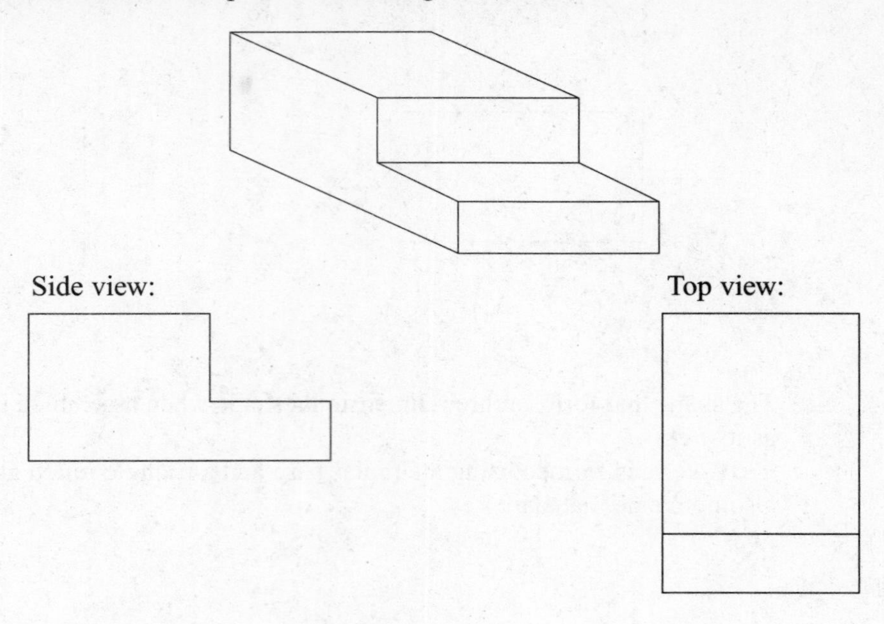

PRACTICE SET

★ **1.** How many smaller cubes are used to make the figure below?

★ A star next to a question indicates that the question is more difficult than ones on the actual HSPA. However, they're worth practicing.

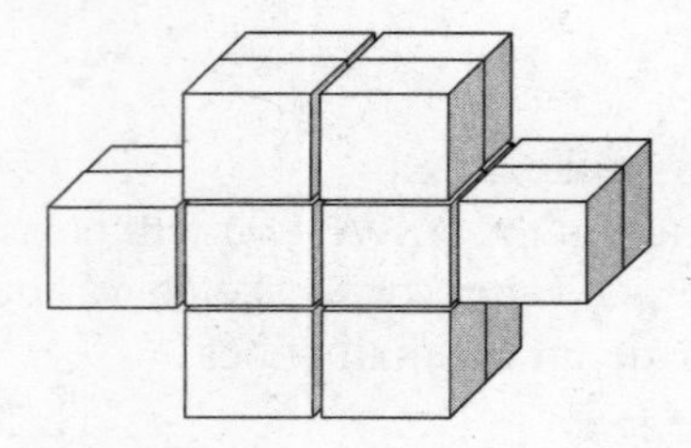

2. Draw the side and top views of the figure below.

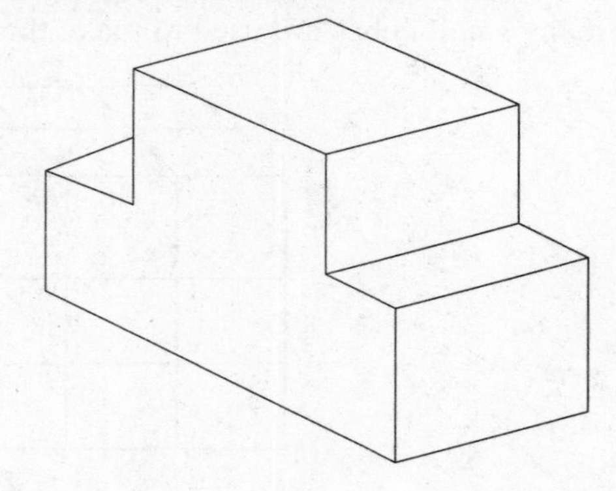

MIXED PRACTICE, CLUSTER II, MACRO A

1. Which of the following statements is true?

A. Every rhombus is a square.
B. Every square is a rhombus.
C. Every parallelogram is a rectangle.
D. Every rectangle is a square.

2. Which of the following are faces of a pentagonal prism?

I pentagon
II rectangle
III triangle

A. I only **B.** I and II **C.** II and III **D.** I and III

3. Given that p = the measure of the vertex angle of an isosceles triangle with base angles measuring 65° each and q = the measure of one of the angles in a right triangle, which statement is correct?

A. $p < q$ **B.** $p > q$ **C.** $p = q$ **D.** The answer can't be determined.

4. What is the measure of the angle formed by the hands of a clock at 10:30?

5. If a quadrilateral is going to be formed using toothpicks that measure 10 inches, 10 inches, 12 inches, and 12 inches, which of the following quadrilaterals *cannot* be made?

A. rhombus **B.** rectangle **C.** parallelogram **D.** kite

6. If you spin this two-dimensional figure about the axis as shown, which three-dimensional solid will result?

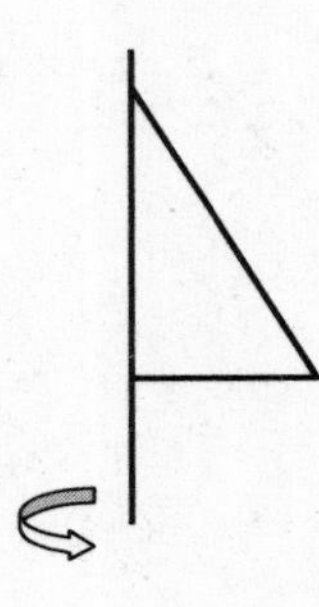

7. If the measures of two supplementary angles are in the ratio 2:8, find the measure of the smaller angle.

8. If the angles of a polygon add up to 2160° how many sides does the polygon have? Show how you determined your response.

MACRO B

CONGRUENCE VERSUS SIMILARITY

Congruent is another term used in geometry to indicate equality. We say, for example, that a line segment is congruent to another line segment if they are equal in length. The symbol for congruent is $\cong$. The symbol looks like an equals sign with a squiggle over it.

Examples

A. Given $\overline{AB} \cong \overline{CD}$, find the value of x.

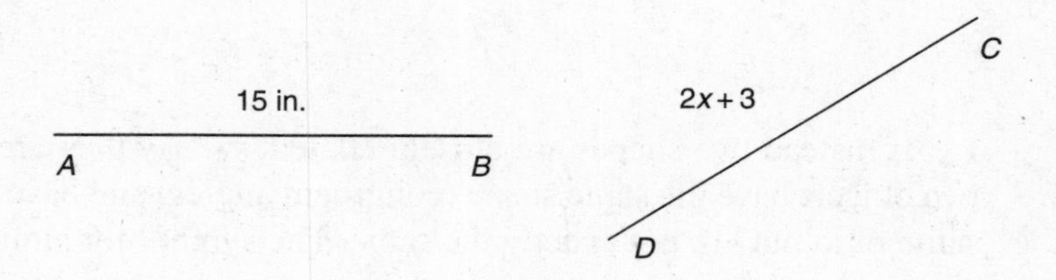

Since the symbol $\cong$ means congruent or equal to, we know that the lengths of the segments are equal. That is, $2x + 3 = 15$. We subtract 3 from both sides to get $2x = 12$ and divide by 2 to get $x = 6$.

B. Given that m$\angle ABC \cong$ m$\angle CBD$, find m $\angle ABD$.

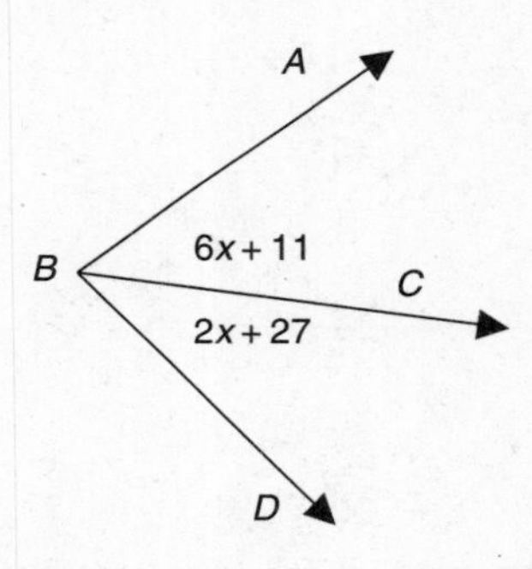

Again, we need to realize the meaning of the symbol $\cong$, which means congruent or equal.

Thus, $6x + 11 = 2x + 27$. If we subtract $2x$ from both sides, we get $4x + 11 = 27$, and then we subtract 11 from both sides to get $4x = 16$ and division gives us $x = 4$.

Therefore, if $x = 4$, $6x + 11 = 6(4) + 11 = 24 + 11 = 35$. The other angle $(2x + 27)$ is also equal to 35, and thus the entire angle is $35 + 35$ or 70.

PRACTICE SET

1. In the diagram below, $\overline{AB} \cong \overline{BC}$. Find the length of $\overline{AC}$.

5x + 9 2x + 15

A B C

2. Given the diagram below, state a statement of congruence among angles. Then find the value of x, m$\angle CED$, and m$\angle BED$.

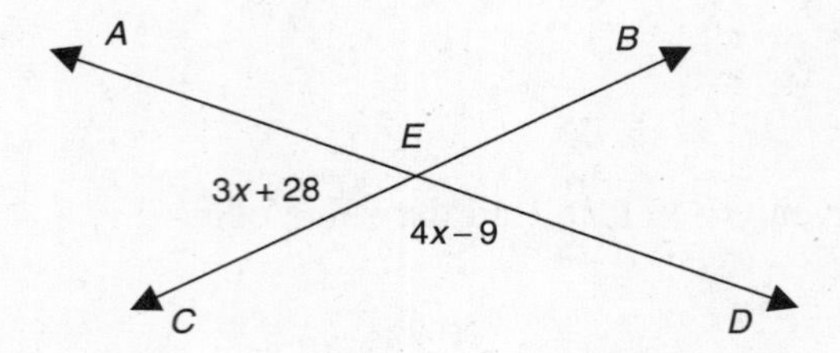

If instead two shapes are closely related, we say they are similar. **Similar** means that two objects have the same shape (congruent angles) and have measurements that are in the same ratio but are not exactly the same. The symbol for similar is ~.

Examples

A. $\triangle RST \sim \triangle UVW$. Find m$\angle RST$ and the length of side $\overline{UV}$.

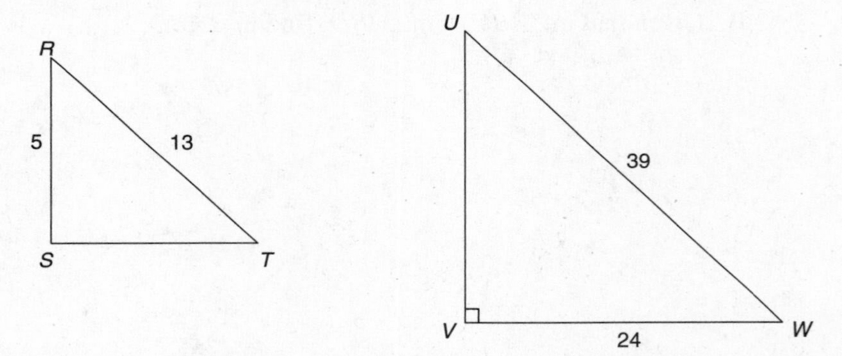

It is very important that you pay attention to the order in which the triangles are named so that you know which sides correspond to one another. The drawings do not always match up nicely.

Since $\triangle RST \sim \triangle UVW$, the sides that match are RS and UV (since these are the first two letters in the name of each triangle), ST and VW (the last two letters in each triangle's name), and RT and UW. Thus, the proportion, should be set up as follows.

$$\frac{13}{39} = \frac{5}{UV}$$

We then cross-multiply, $13UV = 195$, and divide, $195 \div 13 = 15$. The measure of angle RST matches that of angle UVW, which is labeled 90°.

B. Quadrilateral $ABCD \sim$ quadrilateral $EFGH$. Find the missing lengths.

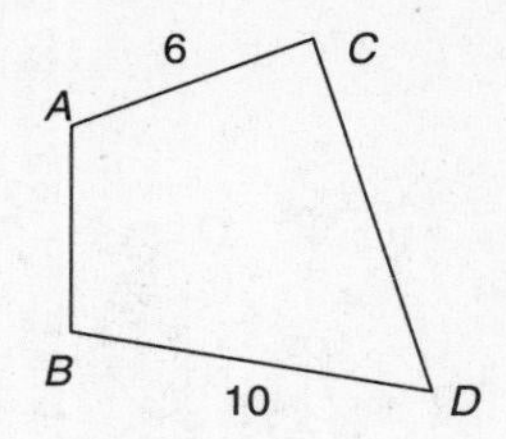

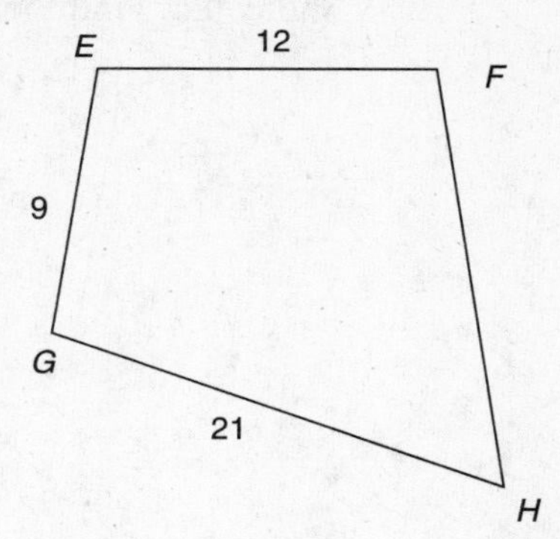

Again, the order of the names of the similar figures is essential in helping to determine which sides correspond. Since the names are given, *ABCD* and *EFGH*, we know that *AB* matches up with *EF*, *BD* with *FH*, *CD* with *GH*, and *AC* with *EG*. So our proportion is as follows.

$$\frac{AB}{12} = \frac{CD}{21} = \frac{6}{9} = \frac{10}{FH}$$

Since $\frac{6}{9}$ reduces to $\frac{2}{3}$, we have the following three proportions to solve:

$$\frac{2}{3} = \frac{AB}{12} \quad \frac{2}{3} = \frac{CD}{21} \quad \frac{2}{3} = \frac{10}{FH}$$

Cross-multiplication gives

$$3AB = 24 \quad 3CD = 42 \quad 2FH = 30$$

Dividing to solve gives

$$AB = 8 \quad CD = 14 \quad FH = 15$$

PRACTICE SET

1. Given: $\triangle GHI \sim \triangle JKL$

$m\angle HIG = 60$, $m\angle JKL = 75$

$GI = 36$ in., $HI = 20$ in., $JK = 18$ in., $JL = 27$ in.

Find the missing angles and lengths.

2. Given quadrilateral $MNOP \sim$ quadrilateral $QRST$ with the lengths of sides as indicated. Find the missing lengths.

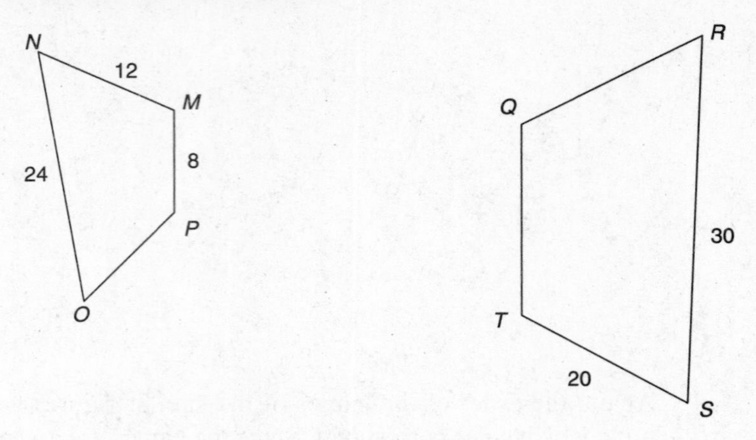

TRANSFORMATIONS

Undoubtedly you noticed that many of the figures given as similar were placed differently on the page so that it was not immediately obvious that they were indeed the same shape but just different sizes. A **transformation** is a change in the orientation of an object that maintains the same size and shape. There are three basic types of transformations:

Translation: The object is slid.
Rotation: The object is turned or spun around a point.
Reflection: The object is flipped or mirrored through a line.

Examples

A. Rotate the letter A clockwise 90°.

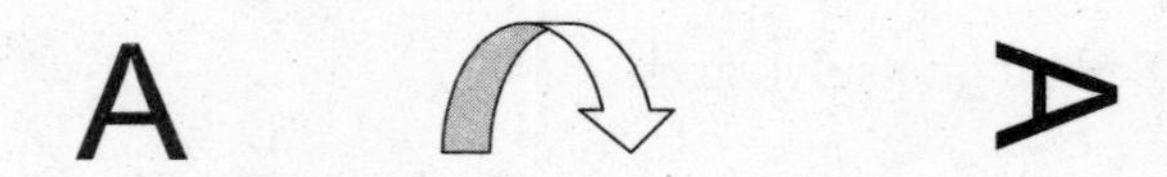

B. Reflect the figure below through the given line.

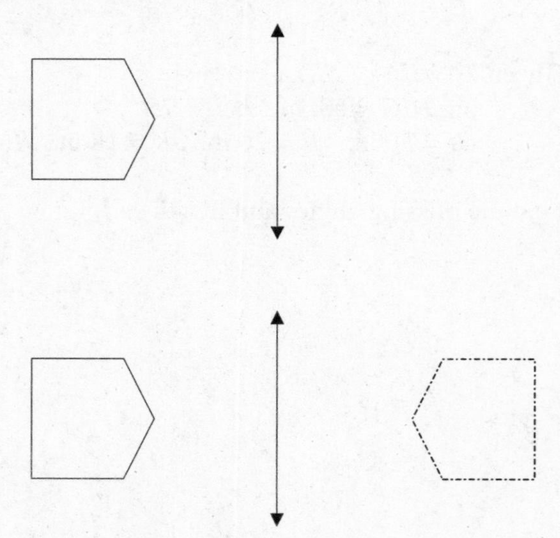

Reflecting is like folding the object over the line—it results in a mirror image of the original figure.

C. Rotate the figure below 45° counterclockwise about its vertex A.

D. Reflect the figure below through the line given.

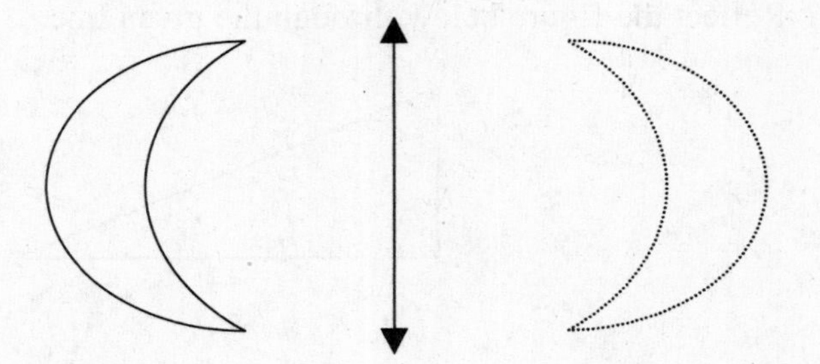

SYMMETRY

A figure has **symmetry** if you can draw a line through it such that each half is a mirror image of the other. When a symmetric figure is reflected through a line parallel to a line of symmetry, the figure remains unchanged.

Examples

A. The figure below has lines of symmetry drawn with dotted lines.

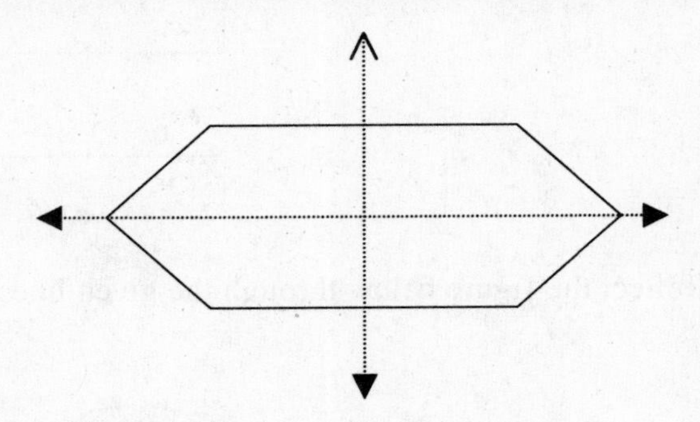

B. Again, the figure below (a square) has lines of symmetry drawn using dotted lines.

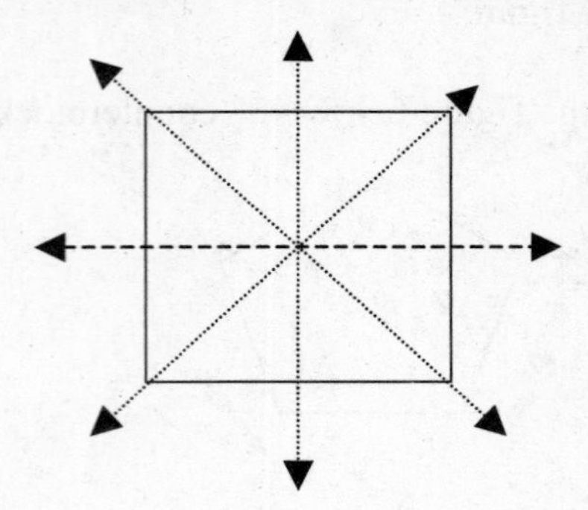

PRACTICE SET

1. Reflect the figure below through the given line.

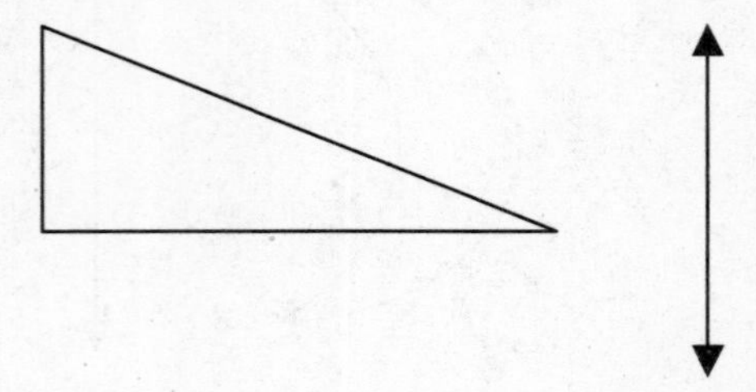

2. Rotate the figure below 90° counterclockwise about the center *O*.

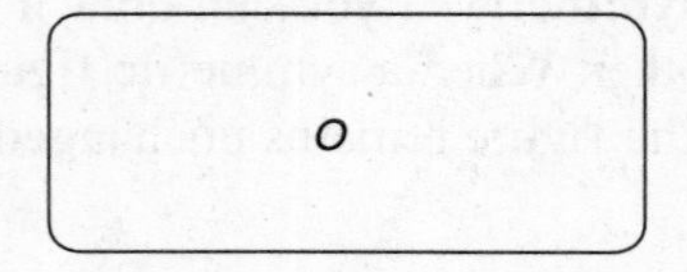

3. Rotate the figure below 60° clockwise about the vertex *A*.

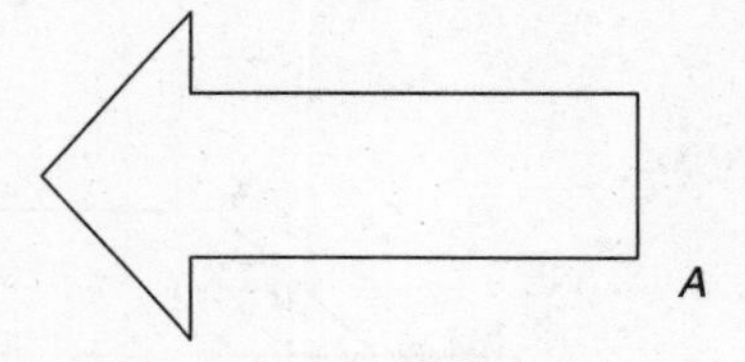

4. Reflect the figure below through the given line.

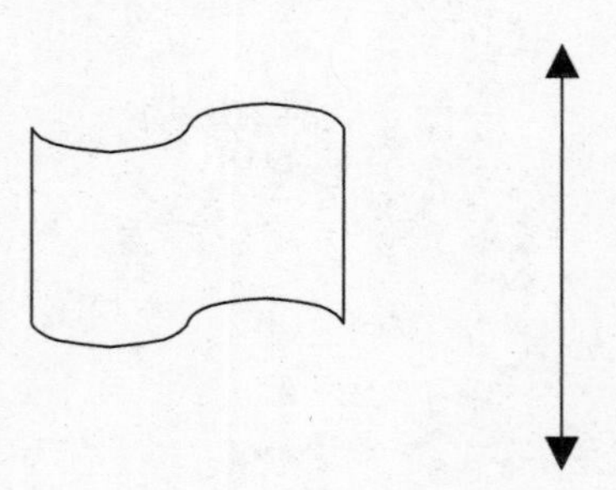

5. Draw all lines of symmetry in the figure below.

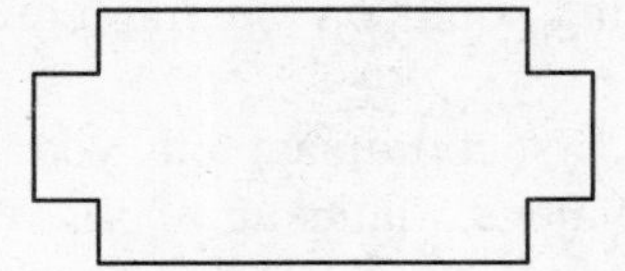

6. Draw all lines of symmetry in the following figure.

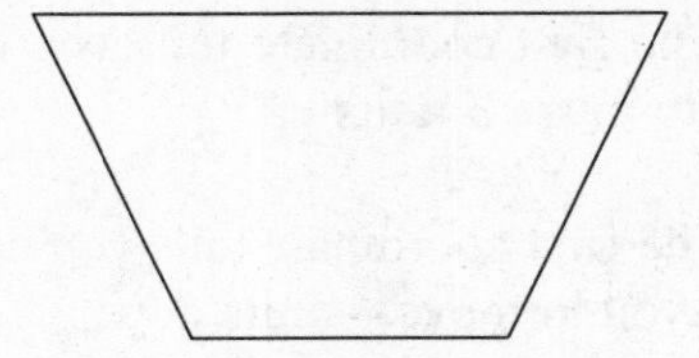

COORDINATE SYSTEM

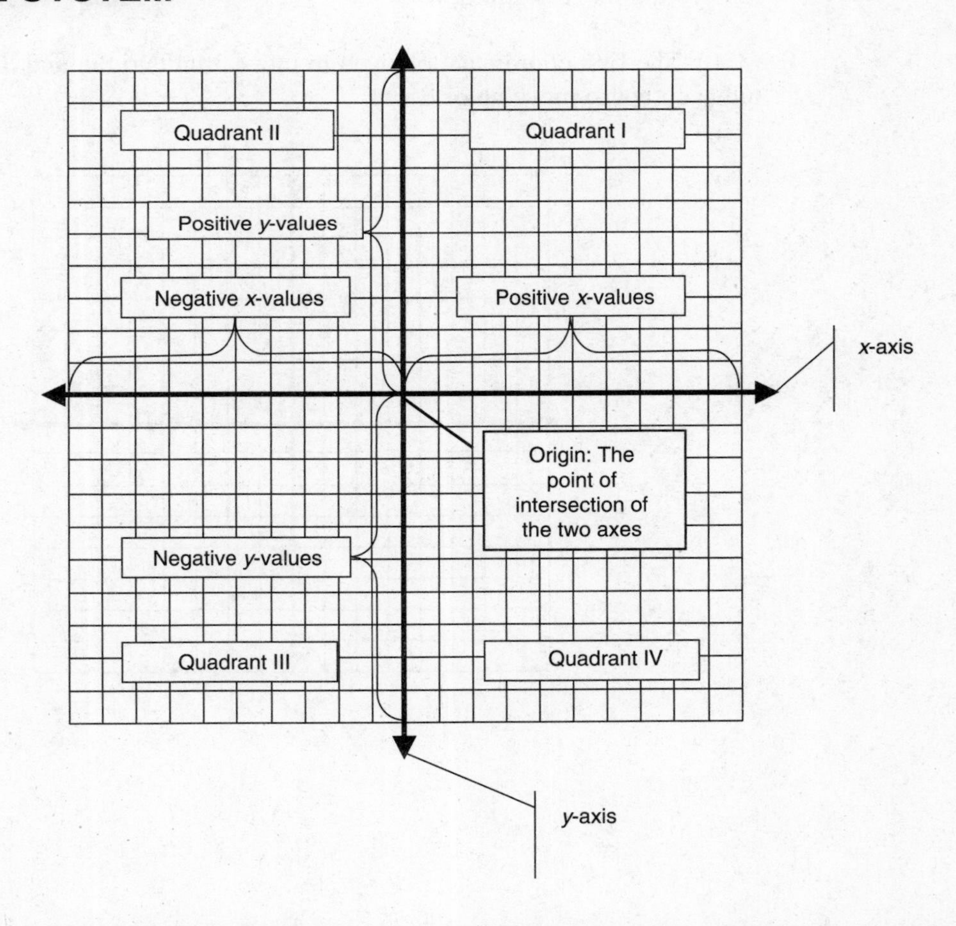

The coordinate system is used to locate points. A coordinate is given as (x, y). The first number tells you the value of the x-coordinate, or how many spaces to move to the right or left from the origin. The second number tells you the value of the y-coordinate, or how many spaces to move up or down from the origin. Points are plotted by reading the values of x and y from the ordered pair given and then moving from the origin to that point.

Examples

Plot the following points on the graph given:

A. (2, 5). The first coordinate tells you to move 2 units right, and the second coordinate tells you to move 5 units up.

B. (−3, −2). The first coordinate tells you to move 3 units left, and the second coordinate tells you to move 2 units down.

C. (−1, 3). The first coordinate tells you to move 1 unit left, and the second coordinate tells you to move 3 units up.

D. (0, −4). The first coordinate tells you not to move right or left, and the second coordinate tells you to move 4 units down.

E. (4, −1). The first coordinate tells you to move 4 units right, and the second coordinate tells you to move 1 unit down.

F. (3, 0). The first coordinate tells you to move 3 units right, and the second coordinate tells you not to move up or down.

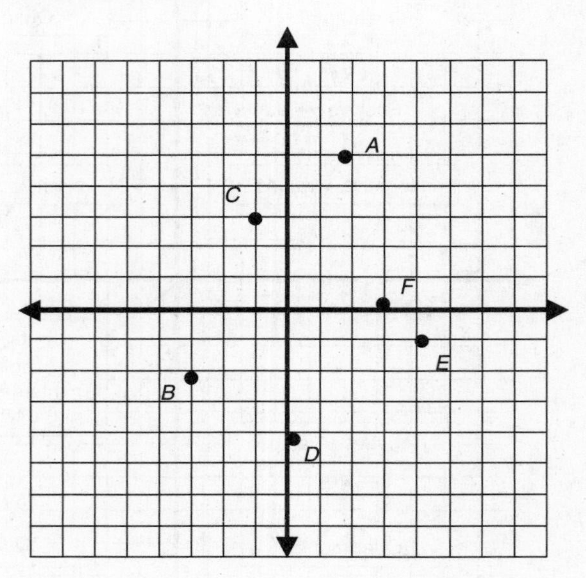

PRACTICE SET

1. Give the coordinates of each point plotted on the graph below.

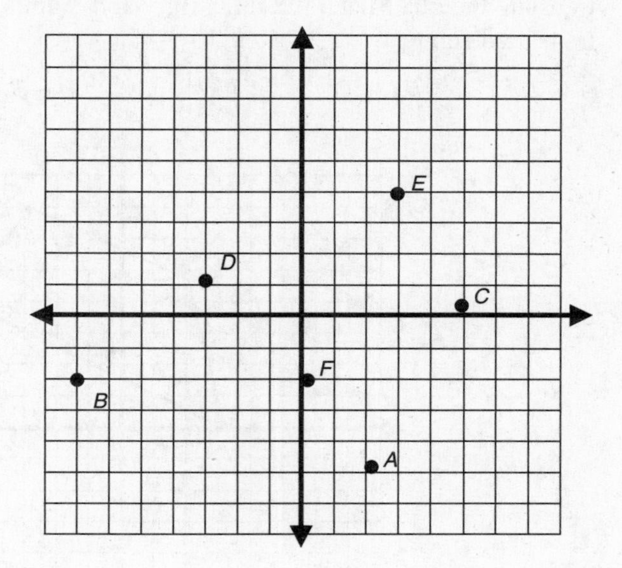

2. Plot and label the given points on a graph.

   R (−3, 5) *S* (0, 6) *T* (4, −1) *U* (−2, −3) *V* (−1, 0) *W* (2, 5)

3. If the following coordinates are plotted and we need a fourth coordinate, *P*, to complete a square, what would the coordinates be?

   M (−2, 5) *N* (−2, −3) *O* (6, 5) *P* (?, ?)

4. What figure do the following points make? Be specific in your description.

   A (−3, 5) *B* (2, 5) *C* (2, 0)

5. Add a fourth point to the following so that you create a trapezoid that has a vertical line of symmetry.

   R (−3, −1) *S* (7, −1) *T* (4, 5) *U* (?, ?)

USING THE COORDINATE PLANE

Many times on the HSPA, you will use a coordinate plane to work with questions involving transformations, including the last type of transformation, which is a translation. The easiest way to describe a translation is that it is a slide. A figure is slid or moved a certain number of units up, down, right, and/or left.

To perform a translation, simply move one point or corner of the figure at a time. Typically, we label the new coordinates using the same letter but with a ′ afterward. For instance, the point *A* would be called *A*′ in the new translated shape.

Examples

A. Plot the following coordinates and connect them to form a quadrilateral. Then translate the quadrilateral 2 units right and 3 units down. Give the coordinates of the translated figure.

$A\ (-2, 5)$ $B\ (3, 7)$ $C\ (5, 4)$ $D\ (-3, 0)$

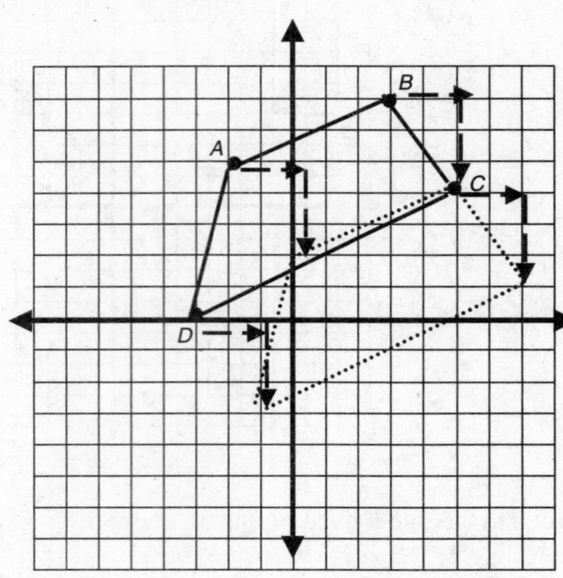

As indicated, each of the four points was moved 2 units right and 3 units down (as given in the translation instructions). The new coordinates are

$A'\ (0, 2)$ $B'\ (5, 4)$ $C'\ (7, 1)$ $D'\ (-1, -3)$

B. Translate the given triangle 4 units left and 5 units up. Then reflect the translated triangle through the *y*-axis. Give the coordinates of the final figure.

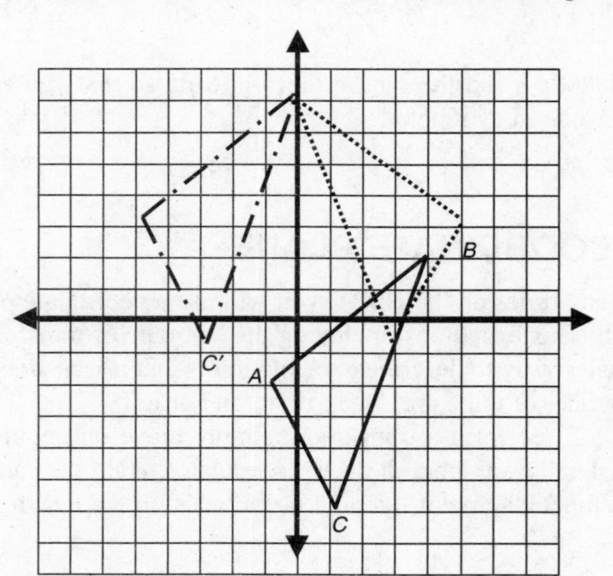

Once again, each point was moved according to the translation—in this case, four units left and 5 units up. The resulting figure was then flipped over the y-axis (the vertical axis). When reflecting, be sure to keep the figure the same distance from the axis you're reflecting through and just put it on the opposite side. For example, C' is 3 units to the left of the y-axis, so when we reflect it, we keep it 3 units from the y-axis but on the right side of the y-axis. The new coordinates of the final figure are:

A'' (5, 3) B'' (0, 7) C'' (3, −1)

PRACTICE SET

1. Plot the following points and then connect them. Then, translate the figure 4 units right and 2 units down. Give the coordinates of the new figure.

   P (−2, 1) Q (−4, 4) R (1, 5) S (1, −3)

2. Plot a right triangle such that one side is on the y-axis and the triangle falls into the fourth quadrant. Give the coordinates of your original triangle. Then, translate the triangle 3 units left and 5 units up. Finally, reflect the translated triangle through the y-axis. State the coordinates of the final figure.

3. Plot the given points and then reflect the figure through the x-axis. Finally, translate the figure 6 units right and 2 units up. Give the coordinates of the final figure.

   A (−7, 3) B (−5, 7) C (−1, 7) D (−1, 3)

One final skill you should learn with regard to transformations is to be able to write a set of transformations that would result in a given figure transforming to what is shown. It is important to realize that more than one set of instructions for performing transformations may be correct.

Examples

A. State a transformation that would result in the figure moving to the figure indicated by the dashed lines.

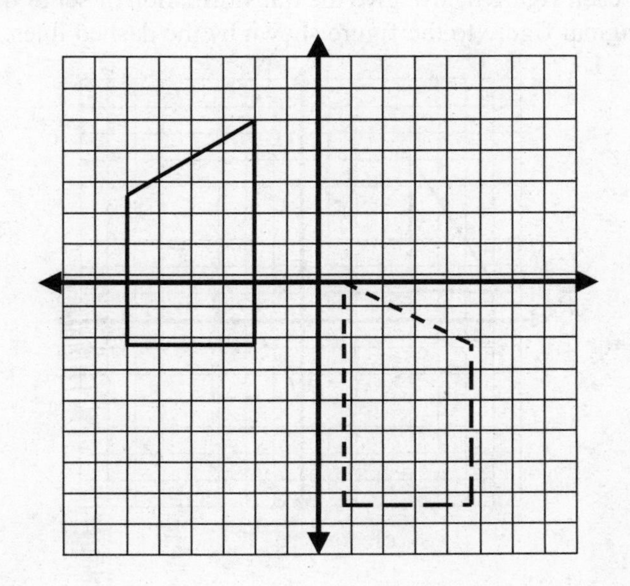

When asked to give transformations to obtain a given result, first determine if there's been a reflection. Then determine if there has been any sliding or translating.

When you look at this shape, you can see that its orientation is different; it has been reflected through the y-axis. If you backtrack and reflect it, you will see that it still has to be shifted left and up. This means the original shape had to be translated right and down. The final result is that the figure needs to be translated 1 unit right and 5 units down and then reflected through the y-axis.

B. Again, give a transformation or transformations that would map the original figure to the figure shown by the dashed lines.

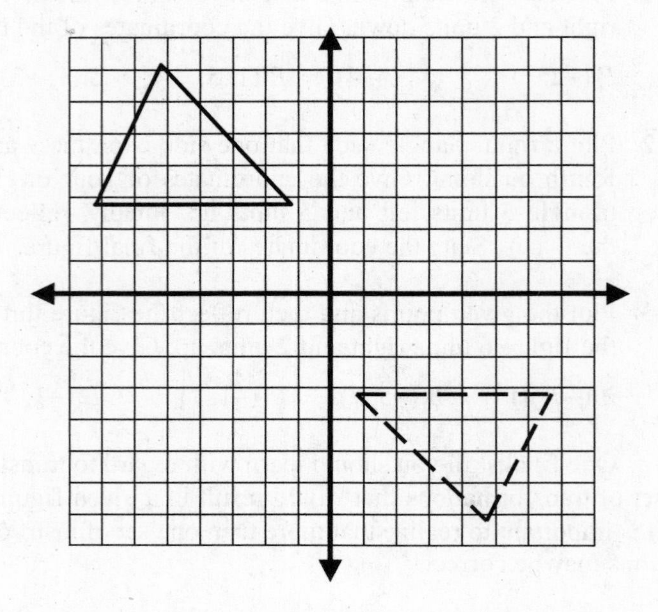

This translation requires a double reflection through the x- and y-axes (in either order).

PRACTICE SET

In each figure below, give the transformation or set of transformations that would map the original figure to the figure shown by the dashed lines.

1.

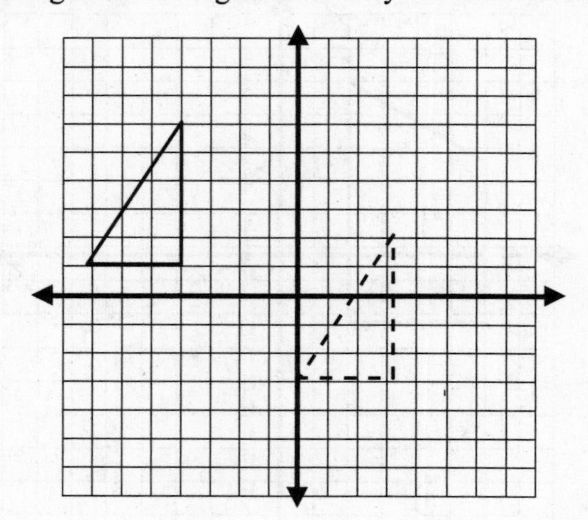

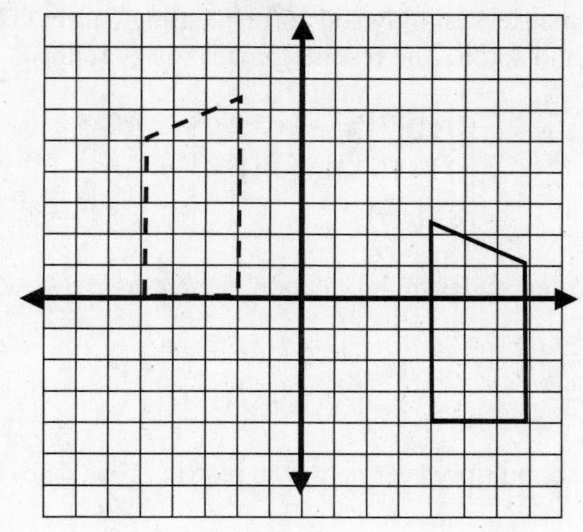

PRACTICE SET

Give a transformation that would map the original figure to the second figure.

1. $A\ (-5, -2)$ $B\ (-3, 0)$ $C\ (-1, -5)$
   $A'\ (0, -1)$ $B'\ (2, 1)$ $C'\ (4, -4)$

2. $R\ (2, 3)$ $S\ (-3, 4)$ $T\ (-3, -1)$ $U\ (3, -1)$
   $R'\ (0, 1)$ $S'\ (-5, 0)$ $T'\ (-5, 5)$ $U'\ (1, 5)$

MATRICES

A **matrix** is a rectangular arrangement of numbers enclosed in parentheses or square brackets. The plural of *matrix* is *matrices*.

The following is an example of a matrix A:

$$A = \begin{bmatrix} 3 & 2 & 6 \\ -4 & 7 & 5 \end{bmatrix}$$

Matrix A contains 2 rows and 3 columns. We sometimes refer to this as a 2×3 matrix (read "2 by 3 matrix"). Each number or entry in a matrix can be referred to by its location (row, column). For example, $a_{23} = 5$ (row 2, column 3).

If two or more matrices are the same size, they can be added. To *add* matrices, simply add the corresponding entries. We can also perform what is known as scalar multiplication on any matrix. *Scalar multiplication* simply means that each entry in a matrix is multiplied by the same number.

Examples

A. Add the given matrices.

$$\begin{bmatrix} 5 & -3 \\ -1 & 4 \\ 8 & 2 \end{bmatrix} + \begin{bmatrix} -7 & 5 \\ 10 & -4 \\ -5 & 6 \end{bmatrix}$$

To add these two matrices, simply add corresponding entries. That is, add 5 + (–7) and then add –3 + 5, and so on. The resulting matrix is as follows.

$$\begin{bmatrix} 5+(-7) & -3+5 \\ -1+10 & 4+(-4) \\ 8+(-5) & 2+6 \end{bmatrix} = \begin{bmatrix} -2 & 2 \\ 9 & 0 \\ 3 & 8 \end{bmatrix}$$

B. Perform the indicated scalar multiplication. Given matrix B, what is the value of $3B$?

$$B = \begin{bmatrix} 6 & 7 & -1 & 4 \\ -5 & 3 & 2 & -7 \end{bmatrix}$$

This means that we multiply every entry in matrix B by 3. So, the result is as follows.

$$\begin{bmatrix} 6\times 3 & 7\times 3 & -1\times 3 & 4\times 3 \\ -5\times 3 & 3\times 3 & 2\times 3 & -7\times 3 \end{bmatrix} = \begin{bmatrix} 18 & 21 & -3 & 12 \\ -15 & 9 & 6 & -21 \end{bmatrix}$$

PRACTICE SET

Given these matrices, evaluate the expressions that follow.

$$A = \begin{bmatrix} -5 & 7 \\ 2 & 3 \\ -4 & 1 \end{bmatrix} \quad B = \begin{bmatrix} 6 & -2 & 9 \\ -3 & 4 & 10 \end{bmatrix} \quad C = \begin{bmatrix} 1 & 5 & -13 \\ -2 & 1 & -7 \end{bmatrix} \quad D = \begin{bmatrix} 2 & -1 \\ 3 & 4 \\ -1 & -6 \end{bmatrix}$$

1. $A + D$ **2.** $2B$ **3.** $-3D$ **4.** $B + C$ **5.** $-4A + D$

TESSELATIONS

A **tesselation** is a geometric design formed by using one or more figures to completely cover an area with no gaps or overlapping. You can think of it as a jigsaw puzzle in which either all the pieces are the same shape and size or there are only two different-shaped pieces. The pieces all lock together to cover the area.

A tesselation is called *regular* if it is made using *one* regular polygon (recall that a regular polygon has all sides equal and all angles equal). A tesselation is called *semiregular* if it is made using two regular polygons. The only regular polygons that form a regular tesselation are triangles, squares, and hexagons.

Examples

Below are some examples of what tesselations look like.

A. A tesselation formed by equilateral triangles:

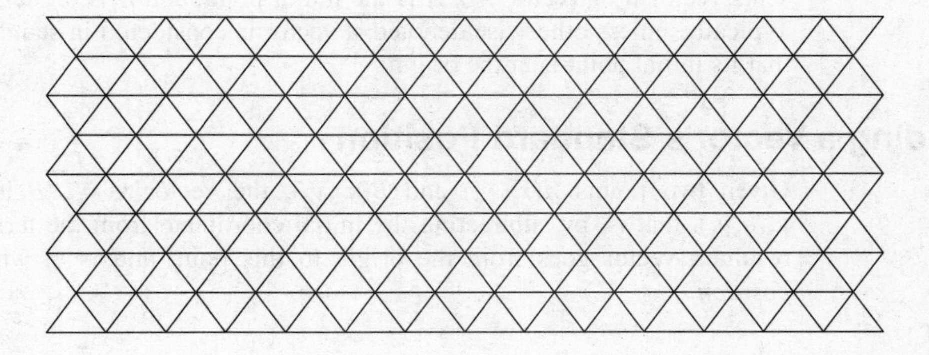

B. A tesselation formed by regular hexagons:

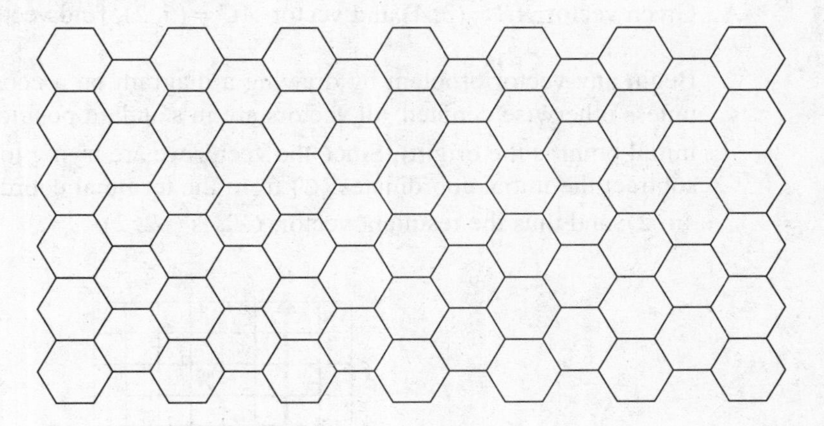

C. A semiregular tesselation formed by pentagons and rhombi:

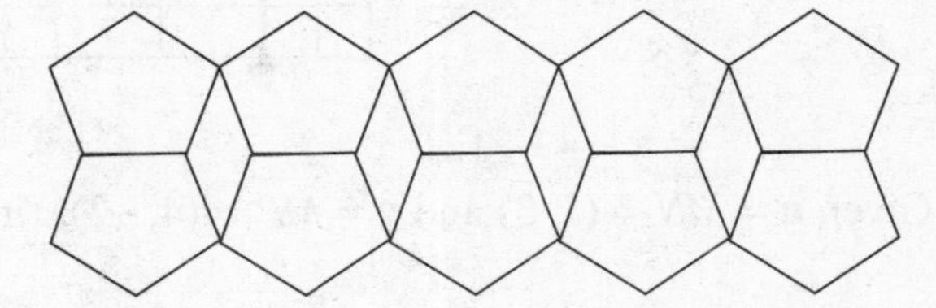

D. A tesselation formed using an irregular shape:

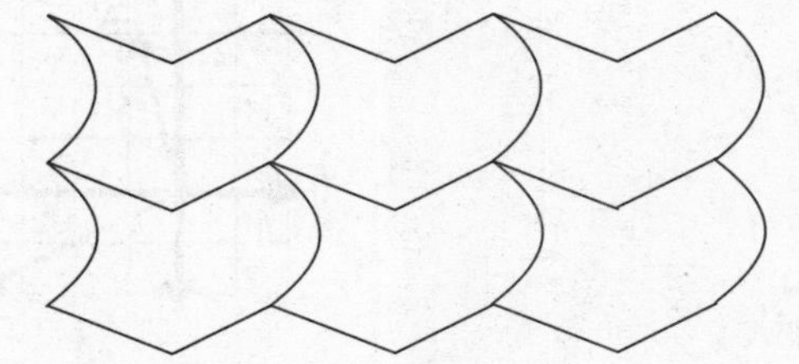

VECTORS

A **vector** is a line segment or ray that has both direction and magnitude (or strength). Vectors are commonly used in physics.

We use a boldface, lowercase letter to represent a vector. For example, $\mathbf{u} = \overrightarrow{AB}$ represents vector **u** or vector $\overrightarrow{AB}$. A is the initial point, and B is the terminal point. Typically, unless otherwise denoted, a vector is considered in standard position, meaning that its initial point is at the origin.

Finding a Vector's Standard Position

Given two points $A(x_1, y_1)$ and $B(x_2, y_2)$, the vector $\mathbf{v} = \overrightarrow{AB}$ is found by $(x_2 - x_1, y_2 - y_1)$, that is, by subtracting the initial coordinate from the terminal coordinate. The resultant vector goes from the origin to this point and is in what is called *standard position*.

Examples

A. Given vector $\overrightarrow{AB} = (3, 4)$ and vector $\overrightarrow{AC} = (5, 2)$, find vector $\overrightarrow{CB}$.

Begin any vector problem by drawing a diagram on a coordinate plane. Remember, unless otherwise denoted, all vectors are in standard position (which means that their initial point is the origin). Since the vector we are trying to represent is $\overrightarrow{CB}$, we must subtract the initial coordinates (C) from the terminal coordinates (B). That is, $(3 - 5, 4 - 2)$, and thus the resultant vector, $\overrightarrow{CB}$, is $(-2, 2)$.

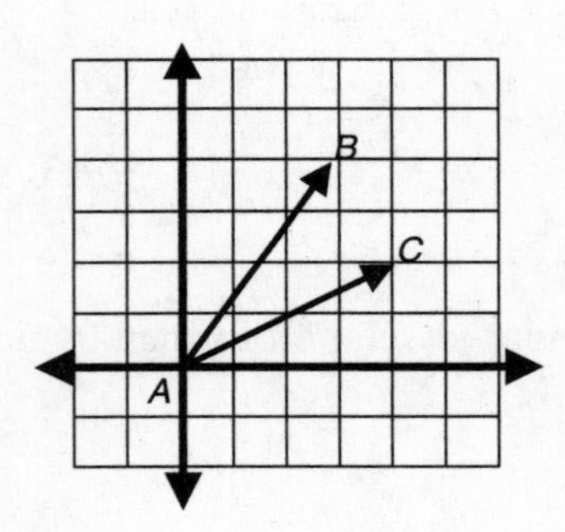

B. Given $\mathbf{u} = \overrightarrow{MN} = (1, 3)$ and $\mathbf{v} = \overrightarrow{MP} = (4, -2)$, find the vector $\overrightarrow{NP}$.

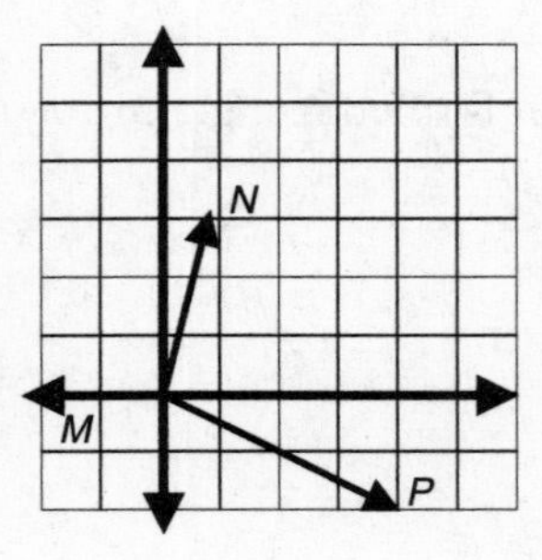

Remember, to find a representation in standard form for a vector between two points, subtract the coordinates of the initial point from the coordinates of the terminal point.

So, in this case, since we are asked for vector $\overrightarrow{NP}$, we subtract the coordinates of N from the coordinates of P: $(4 - 1, -2 - 3)$ or $(3, -5)$.

Adding Two Vectors

To add two vectors, simply add the coordinates. The resultant vector is the diagonal of the parallelogram formed using the two original vectors as two of its sides.

Example

Add $\mathbf{u} = (-2, 3)$ and $\mathbf{v} = (4, 2)$ and draw a diagram showing the resultant vector.

To add the two vectors, simply add the coordinates: $(-2 + 4, 3 + 2)$ to get $(2, 5)$. The diagram looks like this.

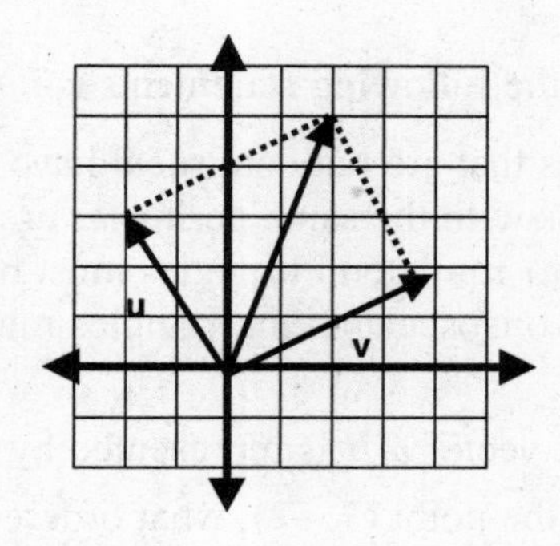

PRACTICE SET

Add the given vectors and draw a diagram showing the original two vectors and the resultant sum.

1. $\mathbf{u} = (1, 5)$ and $\mathbf{v} = (3, -2)$

2. $\mathbf{u} = (-2, -4)$ and $\mathbf{v} = (4, -1)$

3. $\mathbf{u} = (0, 4)$ and $\mathbf{v} = (-3, -5)$

For the following problems find the missing vector $\mathbf{v}$ that would have to be added to the given vector $\mathbf{u}$ to get the specified resultant sum.

4. $\mathbf{u} = (2, 4)$ and $\mathbf{u} + \mathbf{v} = (5, 8)$

5. $\mathbf{u} = (-3, 1)$ and $\mathbf{u} + \mathbf{v} = (-2, -4)$

MIXED PRACTICE, CLUSTER II, MACRO B

1. If rectangle $ABCD$ is translated 3 units left and 2 units down, the new coordinates are A' $(-1, 2)$, B' $(3, 2)$, C' $(3, -1)$, and D' $(-1, -1)$. What were the coordinates of A?

A. (2, 4) **B.** (5, 4) **C.** (2, 0) **D.** (4, 2)

2. If the point P (–2, 4) is reflected over the x-axis, then the new point P' is translated 4 units to the right. What are the final coordinates?

A. (2, 4) **B.** (–6, 4) **C.** (2, –4) **D.** (6, 4)

3. A 60-foot long ramp is 24 feet high. If there is a support beam 15 feet from the top of the ramp, how tall is it?

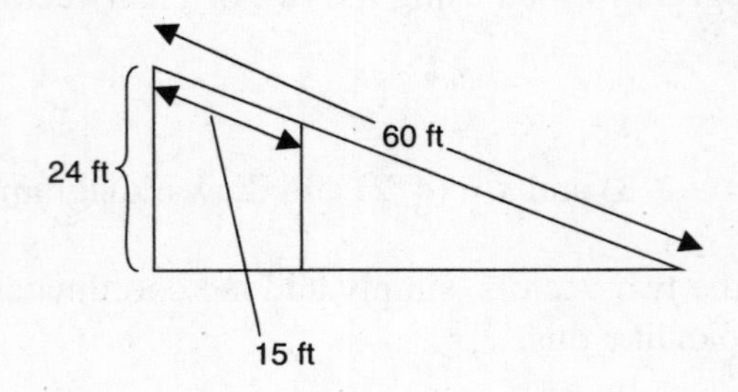

4. Which of the following statements is *not* true?

A. Squares that are not congruent have different areas.
B. Squares with the same perimeter are congruent.
C. Any two equilateral triangles must be similar.
D. Any two isosceles right triangles must be congruent.

5. Given that vector $\overrightarrow{AB}$ is represented by the ordered pair (2, 5). If vector $\overrightarrow{AC}$ is represented by the point (7, –2), what ordered pair represents vector $\overleftrightarrow{BC}$?

A. (9, 3) **B.** (9, –3) **C.** (5, –7) **D.** (7, –2)

6. If right triangle ABC is reflected about the y-axis, which of the following properties of the triangle is changed as a result of the reflection?

A. the measure of $\angle C$
B. the length of the hypotenuse
C. the triangle's perimeter
D. the orientation of the triangle

7. Which of the following points do NOT represent three consecutive vertices of a square?

A. (–2, 5) (–2, 8) (1, 8)
B. (–4, 3) (–4, 7) (–1, 7)
C. (5, 8) (–1, 8) (–1, 14)
D. (0, 2) (0, 7) (5, 7)

8. Plot the following points and connect them to form a polygon.

$$(-5,-2)\ (5,-2)\ (-2,6)\ (2,6)$$

What type of figure is formed?

9. The end points of a circle's diameter are (–3, 2) and (7, 2). What are the coordinates of its center?

10. If two vertices of a rectangle are R (0, 0) and S (5, 0) and the rectangle's perimeter is 26, find the other two coordinates.

MACRO C

MEASURING ANGLES

Angles are measured and drawn using a tool called a **protractor**.

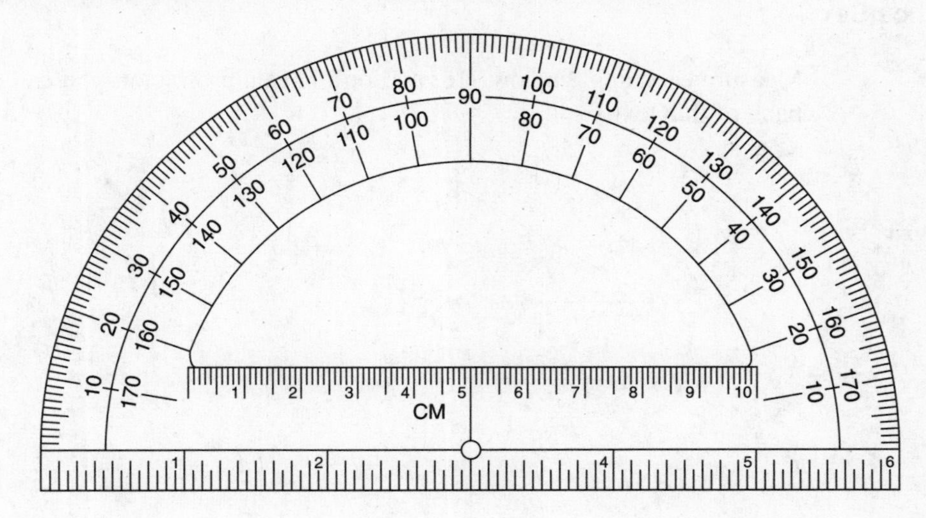

A protractor either has a hole at the center of its bottom or has a center mark. The hole or mark is placed over the vertex of the angle, and the horizontal line is placed along one ray of the angle as shown below.

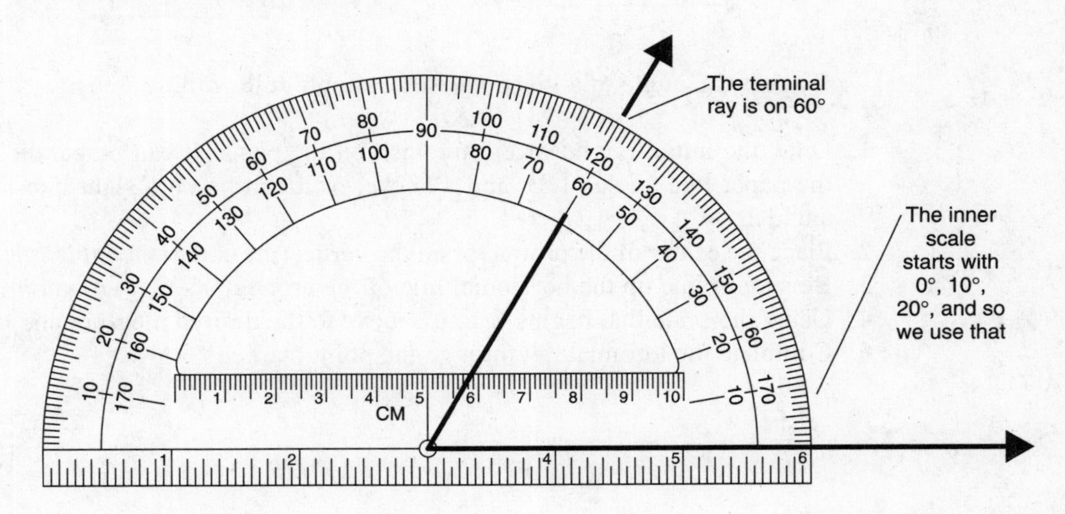

If you know what type of angle you have, you will know which scale to read from. Or, use the scale that starts with 0° along the horizontal ray. In this case, we have an acute angle, so we will use the measurement of 60°.

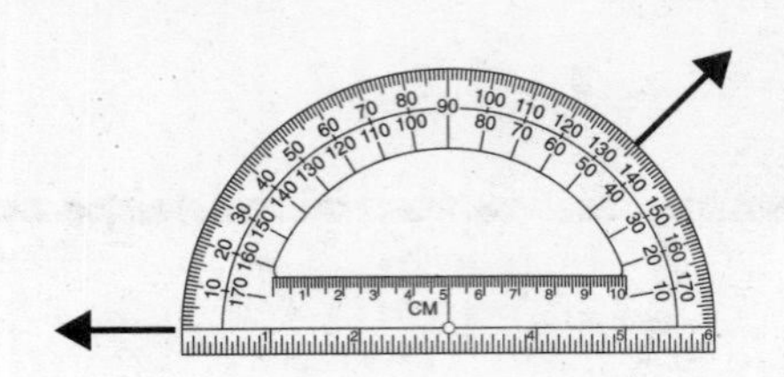

The angle to the left is obtuse, so we know that its measure should be 135° (not 45°). Also, this time the outer scale is the one marked 0°, 10°, 20°, and so on.

Practice Set

Measure each angle below. (If you don't have a protractor, you can cut out the one in the back of this text.)

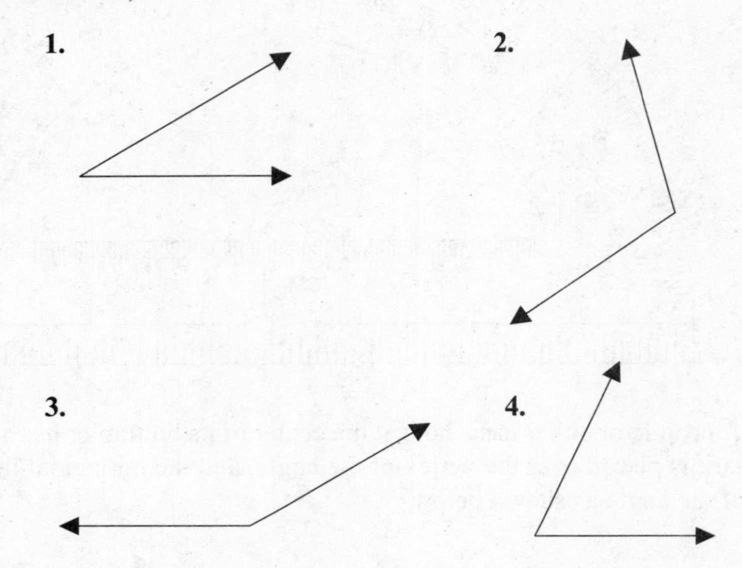

To draw an angle of a given measure, simply follow these steps:

1. Draw the initial ray however you wish on the paper. It can be parallel to the edge of the paper like angles 1, 3, and 4 above, or it can be on a slant like the initial ray in angle 2.
2. Place the center of the protractor on the vertex (the end of the initial ray you just drew).
3. Be sure to line up the horizontal line of the protractor's bottom with the initial ray.
4. Using the scale that begins with 0°, move to the desired measure and make a mark.
5. Complete the terminal ray through the point marked.

Example

Sketch an angle measuring 120°. Below, one has been done for you.

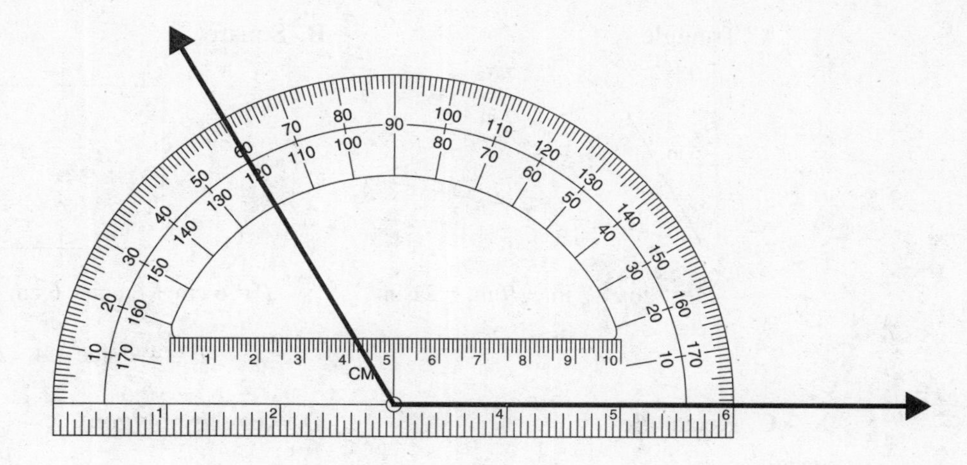

Now we just need to connect the vertex to the 120° mark and our angle is finished:

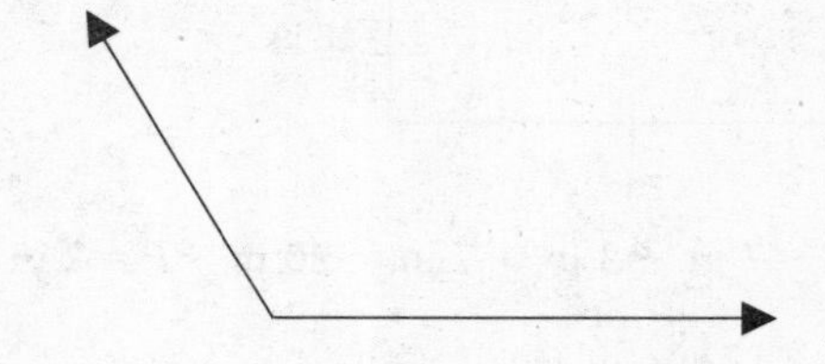

Practice Set

1. Draw an angle measuring 75°.

2. Draw a 160° angle.

3. Draw an angle that has a measure of 115°.

PERIMETER

Perimeter refers to the distance around a figure. To find the perimeter of a figure, you simply *add* up the lengths of all the sides. Perimeter is measured in linear units (line segments). The main application of perimeter is fencing around an area.

Certain figures have a formula you can follow, which saves a bit of time.

Examples

Find the perimeter of each figure shown below.

A. Triangle

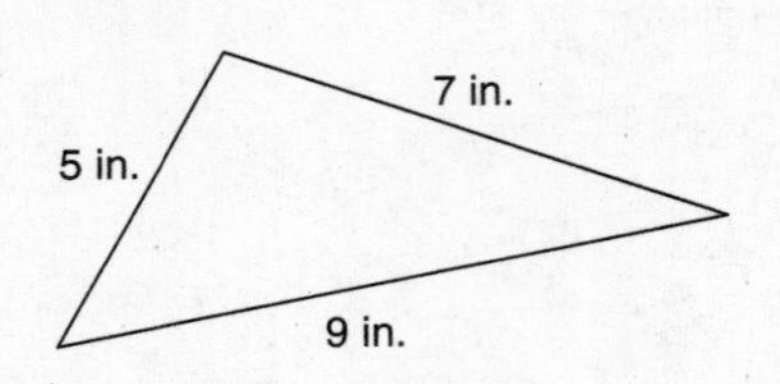

$P = 5 \text{ in.} + 7 \text{ in.} + 9 \text{ in.} = 21 \text{ in.}$

B. Square

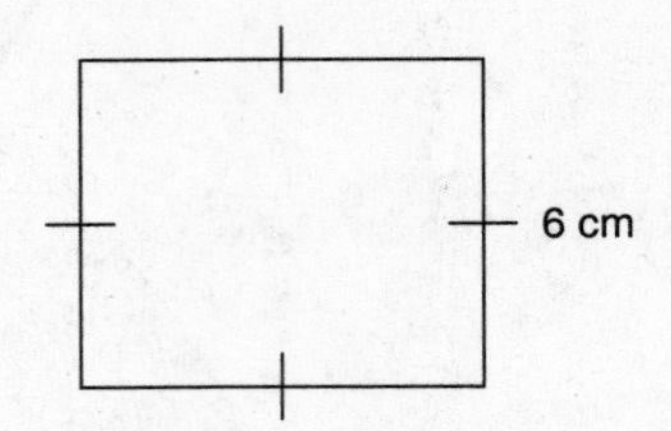

$P = 6 \text{ cm} + 6 \text{ cm} + 6 \text{ cm} + 6 \text{ cm} = 24 \text{ cm}$

or

$P = 4 \times \text{side} = 4 \times 6 = 24 \text{ cm}$

C. Rectangle

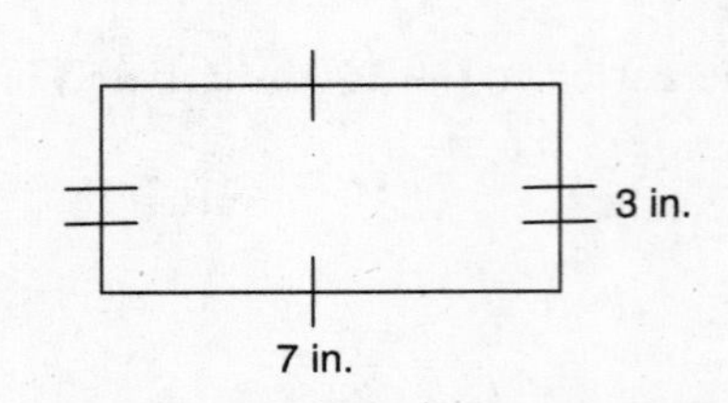

$P = 3 \text{ in.} + 7 \text{ in.} + 3 \text{ in.} + 7 \text{ in.} = 20 \text{ in.}$

or

$P = 2 \cdot \text{length} + 2 \cdot \text{width}$
$= 2 \cdot 7 + 2 \cdot 3 = 14 + 6 = 20$

D. Regular polygon

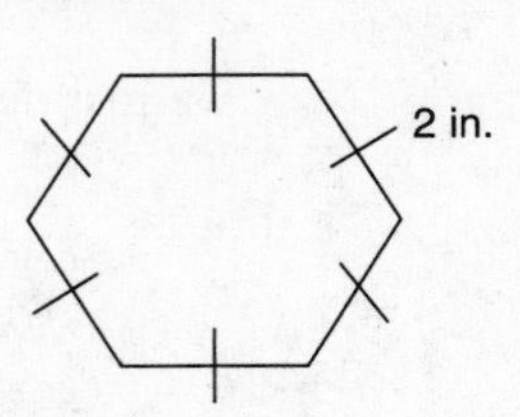

$P = 2 \text{ in.} + 2 \text{ in.} + 2 \text{ in.} + 2 \text{ in.} + 2 \text{ in.} + 2 \text{ in.} = 12 \text{ in.}$

or

$P = \text{no. of sides} \times \text{length of each side}$
$= 6 \times 2 \text{ in.} = 12 \text{ in.}$

E. General polygon

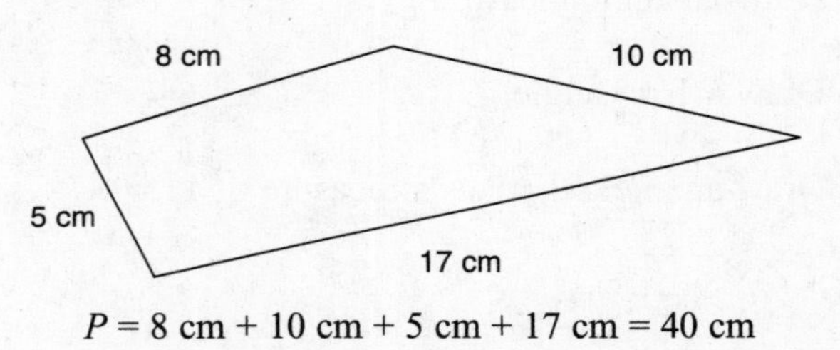

$P = 8 \text{ cm} + 10 \text{ cm} + 5 \text{ cm} + 17 \text{ cm} = 40 \text{ cm}$

PRACTICE SET

Find the perimeter of each polygon below.

1.

12 in.

5 in.

2.

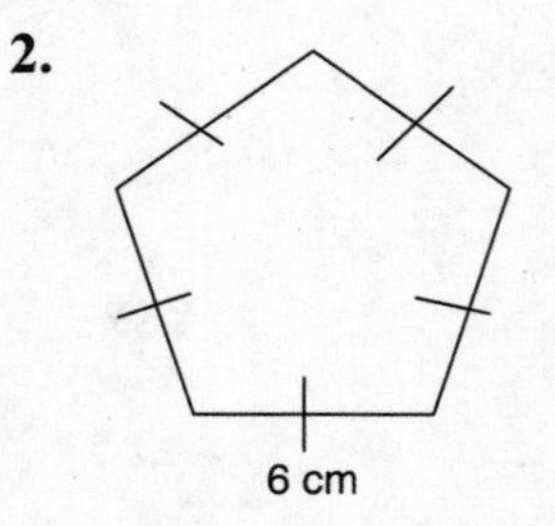

3.

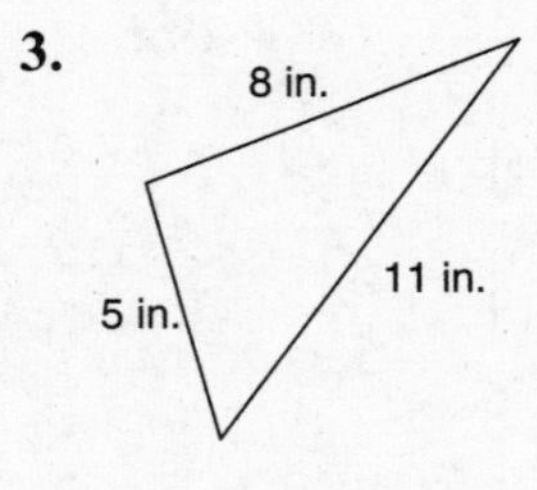

4.

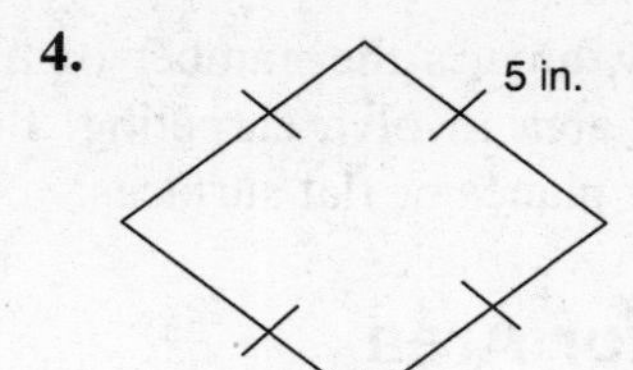

5.

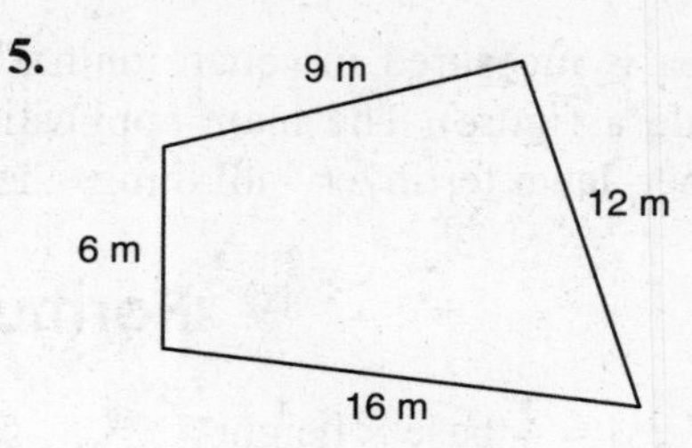

CIRCUMFERENCE

Circumference is the perimeter of a circle. In other words, the circumference is the distance around the edge of a circle. Circumference is calculated using a formula: circumference = pi × diameter = πd.

Examples

Find the circumference of each circle below.

A.

$$\begin{aligned} \text{Circumference} = C &= \pi d \\ &= 8\pi \\ &\approx 8 \times 3.14 \\ &\approx 25.12 \text{ in.} \end{aligned}$$

B.

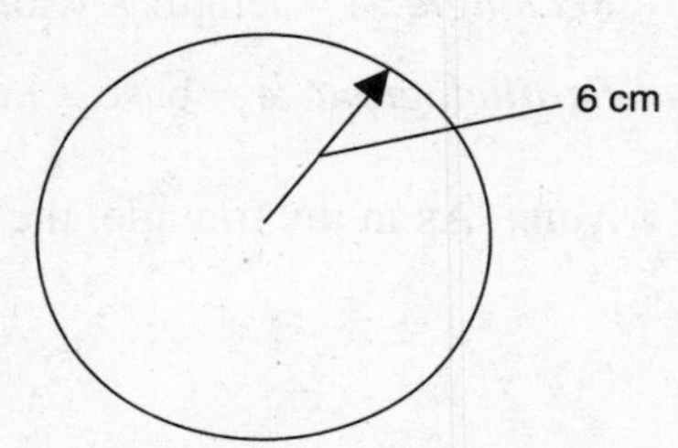

$$\begin{aligned} \text{Circumference} = C &= \pi d \\ &= 12\pi \\ &< 12 \times 3.14 \\ &< 37.68 \text{ cm} \end{aligned}$$

PRACTICE SET

1.

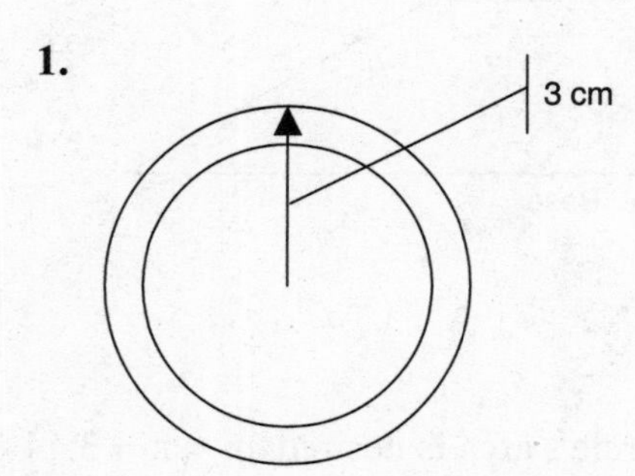

2.

AREA

Area measures the amount of space inside of or enclosed by a figure. There are three main ways to calculate the area of a given figure:

1. Using a formula (if the figure has its own formula).
2. Dividing the figure into two or more figures that have an area formula.
3. Estimating the area by drawing the figure on grid paper and approximating how many squares it covers.

Area is measured in square units (literally, area is the number of tiny squares that fit inside a figure). The main applications of area involve carpeting, floor tiles, paint, wallpaper, lawn fertilizer—all things that cover planes or flat surfaces.

Formulas for Area

Triangle: $A = \frac{1}{2}$ base × height

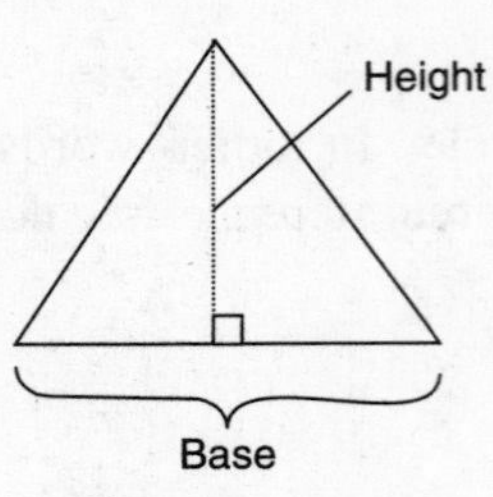

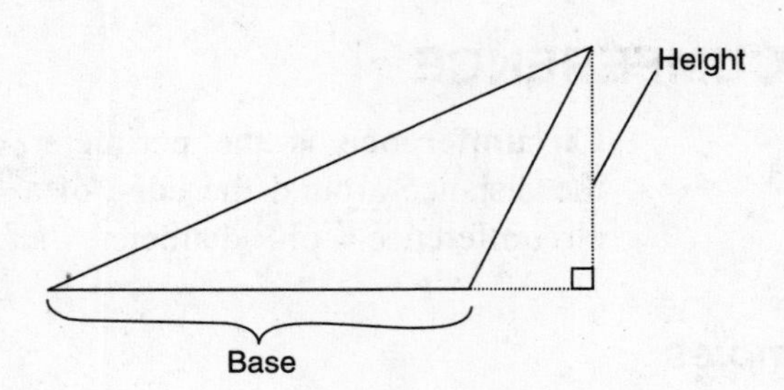

Square: $A = (\text{side})^2$

Rectangle: A = length × width

Parallelogram: A = base × height

Note: As in the triangle, the base and the height form a right angle.

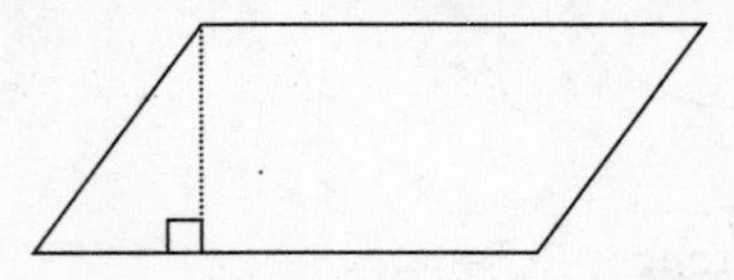

Trapezoid: $A = \frac{\text{base}_1 + \text{base}_2}{2} \times \text{height}$

Note: In the trapezoid, the height is perpendicular to each of the bases

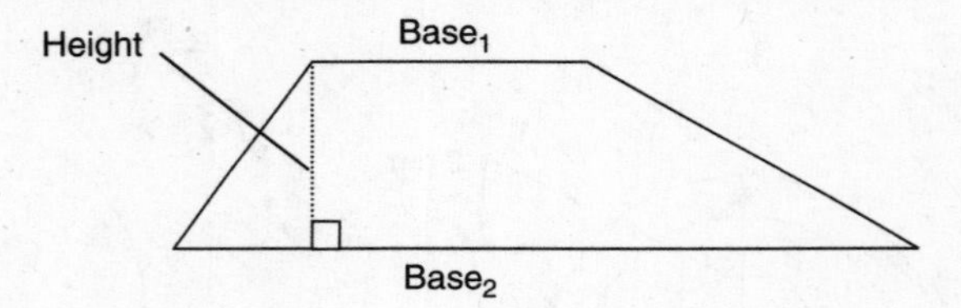

Circle: $A = \text{pi} \times \text{radius}^2 = \pi r^2$

Note: If an approximate value for a circle's area is desired, pi = $\pi \approx 3.14$ or $\frac{22}{7}$.

Examples

Find the area of each figure.

A.

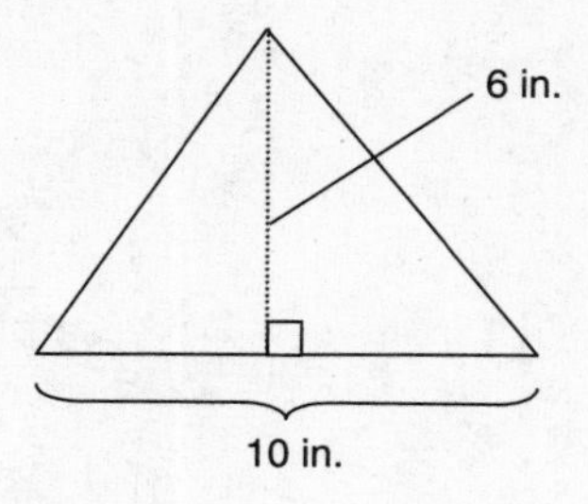

$$\text{Area} = \frac{1}{2}\text{base} \times \text{height}$$
$$= \frac{1}{2}(10 \text{ in.}) \times 6 \text{ in.}$$
$$= 5 \text{ in.} \times 6 \text{ in.} = 30 \text{ in.}^2$$

B.

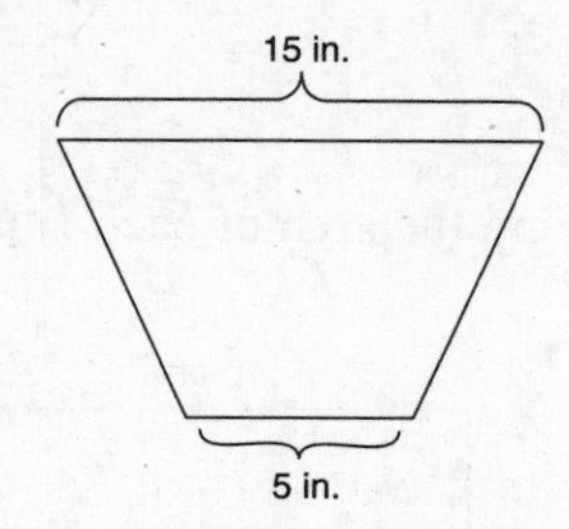

$$\text{Area} = \frac{\text{base}_1 + \text{base}_2}{2} \times \text{height}$$
$$= \frac{15 \text{ in.} + 5 \text{ in.}}{2} \times 8 \text{ in.}$$
$$= 10 \text{ in.} \times 8 \text{ in.} = 80 \text{ in.}^2$$

C.

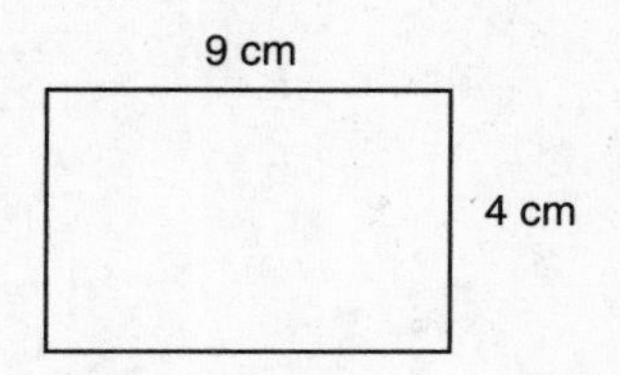

$$\text{Area} = \text{length} \times \text{width}$$
$$= 9 \text{ cm} \times 4 \text{ cm}$$
$$= 36 \text{ cm}^2$$

D.

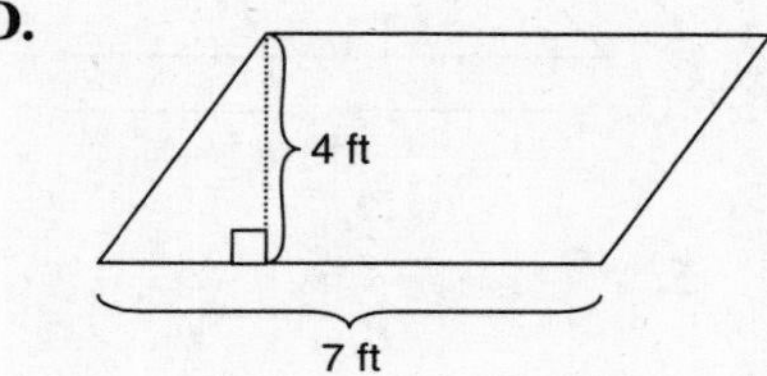

$$\text{Area} = \text{base} \times \text{height}$$
$$= 7 \text{ ft} \times 4 \text{ ft}$$
$$= 28 \text{ ft}^2$$

E.

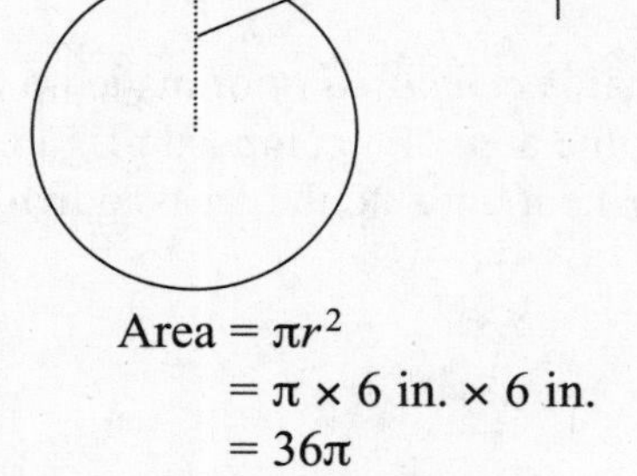

$$\text{Area} = \pi r^2$$
$$= \pi \times 6 \text{ in.} \times 6 \text{ in.}$$
$$= 36\pi$$
$$\approx 36 \times 3.14$$
$$\approx 113.04 \text{ in.}^2$$

F.

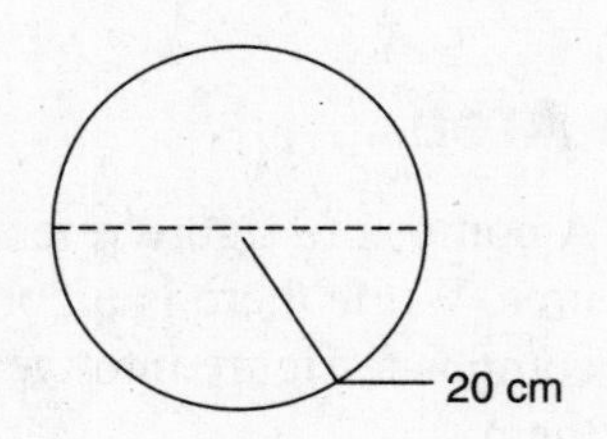

$$\text{Area} = \pi r^2$$
$$= \pi \times 10 \text{ cm} \times 10 \text{ cm}$$
$$= 100\pi$$
$$\approx 100 \times 3.14$$
$$\approx 314 \text{ cm}^2$$

There are a few important things to remember about using the formulas for area. First, in a triangle, you can take half of the base, half of the height, or half of the result when you multiply the base and height. But, remember, take half only once.

Second, for the circle, be sure that if you are given the diameter (the distance across the circle), you take half of it to find the radius. Also, realize there are two formats for the area of a circle. An answer like 64π is in what is called *exact form* or sometimes referred to as "in terms of pi." Pi is actually a nonterminating decimal, so 3.14 is an approximation; thus, when you replace π with 3.14, you are actually getting an estimated or approximate answer.

Finally, be sure that your measurements are all in the same units. That is, if the length of a rectangle is given in inches, then the width must also be in inches or converted to inches before finding the product of the length and width to get the area.

PRACTICE SET

Find the area of each figure.

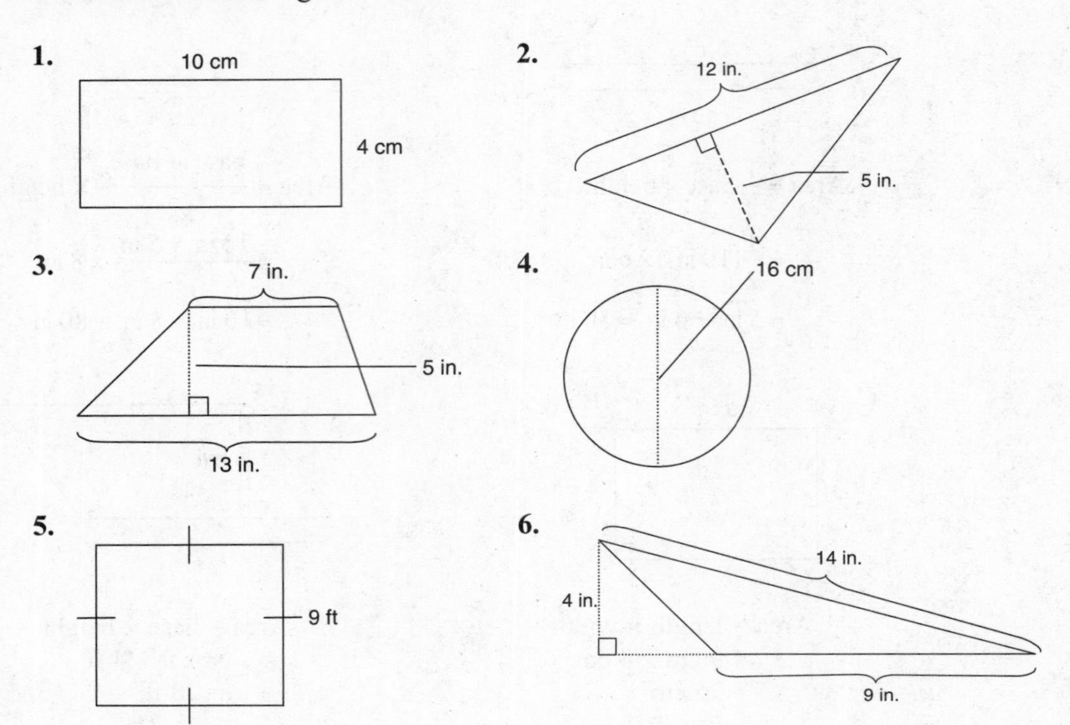

Composite Area

A composite figure refers to a figure that is *composed of* or made up of two or more figures. While there is no one formula for the area of a composite figure, we can divide the composite figure into the shapes that make it up, find the areas individually, and then add them.

Examples

Find the area of each figure.

A.

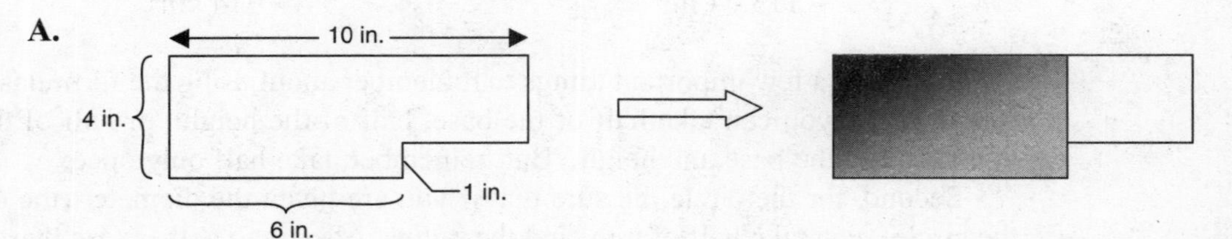

The shaded rectangle measures 4 in. × 6 in., so its area is 24 in^2. The unshaded rectangle measures 3 in. × 4 in. (You figure this out by taking 10 in. [the full length] and subtracting 6 in. [the length of the shaded part] and 4 in. [the full width minus 1 in., which is the amount that the figure is "indented"].) Thus, the area of the unshaded rectangular portion is 12 in^2. The total area is 24 in^2 + 12 in^2 = 36 in^2.

B.

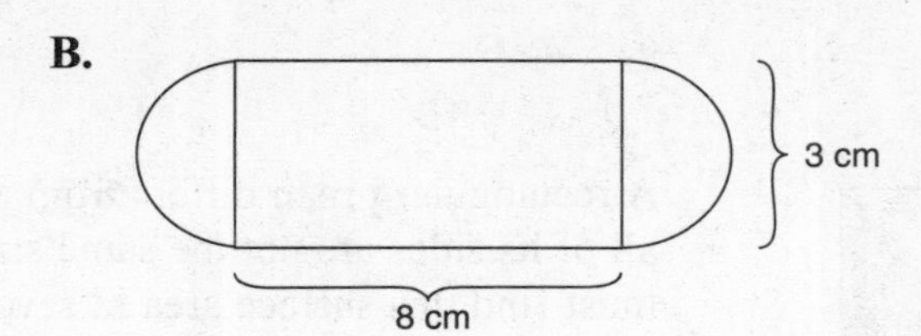

This figure is composed of two half-circles or semicircles, which really means it is one circle in two pieces. The other portion is obviously a rectangle.

Thus, the area is 8 cm × 3 cm = 24 cm^2 for the rectangle. For the circle, the radius is 1.5 cm (half the diameter of 3 cm). Thus, the circle's area is 1.5 cm × 1.5 cm × 3.14 cm or 7.065 cm^2. So, the total composite area is 24 cm^2 + 7.065 cm^2 = 31.065 cm^2.

PRACTICE SET

Find the area of each composite figure below.

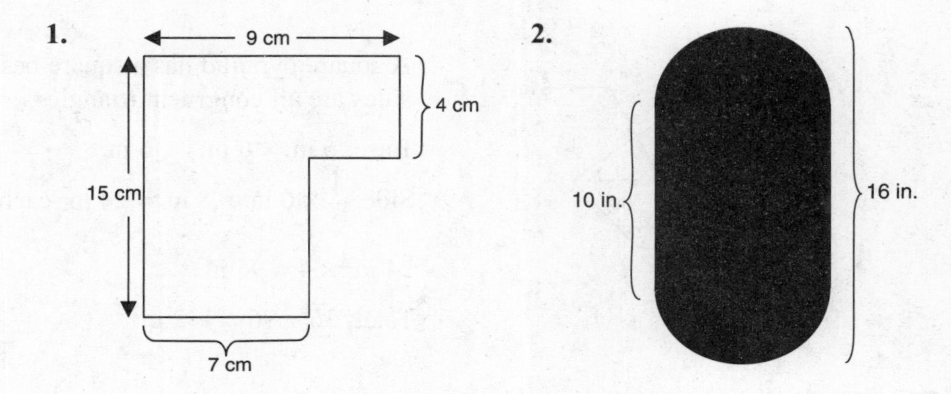

Surface Area

Surface area refers to the total area of all surfaces of a three-dimensional figure. To find surface area, you total up the individual areas of all the sides.

Examples

Find the total surface area of each figure below.

A. Cube:

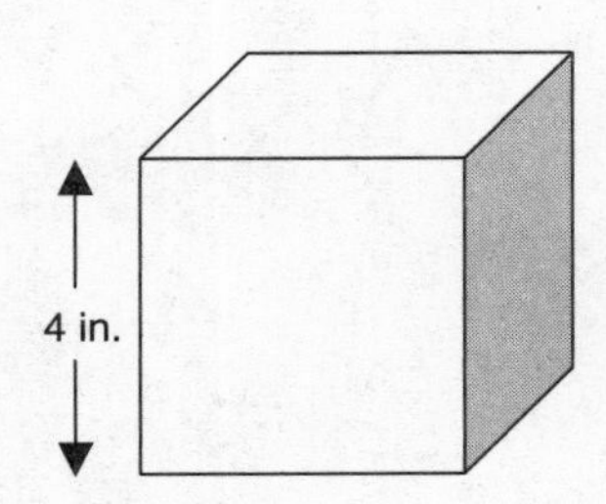

Remember: A cube is a three-dimensional figure in which all sides are squares.

Thus, the area of one side is 4 in. × 4 in. = 16 in^2.

There are six sides—front, back, top, bottom, right, and left. Thus, the total surface area is 16 in^2 × 6 or 96 in^2.

B. Rectangular prism:

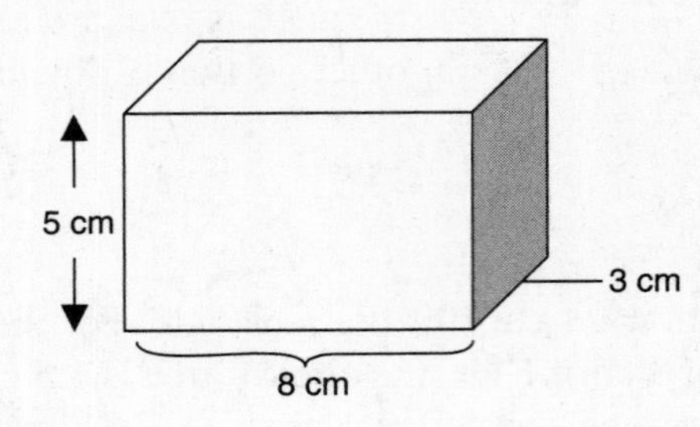

A rectangular prism differs from a cube in that all of its sides are not the same size. So, we must find this surface area in several pieces:

Front/back: 8 cm × 5 cm = 40 cm^2 each, for a total of 80 cm^2

Top/bottom: 8 cm × 3 cm = 24 cm^2 each, for a total of 48 cm^2

Right/left: 5 cm × 3 cm = 15 cm^2 each, for a total of 30 cm^2

Total: 80 + 48 + 30 = 158 cm^2

C. Square pyramid:

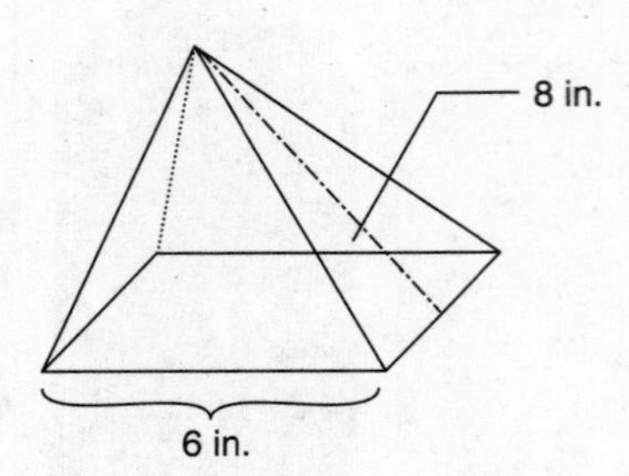

A square pyramid has a square base, and so its sides are all congruent triangles.

Base: 6 in. × 6 in. = 36 in^2

Side: $\frac{1}{2}$ × 6 in. × 8 in. = 24 in^2 each for a total of

24 in^2 × 4 = 96 in^2

Total: 36 + 96 = 132 in^2

Practice Set

Find the total surface area of each figure below.

1.

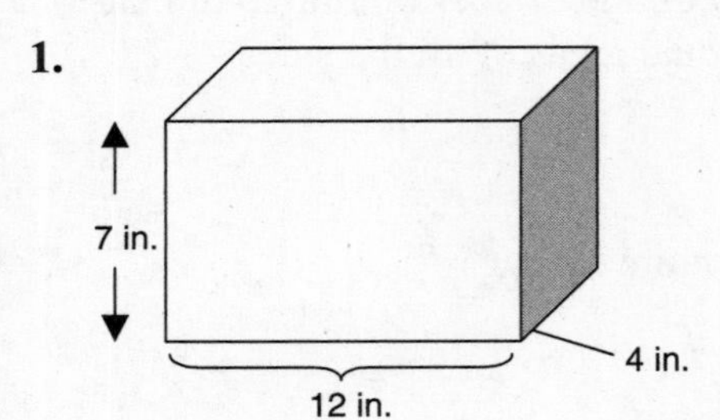

2.

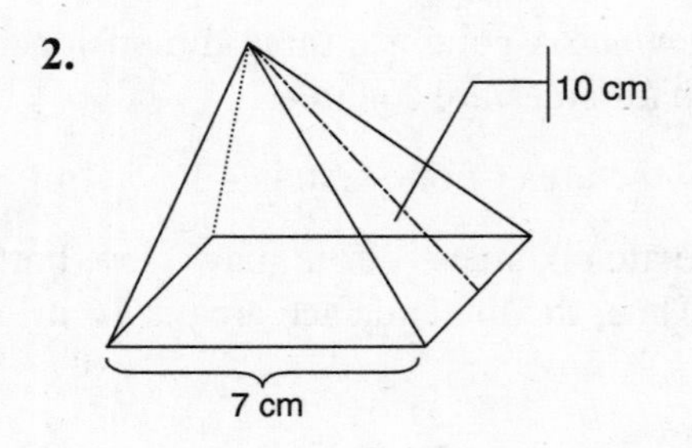

3.

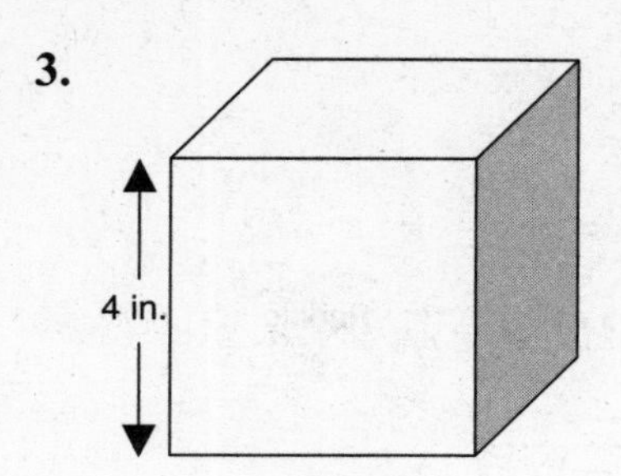

VOLUME

Volume measures the amount a figure can hold or the amount of space inside a three-dimensional figure. Volume is measured in cubic units. This means that we are counting the number of cubes 1 inch or 1 centimeter or 1 foot, and so on, on each side that fit inside the figure. Volume is calculated using formulas.

Formulas for Volume

Rectangular prism: $V = \text{length} \times \text{width} \times \text{height} = lwh$

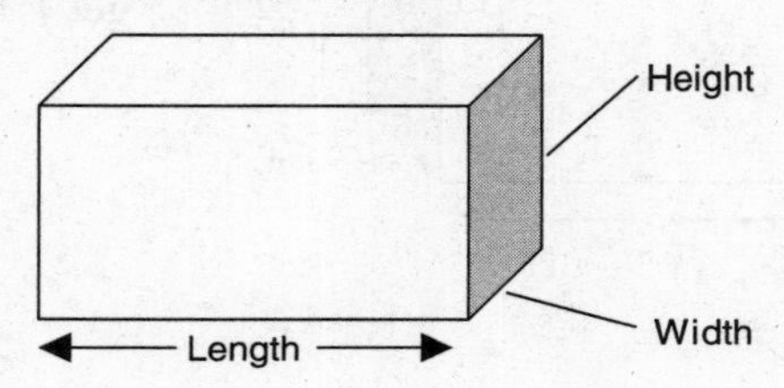

Cylinder: $V = \text{pi} \times \text{radius}^2 \times \text{height} = \pi r^2 h$

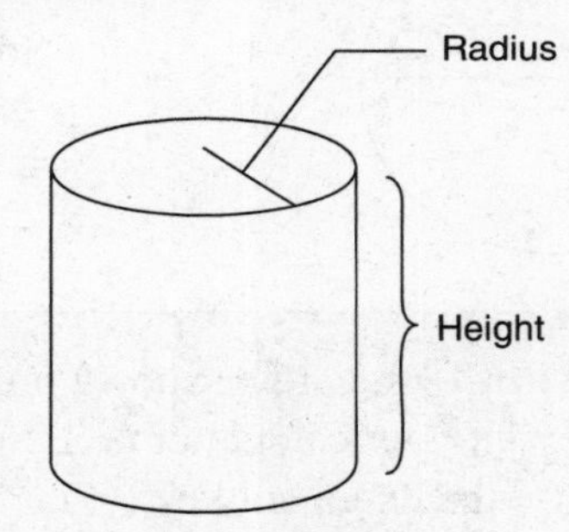

Cone: $V = \frac{1}{3} \times \pi \times \text{radius}^2 \times \text{height} = \frac{1}{3}\pi r^2 h$

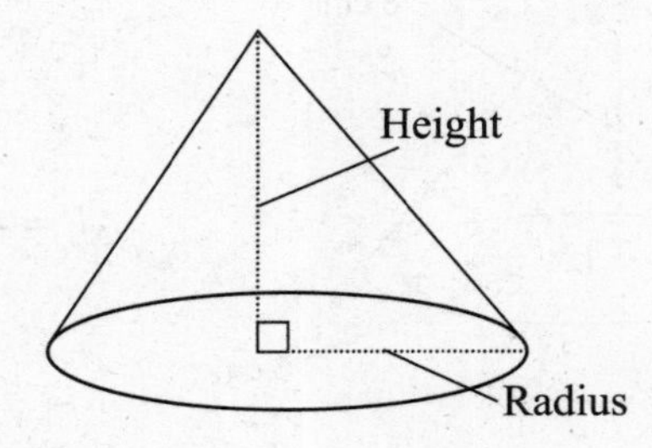

Sphere: $V = \frac{4}{3} \times \pi \times r^3 = \frac{4}{3}\pi r^3$

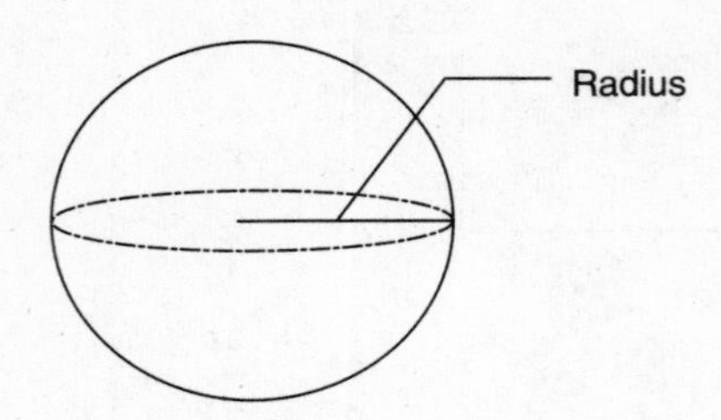

Examples

Find the volume of each figure shown below.

A.

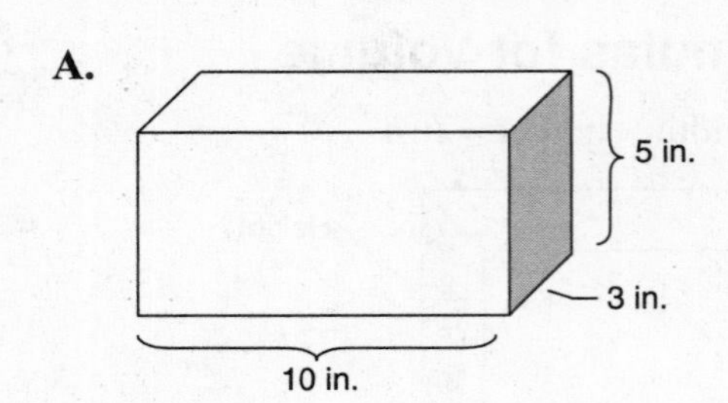

The volume formula for a rectangular prism is $V = lwh$. So, plugging in our values, we get $V = 10 \text{ in.} \times 3 \text{ in.} \times 5 \text{ in.} = 150 \text{ in}^3$.

B.

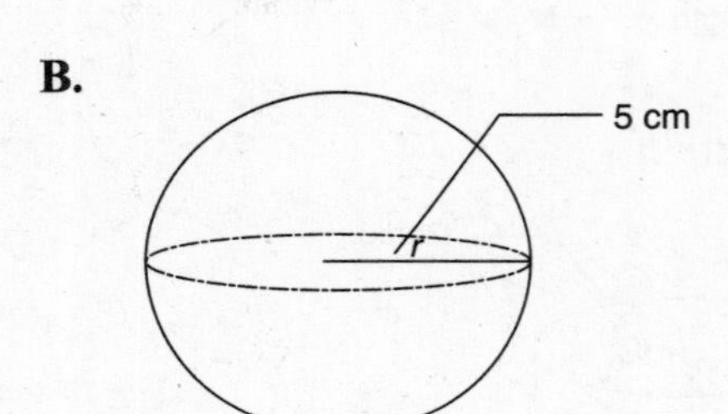

The volume formula for a sphere is $V = \frac{4}{3}\pi r^3$.

So, plugging in our values, we get $V = \frac{4}{3} \times \pi \times 5^3$

$= 166.\overline{6}\pi \approx 166.\overline{6} \times 3.14 \approx 523.\overline{3} \text{ cm}^3$.

> **Calculator Tip:** To square or cube a number: To get r^2, multiply r times r or use the square key on your calculator (x^2). Similarly, to get r^3 you can calculate r times r times r or use the exponent key (y^x or ^). Consult your calculator manual for help.

C.

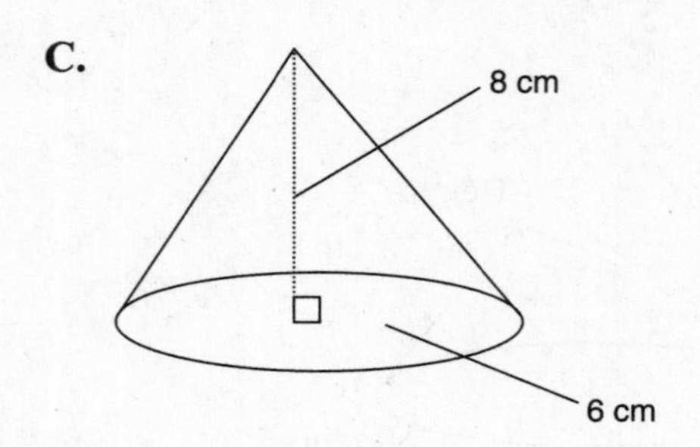

The formula for the volume of a cone is

$V = \frac{1}{3}\pi r^2 h$. So, plugging in our values, we get

$V = \frac{1}{3} \times \pi \times 6^2 \times 8 = 96\pi \approx 301.44 \text{ cm}^3$.

D.

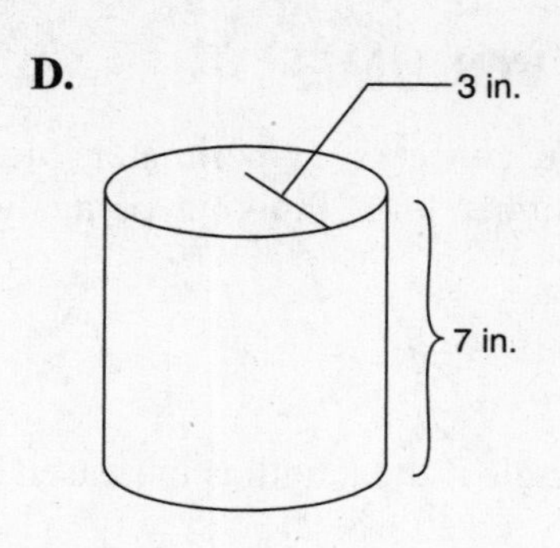

The formula for the volume of a cylinder is $V = \pi r^2 h$. So, plugging in our values, we get $V = \pi \times 3^2 \times 7 = 63\pi \approx 197.82 \text{ in}^3$.

PRACTICE SET

Find the volume of each figure below:

1.

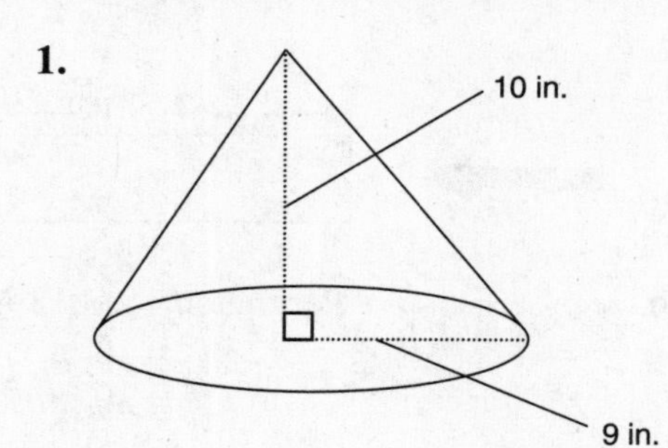

2.

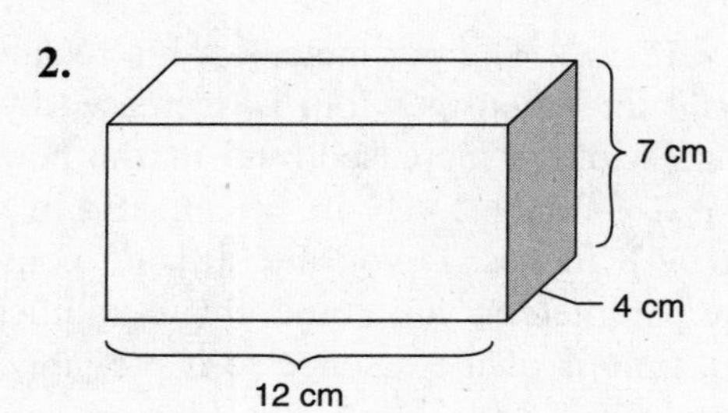

3.

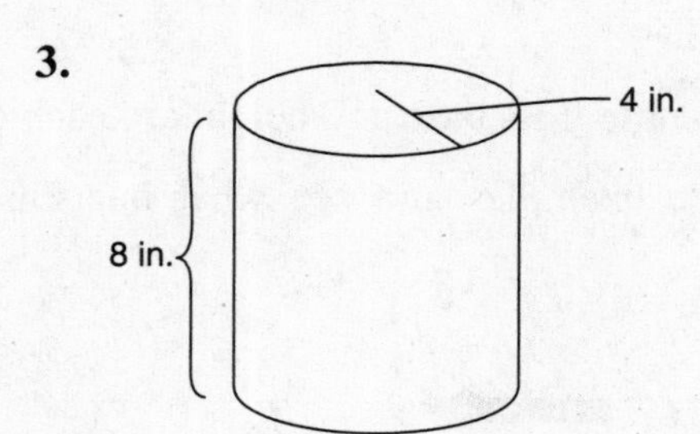

4.

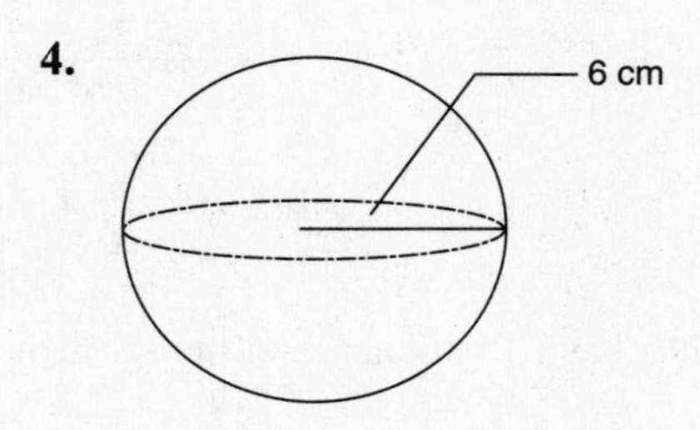

MORE ABOUT PERIMETER, AREA, AND VOLUME

On occasion, you will be asked to use the concepts of perimeter, area, and volume in a context that requires multiple steps, higher level thinking or analysis, and/or the application of more than one formula.

Examples

A. What happens to the perimeter of a rectangle if the length is increased by 5 inches and the width is decreased by 1 inch?

The best way to handle this type of question (What happens if . . . ?) is to do several examples and see if a conclusion can be made. So, assume you have the rectangle below.

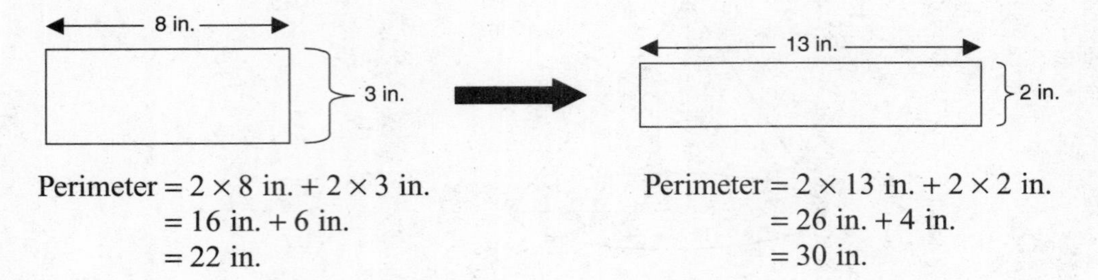

Perimeter = 2 × 8 in. + 2 × 3 in.
= 16 in. + 6 in.
= 22 in.

Perimeter = 2 × 13 in. + 2 × 2 in.
= 26 in. + 4 in.
= 30 in.

Try a second rectangle (5 in. × 12 in.). The perimeter of this rectangle is 2 × 5 in. + 2 × 12 in. = 10 in. + 24 in. = 34 in. When its width is decreased by 1 in., the new width will be 4 in., and when its width is increased by 5 in. the new length will be 17 in. Thus, the new perimeter is 2 × 4 in. + 2 × 17 in. = 8 in. + 34 in. = 42 in. In both cases, the perimeter increased by 8 inches. If you are still not convinced this will always be the case under these parameters, you could create another rectangle, but remember, on the HSPA Exam, time is of the essence so it probably wouldn't be to your benefit to do too many "trials."

B. What happens to the area of a triangle if its base and height are each doubled?

Again your best bet is to create examples and see what happens. Consider the triangle below.

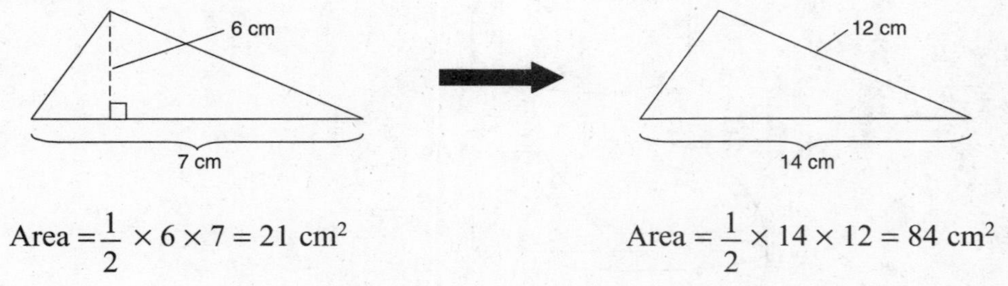

Area $= \frac{1}{2} \times 6 \times 7 = 21$ cm^2

Area $= \frac{1}{2} \times 14 \times 12 = 84$ cm^2

Let's consider a second triangle with base 15 inches and height 8 inches. Its area is $\frac{1}{2} \times 15 \times 8 = 60$ in^2. When its height and base are doubled, the measurements are 16 inches and 30 inches, respectively. Thus, the new area is $\frac{1}{2} \times 16 \times 30 = 240$ in^2.

Thus, the area is multiplied by 4.

C. Find the area of the shaded portion below.

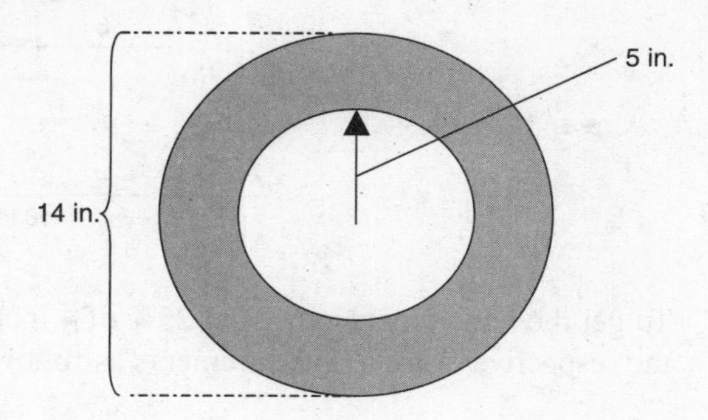

This area can be found by subtracting the area of the smaller circle (unshaded) from the area of the larger circle. So begin by finding the area of the larger circle. The area of a circle is given by the formula $A = \pi r^2$. The radius of the large circle is half of 14 inches (the diameter), or 7 inches. Thus, its area is 49π. The inside circle has the radius given (5 inches). Thus, its area is 25π. So, subtracting gives us an area of $49\pi - 25\pi = 24\pi \approx 24 \times 3.14 \approx 75.36$.

D. Find the perimeter of the figure below.

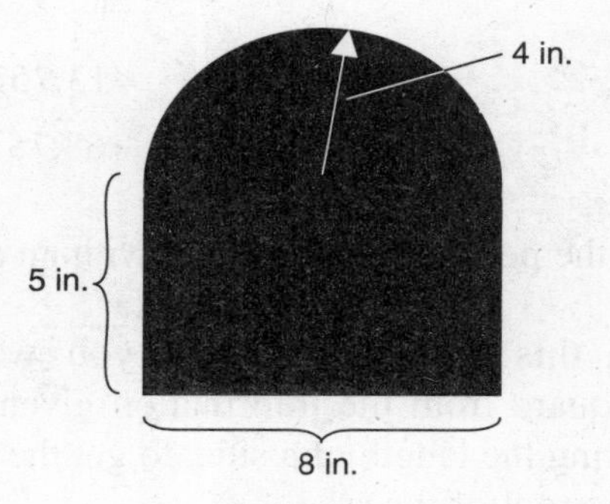

When you are asked to find the perimeter of an irregular-looking shape, you should break the perimeter down into pieces (or even trace it and record each segment). This figure is made up of three segments (5 inches, 8 inches, and 5 inches).

The figure also contains half of a circle. So to find this perimeter, we are really going to use circumference (the perimeter of a circle). After we get the circumference we are going to take half of it since we only have half of a circle (called a *semicircle*).

The radius of the circle is 4 inches. Its circumference is $C = \pi \times d = \pi \times 8$ in. $\approx 3.14 \times 8$ in. ≈ 25.12 in. This circumference is divided in half and then added to the linear perimeter (linear means it is made up of lines). Thus, $25.12 \div 2 = 12.56$. Combined, $12.56 + 5 + 8 + 5 = 30.56$ in.

E. Find the area of the trapezoid below if each side's measurement is increased by 25%.

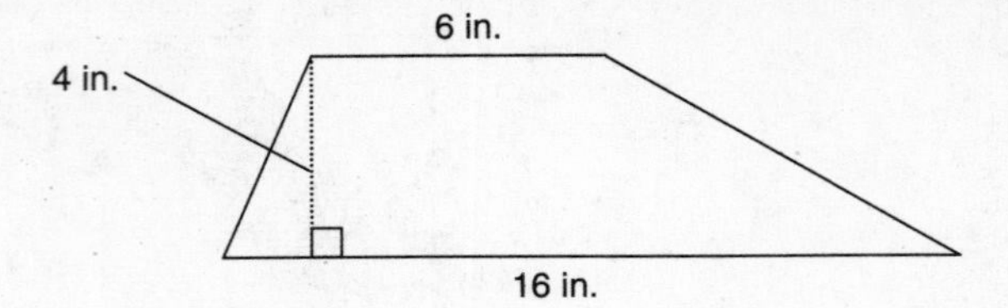

To get the new dimensions, find 25% of 4 inches, 6 inches, and 16 inches and add to the respective original measurements as follows.

$.25 \times 4$ in. = 1 in., so 4 in. + 1 in. = 5 in. (new measurement)
$.25 \times 6$ in. = 1.5 in., so 6 in. + 1.5 in. = 7.5 in. (new measurement)
$.25 \times 16$ in. = 4 in., so 16 in. + 4 in. = 20 in. (new measurement)

Recall the area formula for a trapezoid and plug in your values:

$$\text{Area} = \left(\frac{\text{base}_1 + \text{base}_2}{2}\right) \times \text{height}$$
$$= \left(\frac{20\text{ in.} + 7.5\text{ in.}}{2}\right) \times 5\text{ in.}$$
$$= 13.75 \times 5$$
$$= 68.75\text{ in}^2$$

F. Find the perimeter of a square with an area of 49 cm^2.

With this type of problem you work backward to find the dimensions of the square from the information given. Since the area of a square is calculated by squaring the length of a side, to get the measurement of the side, you have to take the square root of the area.

The square root of 49 is 7, so the square is 7 cm $\times$ 7 cm. Thus, its perimeter is 4×7 cm = 28 cm.

PRACTICE SET

1. Sketch two different rectangles that have an area of 24 in^2. Be sure to label the dimensions.

2. Find the area remaining if a square with perimeter 20 cm is cut from a rectangle measuring 7 cm $\times$ 12 cm.

3. What happens to a rectangle's perimeter if the length is increased by 5 inches and the width is decreased by 3 inches?

4. What is the area of the shaded region in the figure below?

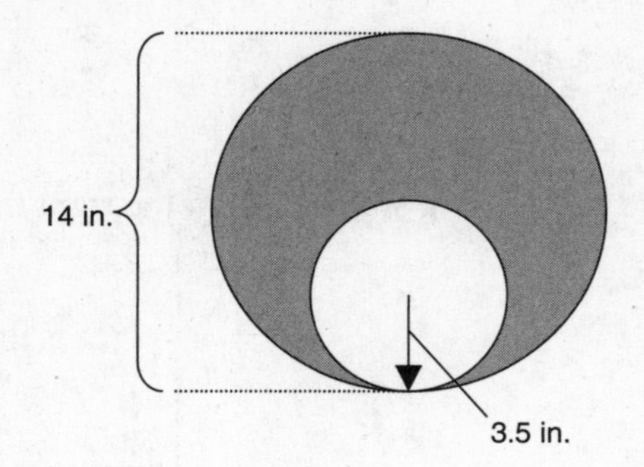

5. Find the area of the figure below.

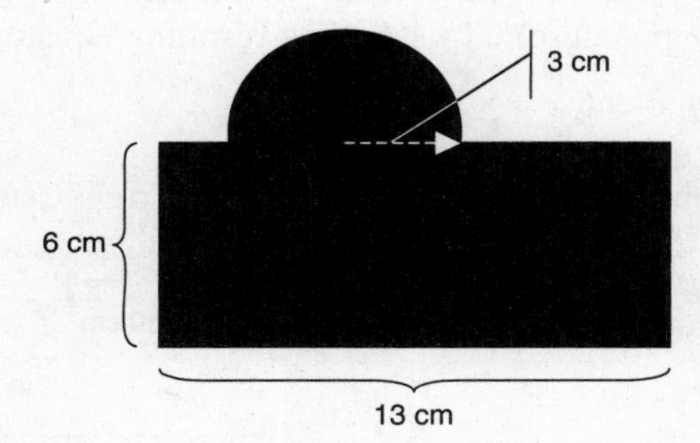

6. Find the area of the trapezoid below if each dimension is increased by 15%.

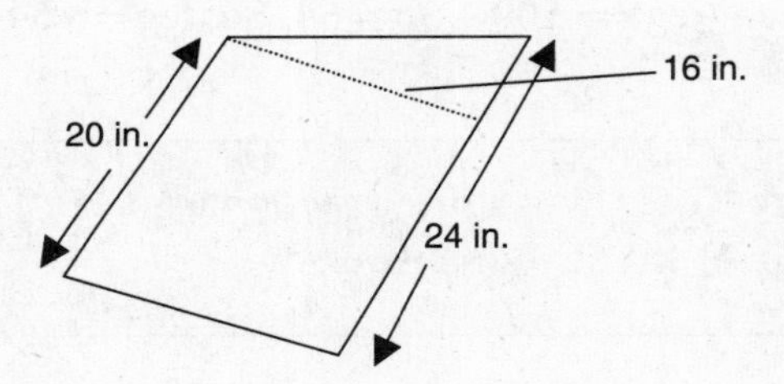

PYTHAGOREAN THEOREM

The **Pythagorean Theorem** is a special rule that applies to right triangles; this rule is named for the Greek mathematician Pythagoras. The theorem states that in any right triangle, the sum of the squares of the legs equals the square of the hypotenuse. To clarify, since this statement involves some vocabulary unique to right triangles, this means

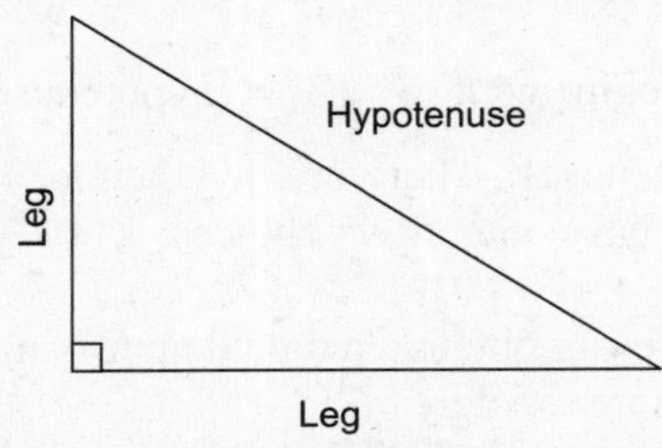

The legs of a right triangle are the two sides of the triangle that meet to form the right angle. The hypotenuse of a right triangle is the side across from the right angle.

The hypotenuse is the longest side. The Pythagorean Theorem says $(\text{leg})^2 + (\text{leg})^2 = (\text{hypotenuse})^2$.

Examples

A. Find the missing side length in the right triangle below.

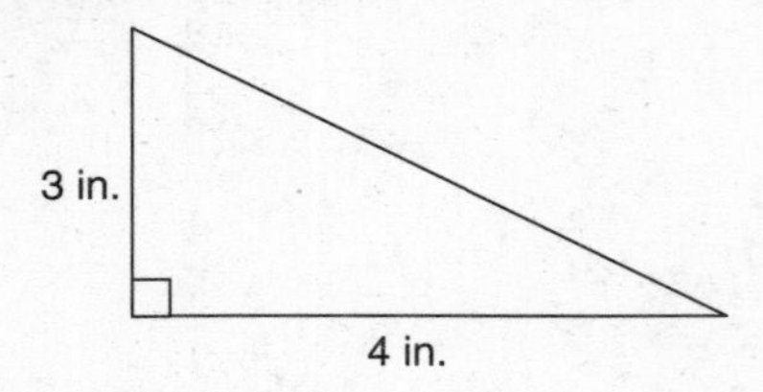

By the Pythagorean Theorem, $3^2 + 4^2 = (\text{hypotenuse})^2$, so $9 + 16 = 25 = (\text{hypotenuse})^2$. To solve the resulting equation, take the square root of both sides; so hypotenuse $= \sqrt{25} = 5$.

B. Find the missing side length in the right triangle below.

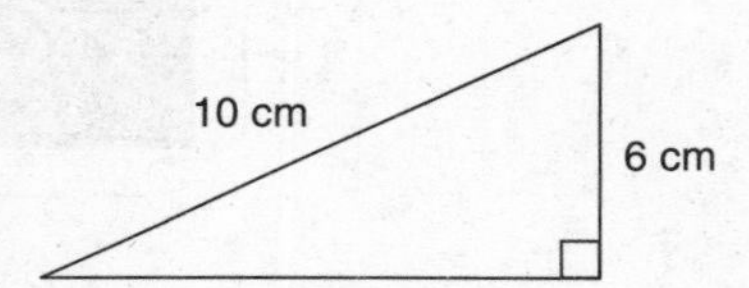

Again, by the Pythagorean Theorem, $6^2 + (\text{leg})^2 = 10^2$. Thus, $36 + (\text{leg})^2 = 100$, and then $(\text{leg})^2 = 100 - 36 = 64$. So, leg $= \sqrt{64} = 8$.

> **Take Note:** When using the Pythagorean Theorem, be sure to take the square root in order to get your answer.

C. Find the length of the missing side in the right triangle given.

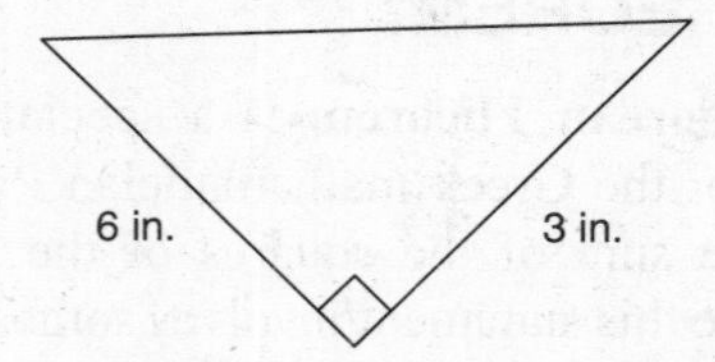

We begin with $6^2 + 3^2 = (\text{hypotenuse})^2$ which simplifies to give $36 + 9 = 45 = (\text{hypotenuse})^2$. Therefore, hypotenuse $= \sqrt{45}$. The square root of 45 is not an integer, so we have one of two choices:

1. We can obtain a decimal approximation for $\sqrt{45}$ using a calculator. This gives us approximately 6.71.
2. We can simplify the square root: $\sqrt{45} = \sqrt{9} \times \sqrt{5} = 3\sqrt{5}$.

D. Find the length of the missing side in the right triangle given below.

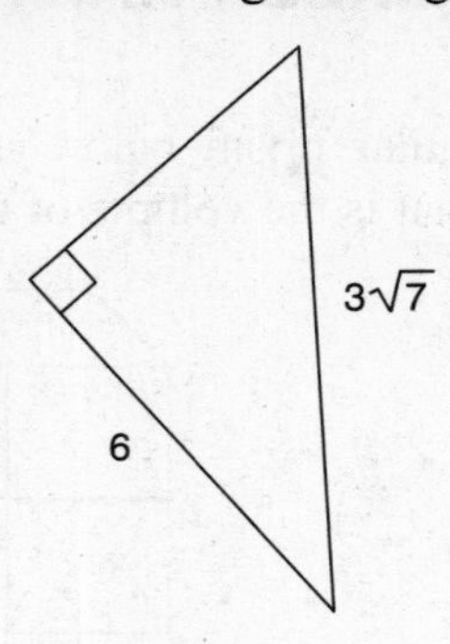

We begin with $6^2 + (\text{leg})^2 = (3\sqrt{7})^2$, which yields $36 + (\text{leg})^2 = (3\sqrt{7}) \times (3\sqrt{7}) = 9\sqrt{49} = 9 \times 7 = 63$. Thus, 36 + $(\text{leg})^2$ = 63 and $(\text{leg})^2$ = 63 − 36 = 18. Finally, taking the square root of both sides gives $\text{leg} = \sqrt{18} = \sqrt{9} \times \sqrt{2} = 3\sqrt{2}$.

PRACTICE SET

Find the missing side length in each right triangle shown below.

1.

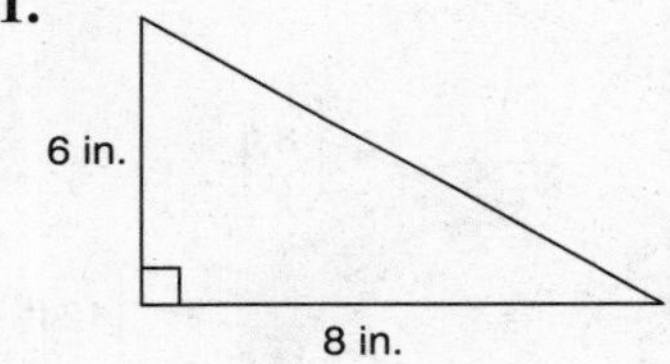

2.

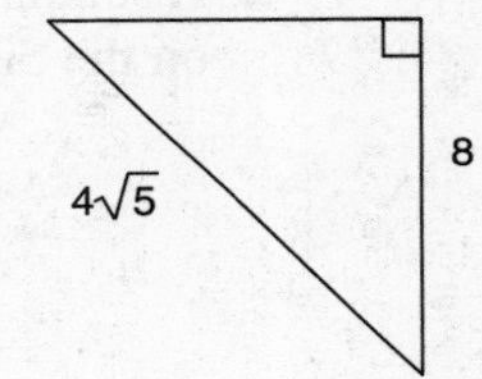

3.

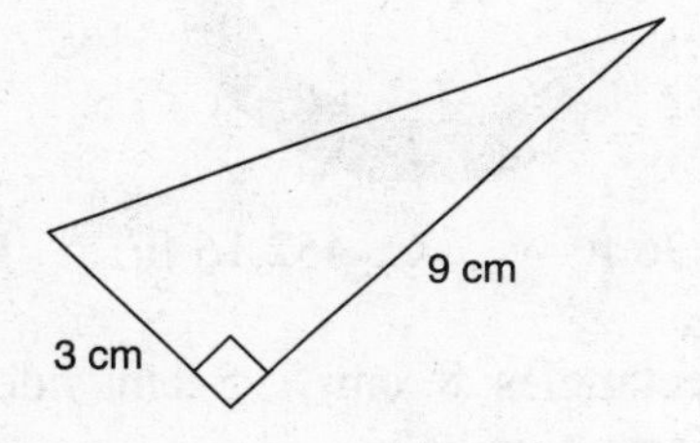

4.

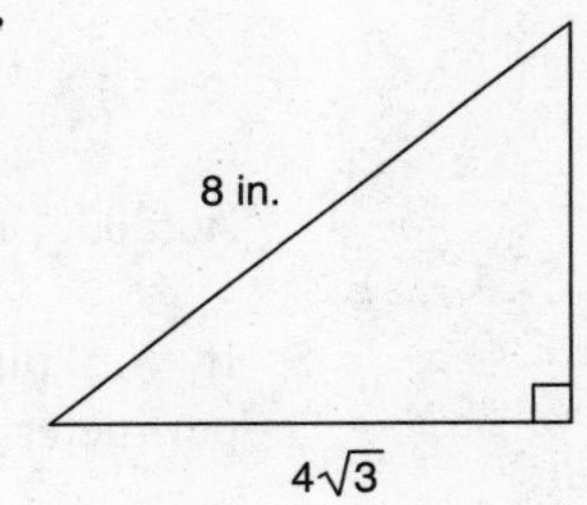

MIXED PRACTICE, CLUSTER II, MACRO C

1. The rectangular prism below is made up of small cubes each 2.5 in. × 2.5 in. × 2.5 in. What is the volume of the rectangular prism?

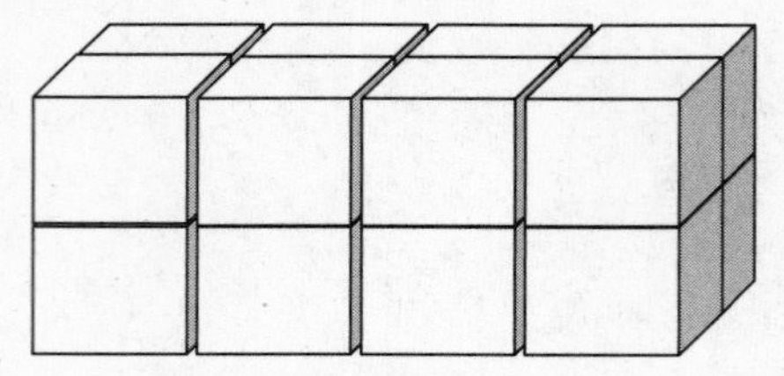

A. 8 in^3 **B.** 16 in^3 **C.** 100 in^3 **D.** 250 in^3

2. If a unicycle's tire has a diameter of 30 inches, about how many times does the tire complete a revolution when the unicycle travels 35 feet?

A. 94 **B.** 35 **C.** 8 **D.** 4.5

3. If Mark has to store 1-in. cubes in boxes that are 6 in. × 8 in. × 3 in., how many cubes will fit in five boxes?

4. A sundial is surrounded by a walkway 8 feet wide. Find the area of the walkway based on the picture below.

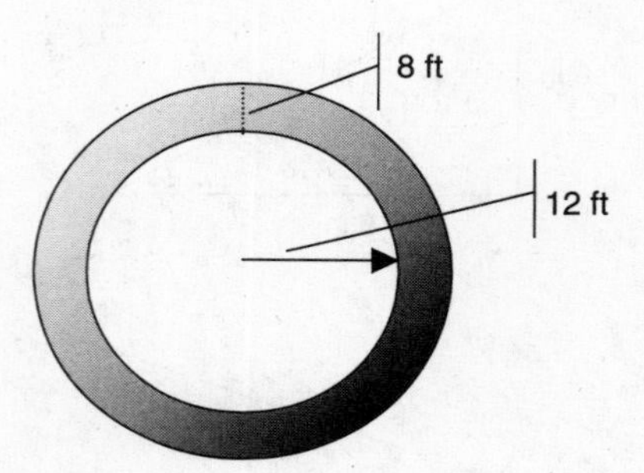

A. 803.4 ft^2 **B.** 200.96 ft^2 **C.** 452.16 ft^2 **D.** 351.68 ft^2

5. If you placed a dozen rectangles 8 cm × 5 cm side by side, what would the perimeter of the resulting formation be?

Example: Three rectangles

6. If the size of a quilt square 6 in. × 6 in. doubles to 12 in. × 12 in., by what percent does the area of the square increase?

7. If a 50-foot ladder is placed $2\frac{1}{2}$ feet from the base of a building, how far up the building will the ladder rest? (Round your answer to the nearest tenth of a foot.)

A. 50 feet **B.** 49.9 feet **C.** 48.7 feet **D.** 47.5 feet

8. The two rectangular prisms below have the same volume. The height of the second prism is 25% greater than the height of the first prism. What is the width of the second prism? Round your answer to the nearest tenth of a meter.

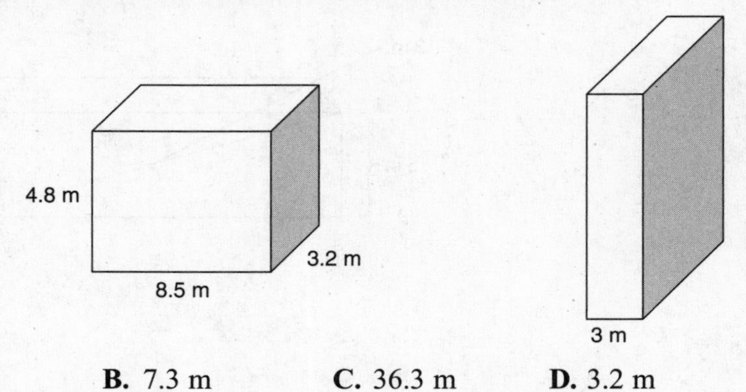

A. 6 m **B.** 7.3 m **C.** 36.3 m **D.** 3.2 m

9. What percent of the area of the rectangle below is shaded?

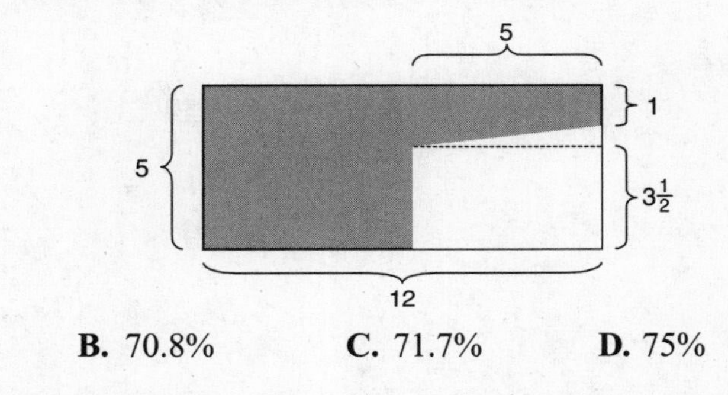

A. 68.75% **B.** 70.8% **C.** 71.7% **D.** 75%

10. If the area of a trapezoid is 60 square inches and the height is 6 inches, what are the lengths of the bases? Is there just one possible answer? Explain.

11. The perimeter of an equilateral triangle is the same as that of a rectangle that measures 5.5 cm × 8 cm. Explain the steps you would follow to find the measurements of the equilateral triangle.

12. Find the approximate area outlined below.

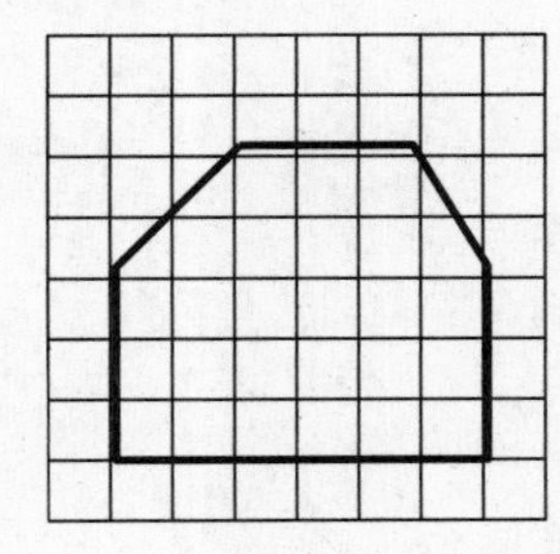

A. 24 square units
B. 26 square units
C. 27 square units
D. 28 square units

13. Find the volume of the prism below and then sketch and label a second prism that has the same volume but less surface area. State the volume and surface area of the prism you create. Show all your work.

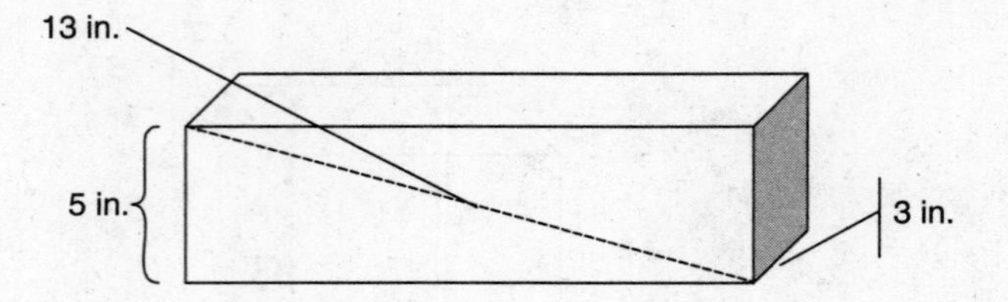

CLUSTER III

DATA ANALYSIS, PROBABILITY, STATISTICS, AND DISCRETE MATHEMATICS

MACRO A

PROBABILITY

Probability is a branch of mathematics dealing with the chances that something will happen. This something is typically referred to as an **event**. The probability that an event will occur is always between 0 and 1.

A probability of 0 means the event is impossible and will not occur.
A probability of 1 means the event is certain to occur.

A probability between 0 and 1 means that an event is possible. To calculate the probability, set up the following ratio and reduce if possible:

$$P(\text{event}) = \frac{\text{no. of favorable outcomes}}{\text{no. of possible outcomes}}$$

$P(A)$ is read "the probability of A." The number of possible outcomes is sometimes referred to as the "size of the sample space." Events are *equally likely* if they all have the same chance of occurring, but if there are, for example, more marbles of a certain color, the events will have different probabilities.

Examples

A. Find the probability of rolling a 2 or a 3 when rolling a standard 6-sided die.

The number of favorable outcomes is 2. (We either roll a 2 or we roll a 3.) The total number of outcomes is obviously 6 since the die has 6 sides. Thus, the probability of rolling a 2 or a 3 is

$$P(\text{rolling a 2 or 3}) = \frac{2}{6} = \frac{1}{3}$$

B. Given a bag containing 15 marbles—4 are blue and 11 are red—what is the probability of choosing a blue marble if you choose one at random from the bag?

There are only 4 blue marbles, so there are only 4 favorable outcomes out of 15 total marbles. So, the probability of choosing a blue is $\frac{4}{15}$.

C. Given the numbers 2, 5, 6, 7, 8, 9, 11, 12, 14, 16, what is the probability of choosing an even number if you choose one number at random?

There are 6 even numbers (2, 6, 8, 12, 14, 16) out of 10 numbers. So the probability is

$$P(\text{no. is even}) = \frac{6}{10} = \frac{3}{5}$$

D. Given the same numbers as in Example C, what is the probability that you will choose a number that is a factor of 12?

There are 3 factors of 12 on the list (2, 6, 12), so the probability is $\frac{3}{10}$.

In all these examples, we found what is called **theoretical probability**. This means it is based on what is expected, not on what is guaranteed to happen. If we conduct an experiment and actually roll a die or choose a number from the list at random, the results represent what is called **experimental probability**. The more trials or times you conduct a given experiment, the closer the experimental probability will be to the theoretical probability. Thus, for example, in Example A we found the theoretical probability of rolling a 2 or a 3 on a standard die to be $\frac{1}{3}$, which means that for every 3 times we roll the die, we should get one 2 or 3. This does not mean that if we actually roll a die 3 times we will get exactly one 2 or 3. However, if we roll the die 60 times, one-third of these rolls (or 20) should be a 2 or a 3. The more times we roll, the closer we will get to the theoretical probability.

PRACTICE SET

1. Find the probability of choosing a vowel given the letters A, B, E, F, H, J, L, M, O, Q, S, T.

2. What is the probability of choosing an odd number given the numbers 1, 4, 5, 9, 11, 13?

3. Find the probability of choosing a factor of 18 from these numbers: 1, 2, 3, 4, 5, 6, 8, 9, 12, 15.

4. If you randomly choose a letter from the word MATHEMATICS, what is the probability that you will choose a consonant?

5. If you have a box that contains 12 red marbles, 8 green marbles, and 4 blue marbles, what is the probability that a randomly chosen marble will be blue?

Multiplication Formula

On some occasions, we use two different sets of outcomes and ask for a probability combining two of these outcomes. For example, a classic instance is rolling a die and tossing a coin. To find the probability of two **independent events**, events that are disconnected or separate, we multiply the two individual probabilities.

Examples

A. If a fair die is rolled and a coin is tossed simultaneously, what is the probability of getting heads and an even number?

$P(A \text{ and } B) = P(A) \times P(B)$, where A is "getting heads" and B is "rolling an even number."

$P(A) = P(\text{getting heads}) = \frac{1}{2}$ since the only possibilities are heads and tails, one of which is our desired outcome.

$P(B) = P(\text{rolling an even number}) = \frac{3}{6} = \frac{1}{2}$ since there are 3 even numbers (2, 4, 6) out of 6.

Thus, $P(A \text{ and } B) = P(\text{getting heads and rolling an even number}) = \frac{1}{2} \times \frac{1}{2} = \frac{1}{4}$.

B. Given cards with the numbers 1, 2, 4, 5, 9, 10, 12, 15, 18 and cards with the letters A, B, E, F, G, I, L, O, what is the probability of choosing a multiple of 3 and a vowel if the cards are in two separate boxes?

$P(\text{choosing a multiple of 3 and a vowel}) = P(\text{choosing a multiple of 3}) \times P(\text{choosing a vowel})$. The probability of choosing a multiple of 3 is $\frac{4}{9}$ since 4 numbers (9, 12, 15, 18) are multiples of 3. The probability of choosing a vowel is $\frac{4}{8}$ since there are 4 vowels on the list (A, E, I, O). This probability reduces to $\frac{1}{2}$. Thus, $P(\text{choosing a multiple of 3 and a vowel}) = \frac{4}{9} \times \frac{1}{2} = \frac{4}{18}$.

If we are faced with events that are not independent (called **dependent events**, which means that the second event is based on or dependent on the first event), then we use a slightly different set of steps.

1. Find the probability of the first event.
2. Reduce the sample space (*sample space* simply refers to the materials you are working with, e.g., numbers, letters, cards, marbles) to reflect the first event occurring.
3. Find the probability of the second event.
4. Multiply and/or add; addition is necessary when there is more than one order of events that can lead to the desired outcome (e.g., if we want an ace of hearts and a queen of spades, this can happen in either order).

Examples

A. If you have 8 gray socks, 12 white socks, and 4 navy blue socks, what is the probability of choosing a pair of white socks if two socks are chosen randomly?

This probability becomes $P(\text{choosing a white sock and then choosing another white sock})$. $P(\text{choosing a white sock}) = \frac{12}{24} = \frac{1}{2}$. Once a white sock is removed, we have a sample space of 8 gray, 11 white, and 4 navy. Thus, $P(\text{choosing a second white sock}) = \frac{11}{23}$.

So, P(choosing a white sock and then choosing another white sock) $= \frac{1}{2} \times \frac{11}{23} = \frac{11}{46}$.

B. Given the letters A, B, E, F, G, H, I, L, O, P, what is the probability of choosing a vowel and a consonant if we choose two letters at random? Since it doesn't matter if we choose a vowel first and then a consonant or a consonant first and then a vowel, we must consider both orders of events:

$$P(\text{vowel and consonant}) = P(\text{vowel}) \times P(\text{consonant}) + P(\text{consonant}) \times P(\text{vowel})$$

$$= \frac{4}{10} \times \frac{6}{9} + \frac{6}{10} \times \frac{4}{9}$$

$$= \frac{2}{5} \times \frac{2}{3} + \frac{3}{5} \times \frac{4}{9}$$

$$= \frac{4}{15} + \frac{4}{15}$$

$$= \frac{8}{15}$$

This probability can also be obtained by looking at what is called a **tree diagram** or at least a partial tree diagram. A tree diagram shows the outcomes written out as follows.

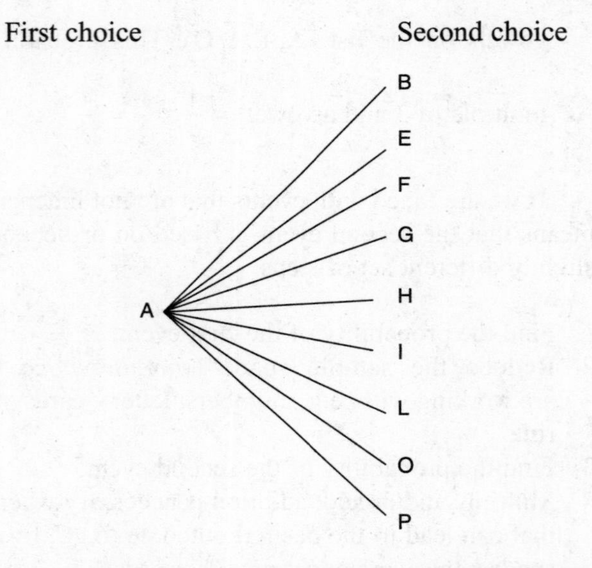

Since each letter will pair up with every letter other than itself, all 10 letters will pair up with 9, for a total of 90 combinations. For the letter A, there are 6 combinations that pair it with a vowel. So, for each of the other vowels, there are 6 combinations that pair a vowel with a consonant. All together there are 24 combinations based on a vowel-consonant combination.

Similarly, we could show a sample of a tree diagram with a consonant first:

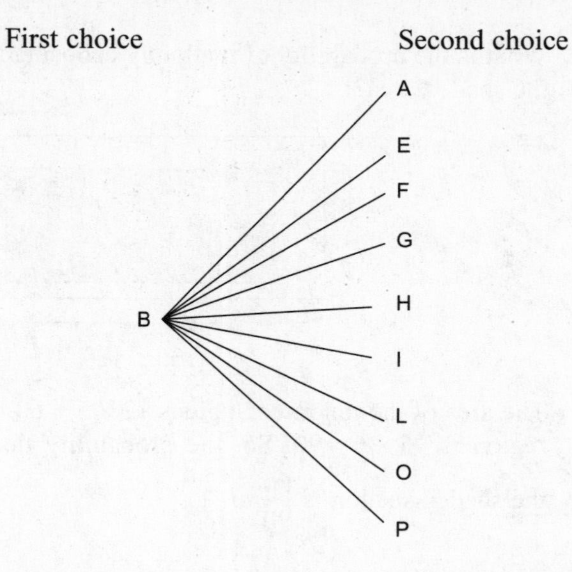

Here there are only 4 combinations that pair a consonant with a vowel. So, there are a total of 24 consonant-vowel combinations. There are, therefore, a total of 48 out of 90 combinations that pair a vowel with a consonant. This fraction, $\frac{48}{90}$, reduces to $\frac{8}{15}$, which agrees with what we calculated earlier.

Practice Set

1. If you choose a color from red, blue, green, yellow, and purple and a number from 1, 2, 5, 9, 10, 11, 12, and 18, what is the probability that you will choose red and an odd number?

2. If you have a list of numbers—2, 6, 7, 10, 12, 13, 14, 16, 17, 19, 20, 22—what is the probability of choosing two even numbers?

3. Given 3 orange beads, 5 yellow beads, and 2 aqua beads, what is the probability of choosing an orange bead and an aqua bead?

4. If we roll a die and choose a letter from the alphabet, what is the probability we will roll a number that is a factor of 6 and choose a letter from the word MATHEMATICS?

5. Which event below is more likely?
   A. Choosing a 2-digit number randomly and choosing a multiple of 12.
   B. Choosing an ace from a standard 52-card deck.

6. If you randomly choose a number between 1 and 30, what is the probability that it will contain at least one 2?

Probability can also be combined with the idea of area as follows.

Examples

A. What is the probability of randomly choosing a point in the rectangle below that is in the shaded area?

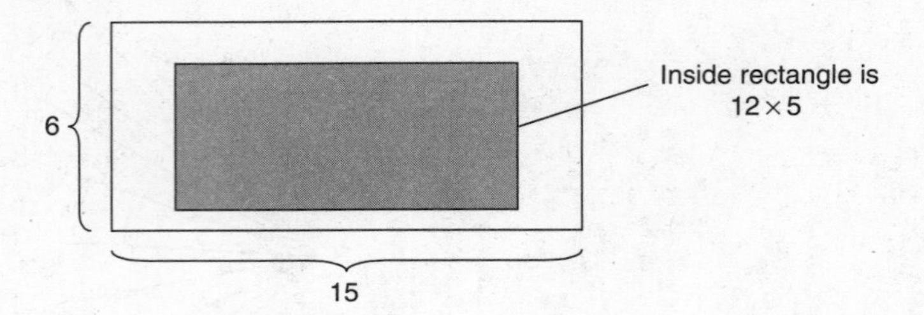

The area of the inner rectangle is $12 \times 5 = 60$, while the area of the entire rectangular region is $15 \times 6 = 90$. So, the probability that a randomly selected point is inside the shaded section is $\frac{60}{90}$ or $\frac{2}{3}$.

B. What is the probability that a randomly chosen point in the figure below will be outside the circle?

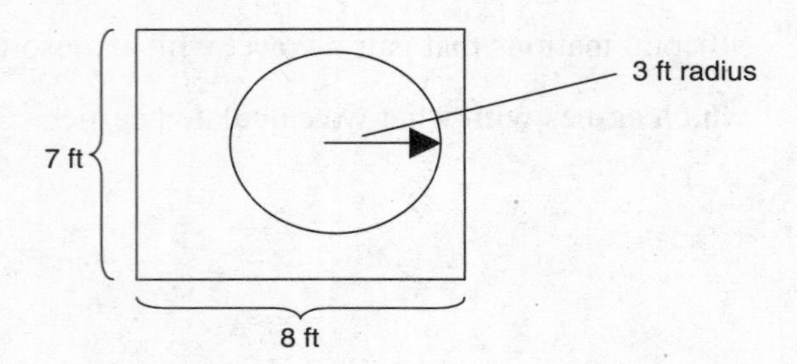

The area of the circle: $\pi r^2 = 9\pi \approx 28.26$.
The area of the rectangle: $l \times w = 8 \times 7 = 56$.
The area of the rectangle outside the circle: $56 - 28.26 = 27.74$
Thus, the probability of landing outside the circle is: $\frac{27.74}{56} \approx .4954$. As a percent, $.4954 \approx 49.5\%$.

In the last example, you saw how sometimes it is easier to write the resulting fraction for probability as a decimal or a percent.

Practice Set

1. What is the probability of choosing a random point in the trapezoid below and having it be in the shaded section?

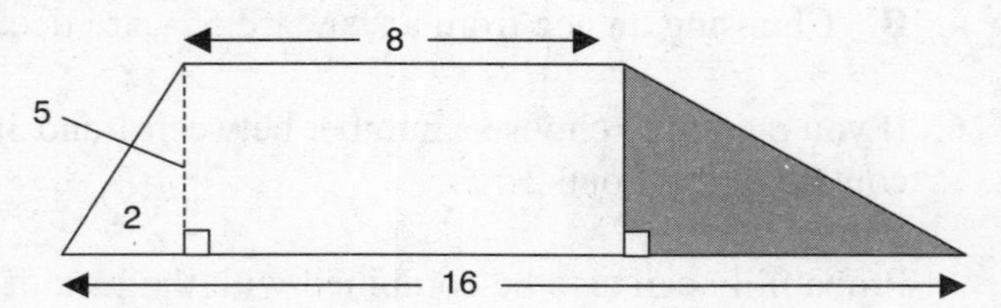

2. What is the probability of each of the following based on the dartboard given? Answers may be given as decimals rounded to the nearest thousandth.

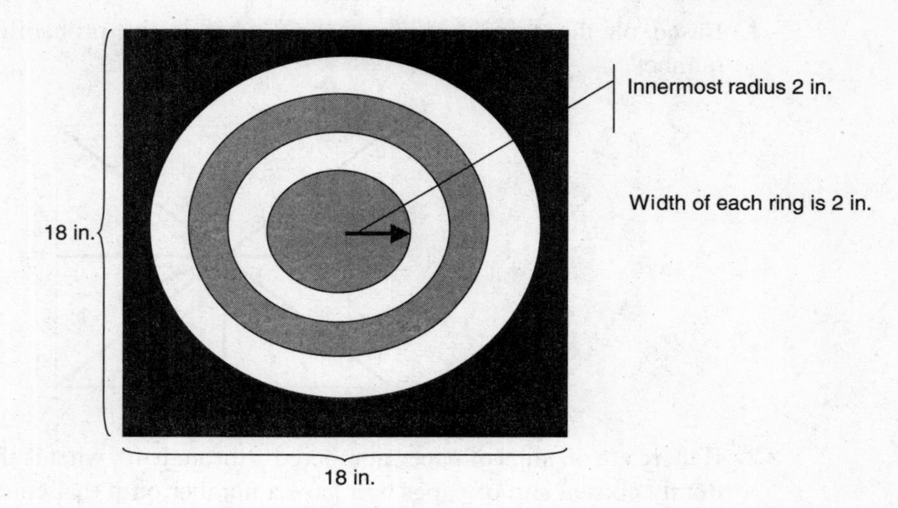

(a) What is the probability of hitting the target if the dart lands somewhere on the board?
(b) What is the probability that a dart that hits the target will land on the bullseye?
(c) What is the probability of throwing two darts consecutively that will land on the board but miss the target?
(d) What is the probability that a dart that hits the target will land on either of the two outermost rings?

For certain probability situations, it is essential to write out the resulting sample space. A classic example of this is rolling two dice and adding the two numbers rolled. It is not as easy to count the possible outcomes, so a chart to organize the outcomes helps. The chart below shows the possible rolls of 1 through 6, and below these the sum is recorded.

	1	**2**	**3**	**4**	**5**	**6**
1	2	3	4	5	6	7
2	3	4	5	6	7	8
3	4	5	6	7	8	9
4	5	6	7	8	9	10
5	6	7	8	9	10	11
6	7	8	9	10	11	12

Examples

Based on the results shown in this chart, we can find various probabilities:

A. $P(\text{sum of } 5) = \frac{4}{36} = \frac{1}{9}$

B. $P(\text{sum greater than } 9) = \frac{6}{36} = \frac{1}{6}$

C. $P(\text{sum of 5 or 9}) = \frac{8}{36} = \frac{2}{9}$

MIXED PRACTICE, CLUSTER III, MACRO A

1. Based on the spinner shown below, what is the probability of spinning an odd number?

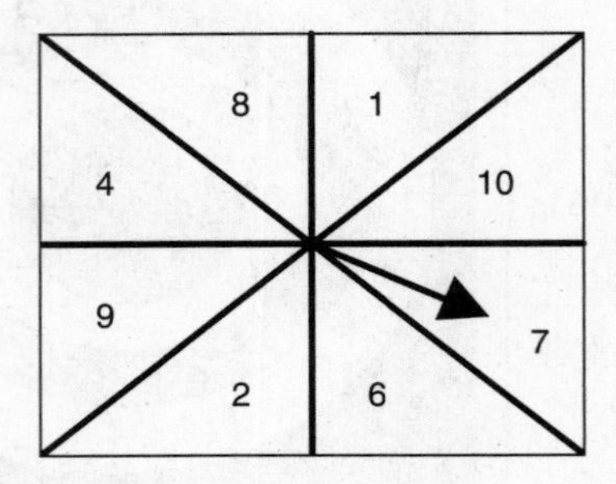

2. If there are 40 slips of paper numbered 1 through 40, what is the probability that a randomly chosen slip of paper will have a number on it that contains at least one 3?

A. .1 **B.** .25 **C.** .35 **D.** .3

3. In the circle spinner shown below, what is the probability of spinning a 2?

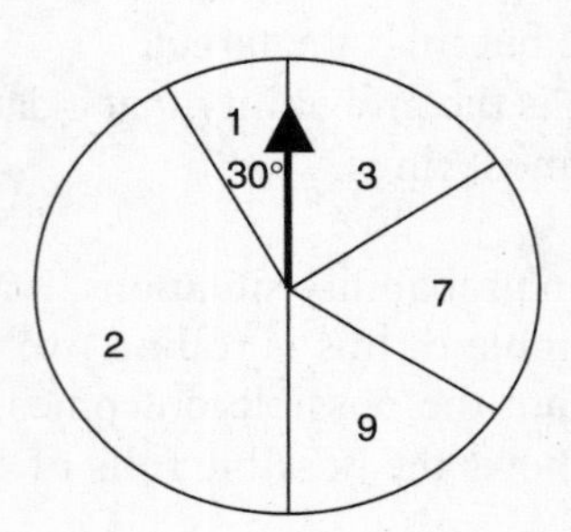

A. $\frac{1}{5}$ **B.** $\frac{2}{5}$ **C.** $\frac{3}{8}$ **D.** $\frac{5}{12}$

4. During a visit to the zoo, Nathaniel notices that the tiger is in a fenced rectangular area measuring 100 ft × 40 ft. What is the probability that the tiger is within 5 feet of the longest side toward the front of the exhibit?

A. $\frac{1}{20}$ **B.** $\frac{5}{8}$ **C.** $\frac{1}{8}$ **D.** $\frac{3}{20}$

5. While scouting, a baseball team's manager tries to base a player's performance on his performance under the same or very similar previous conditions. If Tyler has had 120 hits in 180 similar batting situations, what is the best estimate of the probability of a hit during this at-bat based on this method of prediction?

A. 12% **B.** 25% **C.** 60% **D.** 67%

6. Which of the following spinners has the best probability of spinning an E?

A.

B.

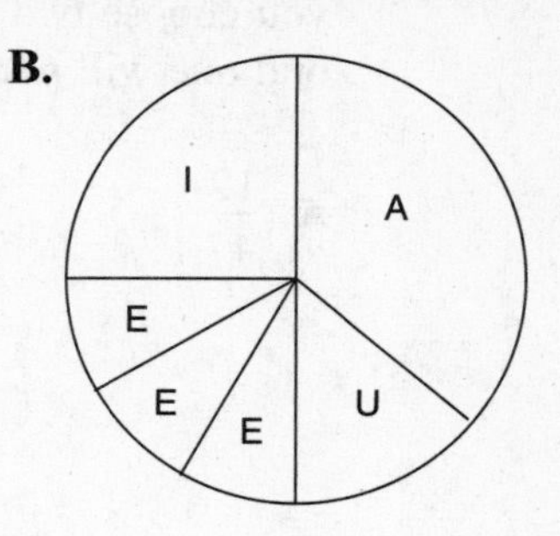

C.

D.

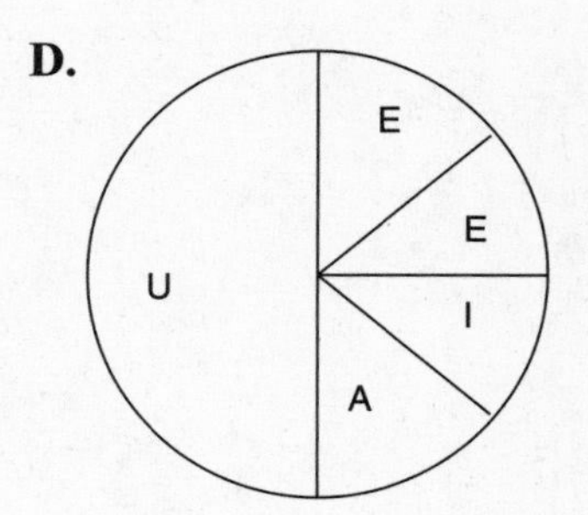

7. If a bumblebee is buzzing around a room 12 ft × 24 ft × 10 ft, at a given second what is the probability that the bee will be less than 4 ft from the ceiling? Show how you approached this problem and explain your answer.

8. If Miranda reaches into a velvet bag containing 6 blue marbles, 4 green marbles, and 8 yellow marbles and chooses 3 marbles, what percent of the time (on average) should we expect that she will choose 3 yellow marbles?

9. Based on the results of a computer simulation of tossing three coins 500 times, answer the questions that follow.

HHH	64	TTT	57
HHT	59	TTH	63
HTT	65	THH	61
HTH	80	THT	51

(a) What is the experimental probability of tossing two tails and a head according to the results obtained during this 500-toss simulation? Round your answer to the nearest thousandth.

(b) Find the theoretical probability of tossing two tails and a head. Again, round your result to the nearest thousandth.

(c) Compare and contrast your results for **(a)** and **(b)**.

10. If you roll two dice numbered the standard 1 through 6 and add the two resulting rolls, what sum is most likely to occur?

A. 5 **B.** 6 **C.** 7 **D.** 8

11. There are 12 of each of the following colored marbles: blue, red, green, and white. If you choose two marbles and the first one is red, what is the probability that the second one will *not* be green?

A. $\frac{12}{47}$ **B.** $\frac{3}{4}$ **C.** $\frac{1}{4}$ **D.** $\frac{35}{47}$

MACRO B

SCATTERPLOTS

A scatterplot is one means of graphically representing data. In a **scatterplot**, two sets of information are plotted for comparison. There is a scale along the side and along the bottom. Data are plotted much like an ordered pair. Once the data are plotted, we can observe a trend or general path that the data fall along. This is called the **line of best fit**. The line may be close to a straight line or it may be a curve. Or, there may be no line of best fit, meaning the data are too scattered.

If the line of best fit goes from the bottom left toward the upper right, we say we have positive correlation. If the line of best fit goes from the upper left toward the lower right, we say we have a negative correlation. A **positive correlation** indicates that as one variable increases or decreases, the other has the same behavior. A **negative correlation** means that as one variable increases, the other decreases, and vice versa. In some cases, the data are so scattered that there is no correlation or it is a weak correlation. Basically, a correlation is a cause-and-effect relationship.

Below is an example of a scatterplot showing grip strength versus arm strength. There's a strong positive correlation between the two.

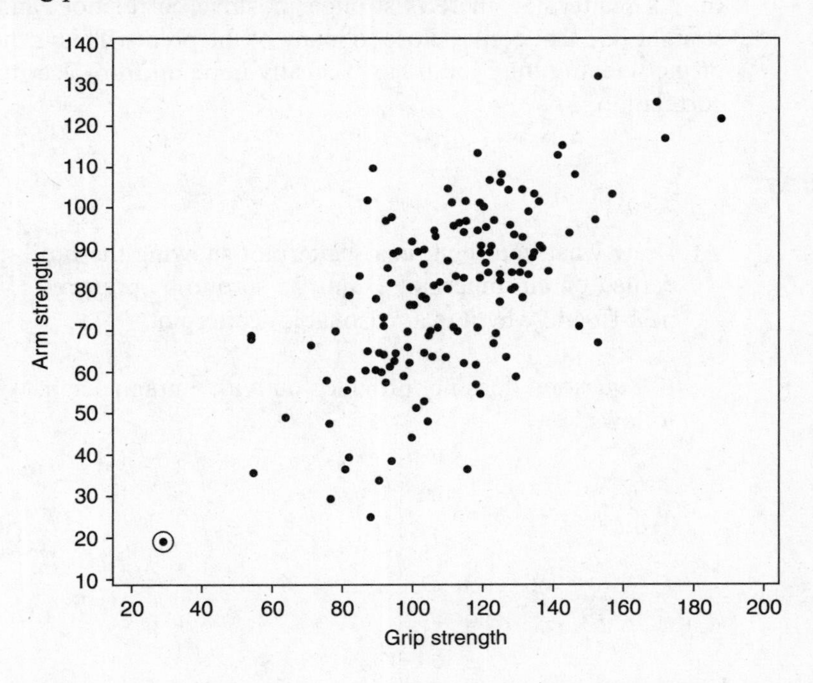

If several pieces of data are extremely high or low compared to the majority of the data, they are considered **outliers.** Closely spaced groups of data are called **clusters**. In the scatterplot above there appears to be an outlier (indicated by the point circled above).

An example of a scatterplot that has stronger correlation is shown below.

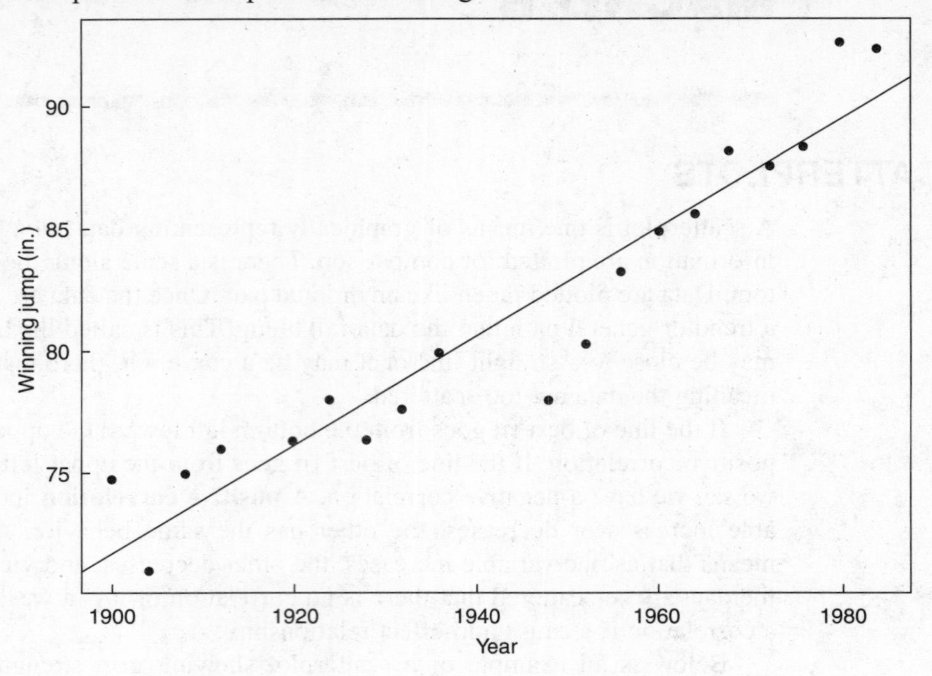

In this scatterplot, there is stronger positive correlation, and as you can see, there is a straight line that comes close to many of the points. This is the line of best fit. The length of the winning long jump has basically gone up from year to year, and this is a positive correlation.

Examples

A. Draw what you think is a scatterplot showing the time spent studying and the grade earned on an important exam. Be sure to be prepared to explain the trend or line of best fit and why it is a reasonable scatterplot.

If a question does not provide you with a graph setup, you will need to set one up as follows.

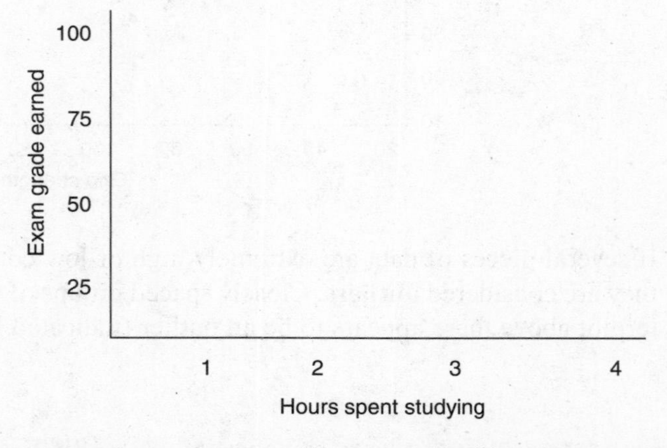

Once you have the graph set up, you will need to decide what you feel is the general trend of the data. Within reason, as long as you explain why you graphed the data the way you did, there is no one correct answer.

In this situation, it is most likely the case that the longer a person studies, the better his or her grade will be. There will probably be one or two outliers: For instance, consider a student with natural ability in a course who can therefore study not at all or very little and still do well. You may also have a student who is just a very poor test taker and, despite considerable studying, still doesn't do well. Once you have a general idea how you think the two items are related to one another, plot some points.

A completed scatterplot for this situation would look similar to the one below.

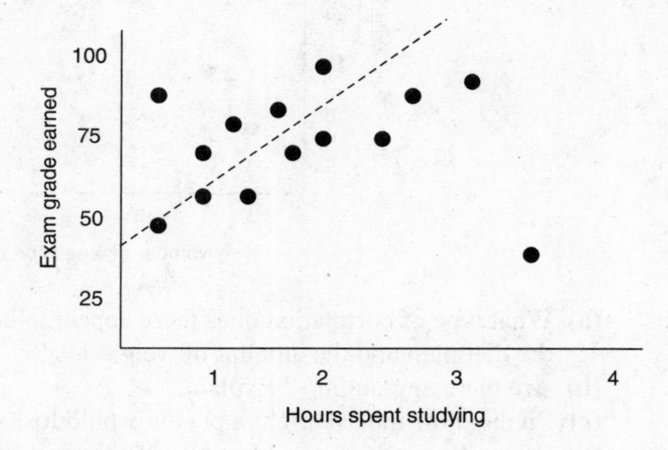

B. Based on the scatterplot between the number of hours worked during a week and the amount of time spent exercising during a week, answer the questions that follow.

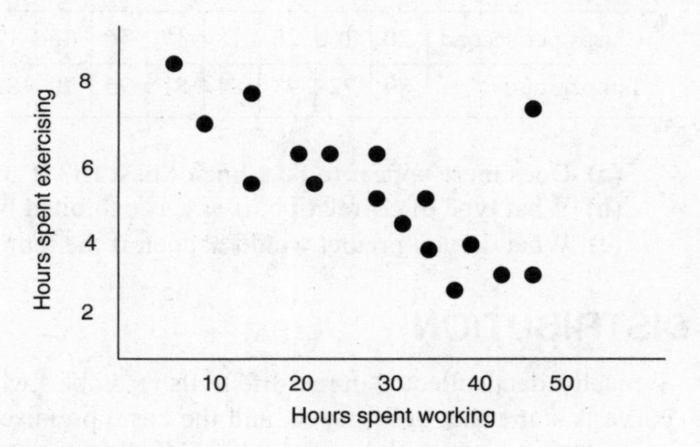

What type of correlation does the scatterplot show?

This scatterplot shows a negative correlation because as the number of hours spent working increases, the number of hours spent exercising does the opposite (decreases).

Are there any outliers on the scatterplot? If so, what explanation can you offer to explain these?

There appears to be an outlier at approximately (50 hours working, 7 hours exercising). This probably represents an "exercise fanatic."

PRACTICE SET

1. Based on the scatterplot, answer the questions that follow.

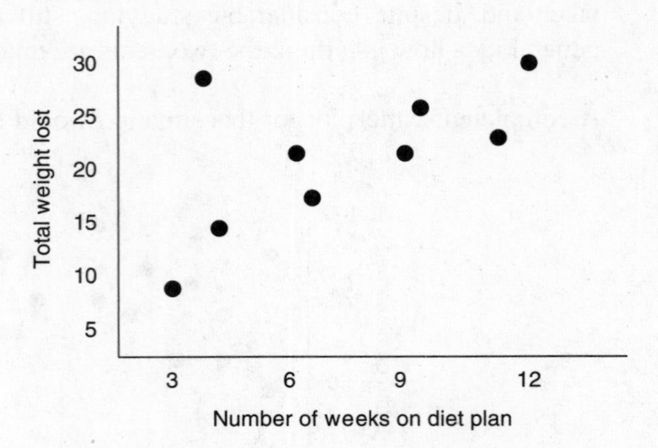

(a) What type of correlation does there appear to be between the number of weeks on the diet plan and the amount of weight lost?
(b) Are there any outliers? Explain.
(c) Predict how much weight a person would lose if they were on the plan 15 weeks.
(d) What do you think would eventually happen to the total weight lost? Why?

2. Using the given data, construct a scatterplot.

Chirps per second	20	16	20	18	17	16	15	17	15	16	15	17	16	17	14
Temperature	89	72	93	84	81	75	70	82	69	83	80	83	81	84	76

(a) Does there appear to be a line of best fit? If so, sketch it.
(b) What type of correlation, if any, is exhibited by the data?
(c) What do you predict would happen if the temperature grew colder? Why?

NORMAL DISTRIBUTION

Typically, data collected in real-life trials resembles what is called the normal **curve**. This curve is somewhat bell-shaped, and the basic premise is that most data should "cluster" around the average and that there should be fewer extreme values (high and low). The normal curve resembles the following.

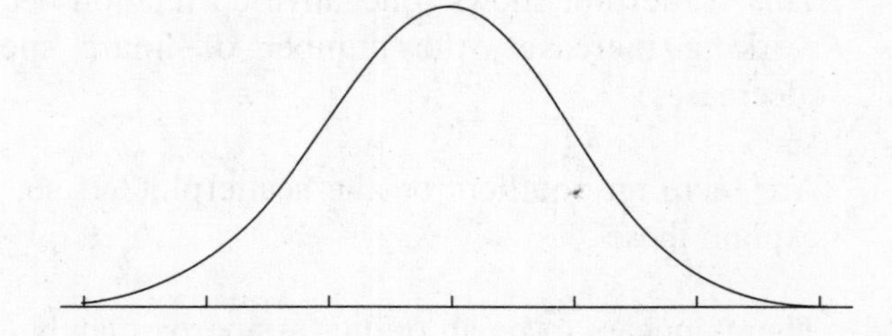

The mean or average of the data points should be where the highest point of the curve falls. The normal curve should be symmetric on either side of the average.

To see if data follow the normal curve, we construct what is called a **histogram**. In a histogram, the outcomes are plotted along the bottom (x-axis) and the frequency (either in amount or percent) of each outcome is plotted along the side (y-axis).

Examples

A.

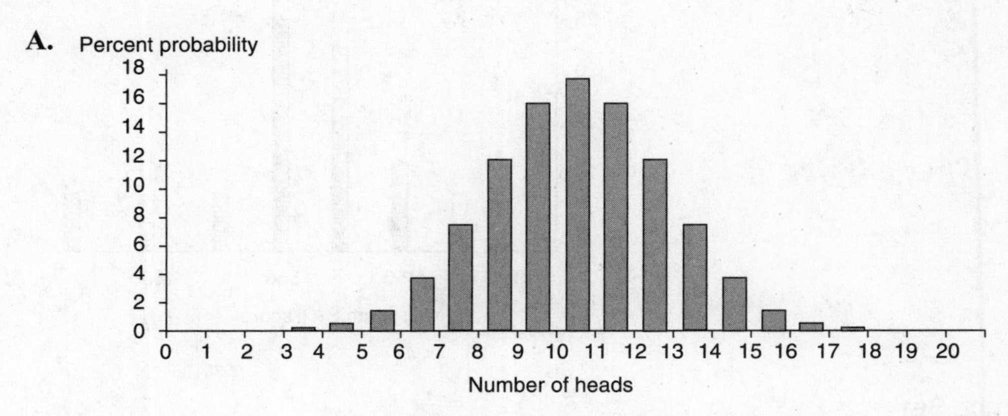

Frequency of heads in 20 coin tosses

Approximately what percent of the time do 7 out of 20 tosses land on heads? Simply look at the bar for 7 heads, and it appears to be at a height of about 8%.

What is the most frequent outcome when tossing a coin 20 times? Why do you think this is the case?

The most frequent outcome is to have 10 heads (and consequently 10 tails). This is the case because the probability of tossing a head is $\frac{1}{2}$, so we would expect that the more times we toss a coin, the more likely we are to get heads 50% of the time and tails the other 50% of the time.

How many heads are about as common an occurrence as 14 heads?

The bar for 6 heads is about the same height as the bar for 14 heads.

B. Make a histogram for the following data.

SAT Math Scores at Anytown College

Score	Percent	Score	Percent
350	4	550	28
400	9	600	10
450	12	650	4
500	24	700	9

To make the histogram, simply construct two axes and then draw bars indicating the corresponding value.

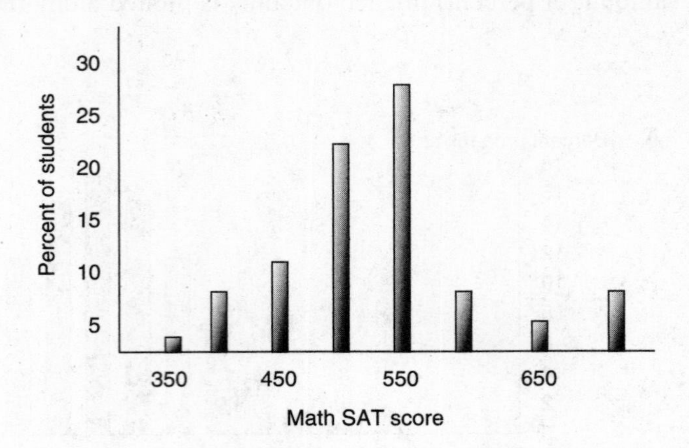

PRACTICE SET

Make a histogram showing the number of heads that would be expected to occur if three coins were tossed 100 times.

With a normal curve, you will also be expected to make use of what is known as **standard deviation**. Standard deviation measures how far away from the mean data falls. We say data is "within 1 standard deviation" if it is less than 1 away from the mean values.

In a situation where data are distributed according to the bell curve:

About 68% of the data should be within 1 standard deviation of the mean.
About 95% of the data should be within 2 standard deviations of the mean.
About 99.7% of the data should be within 3 standard deviations of the mean.

Example

The average sneaker size for men is 9, with a standard deviation of 1.5. What does this indicate about most men's sneaker size?

Assuming the data fall into a normal distribution, $9 + 1.5 = 10.5$ and $9 - 1.5 = 7.5$, so 68% of men wear between a size 7.5 and 10.5 sneaker. Since $9 + 3 = 12$ and $9 - 3 = 6$, we can reasonably say that 95% of men wear between a size 6 and a size 12 sneaker. Since $9 + 4.5 = 13.5$ and $9 - 4.5 = 4.5$, we can safely say that very few men wear smaller than a size 4.5 or larger than a size 13.5.

VARIATION

Very closely related to scatterplots and correlation is a concept called **variation**. Variation has to do with how two quantities are related to one another or how a change in one affects the other.

We said that if one variable quantity increases or decreases and the other increases or decreases (same behavior), this signifies positive correlation. This is also referred to as **direct variation**. When two quantities vary directly, there is a constant ratio between the two numbers. That is, if x and y vary directly, $x/y = c$, where c is a constant.

If one variable decreases while the other increases, or vice versa, this is referred to as **indirect variation**. If x and y vary indirectly, $xy = c$, where c is a constant.

Examples

A. If x and y vary directly and $x = 5$ when $y = 8$, what is y when $x = 20$?

You need to remember that in direct variation there is a constant ratio between the two quantities.

So, in this case, set up a proportion:

$$\frac{5}{8} = \frac{20}{y}$$

Cross-multiplication gives

$$5y = 160$$

Dividing both sides by 5 gives

$$y = 32$$

B. If x and y vary indirectly and $x = 3$ when $y = 8$, what does x equal when $y = 4$?

In indirect variation, you need to remember that the product of the two quantities is always the same value. So, here $xy = 3 \times 8 = 24$, so x times y must always be 24. Thus, the question becomes $4x = 24$ for what value of x? Finally, $x = 6$.

PRACTICE SET

1. If a and b vary indirectly and $a = 5$ when $b = 8$, what does b equal if a equals 4?
2. If m and n vary directly and $m = 3$ when $n = 4$, what does m equal when $n = 20$?
3. If x and y vary directly and $x = 2$ when $y = 5$, what is the value of y when x is 9?
4. If r and s vary indirectly with $r = 4$ when $s = 9$, what does r equal when s is 2?

MIXED PRACTICE, CLUSTER III, MACRO B

1. What type of correlation would you expect between level of education and starting salary?
2. For which of the following situations would you expect there to be a negative correlation?

   A. the cost of a purchase and the amount of tax paid
   B. the age of a computer and its resale value
   C. a car's speed and the distance it travels in 3 hours
   D. the number of people at the beach and the temperature

3. Which of the following situations is *not* an example of direct variation?

A. a used car's value and its age
B. the number of gallons of gas purchased and the total price
C. the area of a square and the length of each side
D. the distance a car travels at a constant speed and the time traveled

4. If it takes 2.4 hours to make a trip traveling at 55 miles per hour, how long will the same trip take at only 40 miles per hour.

5. Given that a normal distribution has a mean of 14 and a standard deviation of 3, complete the following statements.

A. About 68% of the data will be between __________ and __________.
B. About 95% of the data will be between __________ and __________.

6. What percent of the area under the normal curve below is shaded if the mean is 100 and the standard deviation is 40?

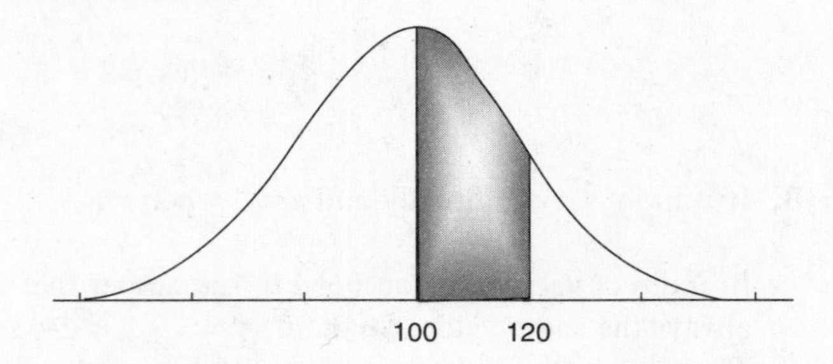

7. Which of the surveys listed below is an example of an unbiased survey?

A. asking visitors to the local zoo whether they support federal funding of zoo programs
B. asking some teens from Alaska and Hawaii what type of music teens between 13 and 15 prefer
C. asking only men between 21 and 30 to determine what type of candy people from 21 to 30 like the most
D. asking some men older than 50 what type of hair care products men over 50 use

8. The heights of American males between 18 and 24 years of age are approximately normally distributed, with a mean height of 5 feet 10 inches and a standard deviation of 3 inches.

A. Find the approximate height of 95% of all American men between 18 and 24 years old.
B. What is the probability that a randomly selected American male between ages 18 and 24 will be taller than 76 inches?

MACRO C

MEASURES OF CENTRAL TENDENCY

There are four main measures of **central tendency**, meaning that these are statistical measures indicative of the tendency of the data to center around a value. The four measures of central tendency are

1. Mean (average)
2. Median
3. Mode
4. Range

Mean

The **mean** is the average of a set of numbers. To find the mean or average of a list of numbers, add the numbers and divide by the quantity of numbers on the list.

Median

To find the **median**, list the numbers in order from lowest to highest. The median is the middle number if there is an odd number of numbers in the list. If there is an even number of numbers, the median is the average of the two middle numbers.

Mode

The **mode** is the number or numbers that occur most often. In some cases, there is no mode; that is, if all the numbers on the data list appear the same number of times, no one number is more common than the others.

Range

The **range** is the highest value minus the lowest value. The range is known as a measure of dispersion.

Examples

A. Find the mean, median, mode, and range for the given data.

$$12, 8, 9, 15, 11, 8, 7, 14, 6$$

Usually, the first thing you should do is put the numbers in order:

$$6, 7, 8, 8, 9, 11, 12, 14, 15$$

Mean: $6 + 7 + 8 + 8 + 9 + 11 + 12 + 14 + 15 = 90$, $90 \div 9 = 10$

Median: The middle number is 9 (there are four numbers lower than 9 and four numbers higher than 9).

Mode: There are two 8's and one of every other number, so 8 is the most common number.

Range: High minus low: $15 - 6 = 9$

B. Find the mean, median, mode, and range for the following data.

52, 47, 60, 48, 50, 46

First, put the numbers in order from lowest to highest:

46, 47, 48, 50, 52, 60

Mean: $46 + 47 + 48 + 50 + 52 + 60 = 303$; $303 \div 6 = 50.5$.

Median: There are two numbers in the middle: 48 and 50. So, we average these numbers: $48 + 50 = 98 \div 2 = 49$.

Mode: Since each number occurs once, there is no mode.

Range: $60 - 46 = 14$.

C. Find the mean, median, mode, and range for these numbers:

$-8, 5, -2, 7, -8, 5, 9, 2, 4, -3, 0$

Again, begin by putting the numbers in order:

$-8, -8, -3, -2, 0, 2, 4, 5, 5, 7, 9$

Mean: $-8 + -8 + -3 + -2 + 0 + 2 + 4 + 5 + 5 + 7 + 9 = 11 \div 11 = 1$.

Median: The middle number is 2.

Mode: The most common numbers are -8 and 5, each of which appears twice, whereas each of the other numbers occurs only once.

Range: $9 - (-8) = 9 + 8 = 17$.

PRACTICE SET

Find the mean, median, mode, and range for each set of data:

1. 25, 37, 34, 21, 31, 25, 23

2. 103, 97, 108, 91, 99, 105

Extending the Ideas of Statistical Measures

More often than not, you will be asked to apply your knowledge of statistical measures rather than just to perform a straightforward calculation.

> **Take Note:** To get an average, use the "formula" AVG = TOTAL / #. Get the total by multiplying AVG by # of numbers.

Examples

A. If Mark's current average after 8 grades is 85%, what grades must he earn on his last two grades if he wants to bring his average up to 87%?

If Mark currently has an 85 average, then his total for the 8 grades must be $85 \times 8 = 680$.

To get an 87 after 10 grades, his total must be $87 \times 10 = 870$, which means he needs 190 ($870 - 680$) points combined. So, he needs two grades that add to 190 (95% and 95% or 97%, and 93%).

B. If the mean of 6 numbers was calculated to be 24 and then it is realized that one number was supposed to be 36 not 24, what happens to the mean?

When faced with a question about how the mean is affected, there are two things to consider: the total of the numbers and the number you are dividing by. In this case, the total changes because one number increases; 24 increases to 36 (an increase of 12). This new increased total is then divided by 6, so there is an increase of $12 \div 6 = 2$ in the average.

C. If the mode of 8 numbers, seven of which are given below, is 54, what is the mean of the 8 numbers?

48, 45, 50, 54, 57, 51, 49

Since one number is missing from the list and the mode is supposed to be 54, the missing number must be 54 (so that there are two 54's).

Thus, take the average of 45, 48, 49, 50, 51, 54, 54, 57.

$$45 + 48 + 49 + 50 + 51 + 54 + 54 + 57 = 408 \div 8 = 51.$$

PRACTICE SET

1. If one number on a list of a dozen numbers was supposed to be a 102 instead of 138, what effect does this have on the mean of the numbers?

2. If the median of 6 numbers, five of which are shown below, is 24, find the mean of the 6 numbers.

15, 16, 20, 29, 30

3. If the mode of 8 numbers, seven of which are shown below, is 35, find the mean and median of the 8 numbers.

25, 35, 28, 26, 30, 23, 32

4. If the 23 on the list below is changed to an 18, what happens to each of the four statistics, if anything?

18, 20, 23, 24, 25, 25, 30

GRAPHS AND OTHER DATA DISPLAYS

There are various ways of displaying data visually. We will discuss each of these and give examples. On the HSPA Exam, you will be expected to interpret and/or construct these types of data displays.

Bar Graph

A **bar graph** has two axes, one of which is labeled with the various categories of data, whereas the other is labeled with quantities. Bars of appropriate length are drawn for each category.

Example

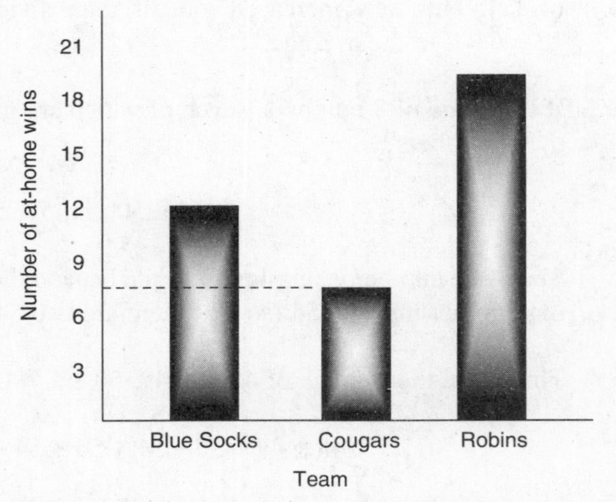

To read off the amount of wins a certain team had at home, simply line up the top of the bar with the scale along the left side. For example, the Cougars won approximately eight at-home games.

Note: A bar graph can also be constructed so that its bars are horizontal rather than vertical, as shown below.

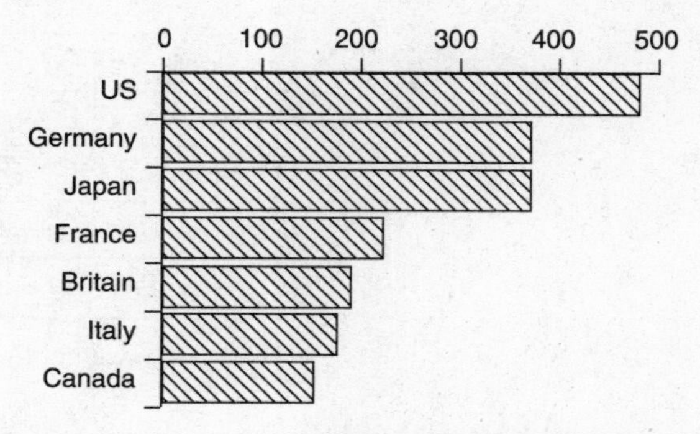

Merchandise exports, 1993 (billions of dollars).

Circle Graph

In a **circle graph**, a circle is divided into sectors or sections by drawing radii and/or diameters.

Since a circle has 360°, we take a certain percent of 360° and construct an angle using one ray as a radius of the circle.

Example

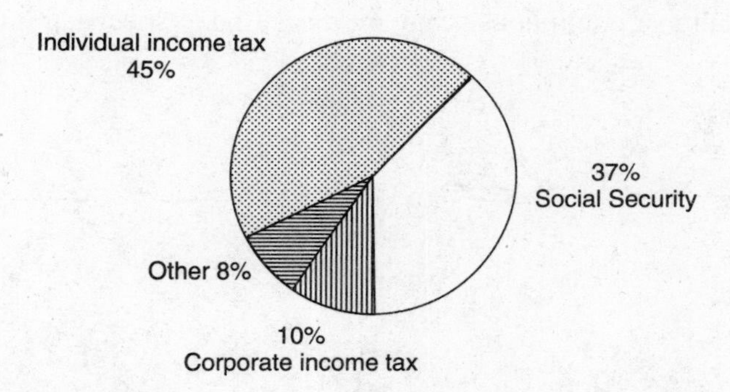

Revenue sources for the federal government.

In the circle graph above, the sector representing corporate income tax (10%) should have an angle measure of 10% of 360°, or 36°.

If we knew that the federal government had an income of \$2 million for a certain period of time, we could figure out that 37% ($.37 \times 2{,}000{,}000$), or \$740,000, of that amount came from Social Security.

Pictograph

In a **pictograph**, a picture or symbol is used to represent a certain quantity.

Walk	⊗ ⊗ ⊗
Ride a Bike	⊗ ⊗ ⊗ ⊗
Ride the Bus	⊗ ⊗ ⊗ ⊗ ⊗
Ride in a Car	⊗ ⊗
Key: Each ⊗ = 10 students.	

How we get to school.

In a pictograph we try to choose a symbol that is related to the data being displayed. Also, we make each symbol equal an appropriate number of people or items so that we do not have to draw too many symbols.

In the pictograph above, you can see in the legend or key that each picture of a wheel stands for 10 students. So, for instance, we can figure out that 20 students ride in a car since two wheels are drawn in that case.

Line Graph

A **line graph** is used when the data are more continuous, rather than just being grouped into categories. For example, a line graph is useful in plotting data having to do with stocks or the economy. In general, a line graph is good for quantities that are changing almost continuously but are measured at specified intervals—yearly, monthly, and so on.

Example

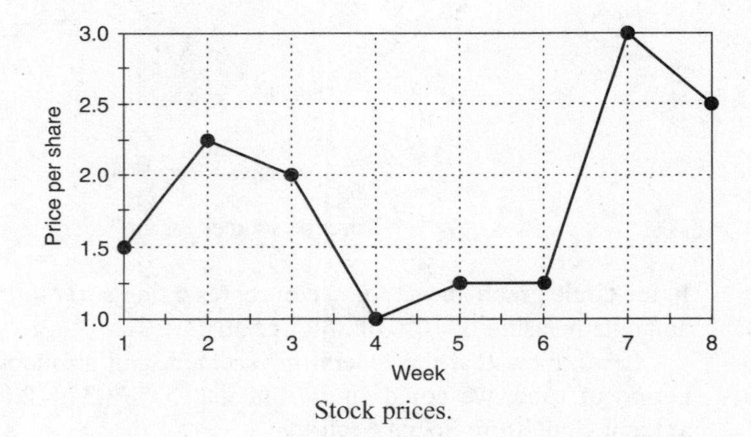

Stock prices.

From the line graph above we can read off any price for one of the marked weeks simply by going to that week and looking to see where the point for that week is marked along the scale on the left. For example, we can see that the price per share for week 3 is $2.00.

We can also approximate values in between marked intervals. For instance, here we can see that after about $3\frac{1}{2}$ weeks, the stock price was about $1.50 per share.

Finally, using a line graph provides a good visual picture of when the sharpest increases or decreases occurred simply by observing the steepness of the lines between points. For example, in this case we can see that the biggest increase in price occurred between the sixth and seventh weeks, as that is where the line is steepest "uphill." You can observe sections of no change by looking for horizontal lines (such as what occurs here between weeks 5 and 6).

Misleading Graphs

Graphs can be misleading if the axis is unevenly marked or if there is a gap in the scale used. This occurs most commonly in bar graphs. In the example below, it appears that the color red is chosen by twice as many people as the color green because the scale begins with 10 and then increases by 2's. Thus, the initial increment of 10 is the same size as the increments of 2 that follow, creating the misleading graph. Actually, when you look at the numbers, 12 liked green while 16 liked red. This is a relatively insignificant difference, yet because of the unevenly marked axis, there is a visually misleading appearance.

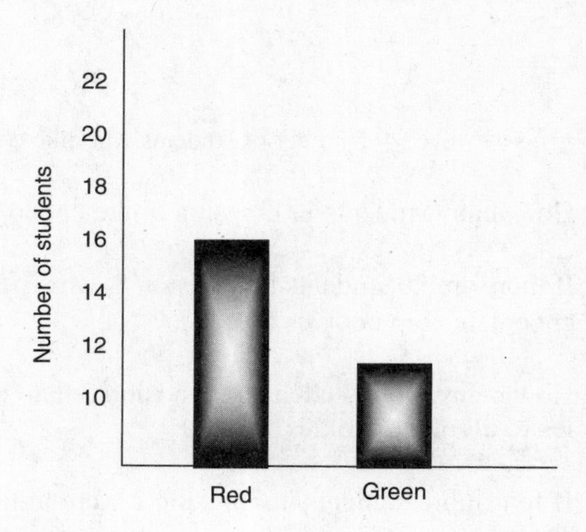

PRACTICE SET

Using the graphs provided, answer the questions that follow each one.

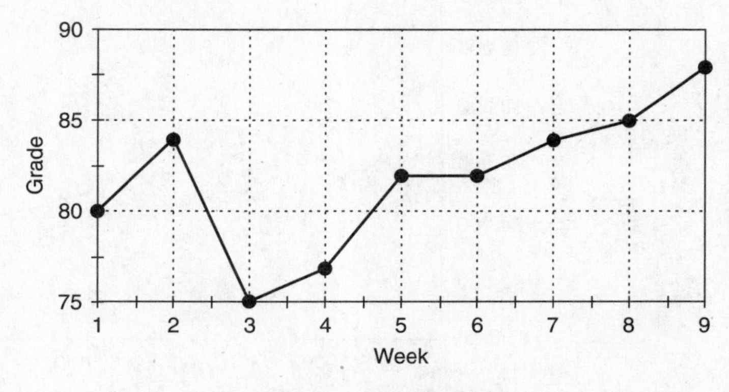

Mark's algebra average.

1. What is Mark's average at the fifth week?

2. Between what two weeks did Mark's grade show the most significant drop?

3. If Mark had two grades between weeks 5 and 6, what do you think his grades were on these two tests? Why?

4. When is Mark's grade about 80%?

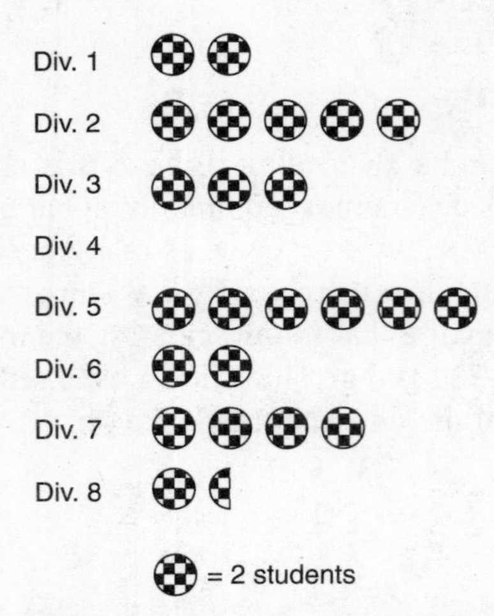

Number of students who like chocolate chip cookies best.

1. How many students in Division 5 like chocolate chip cookies the best?

2. If there are 20 students in Division 8, what percent of the students in that division like chocolate chip cookies best?

3. How many more students in Division 7 than in Division 3 prefer chocolate chip cookies to all other cookies?

4. If four more students in Division 2 were to like chocolate chip cookies the best, 40% of that division would like chocolate chip cookies best. How many students are in Division 2?

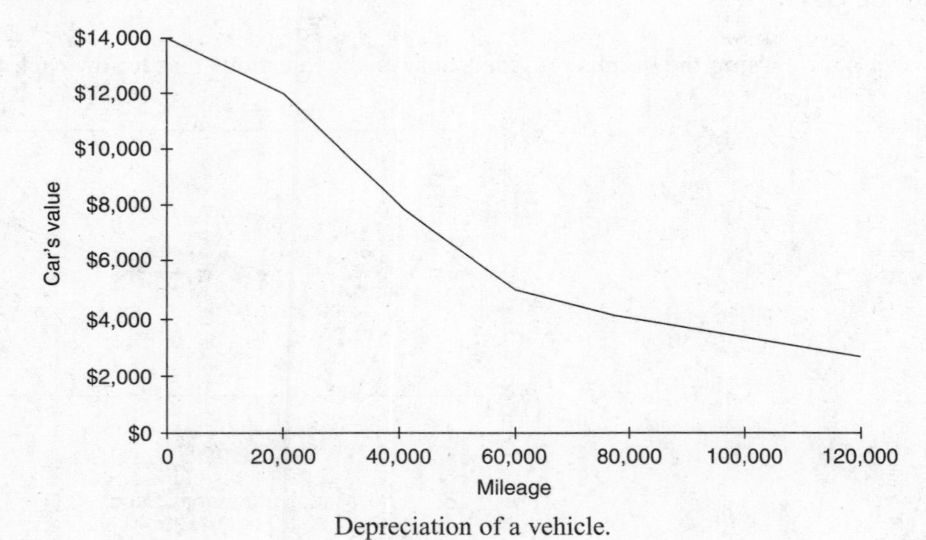

Depreciation of a vehicle.

1. At 40,000 miles, how much has the value of the vehicle decreased?

2. By what percent has the value of the vehicle depreciated after 60,000 miles?

3. Within which mileage interval does the value of the vehicle depreciate the most?

4. Within which mileage interval does the value of the vehicle depreciate least?

5. By what percent does the value of the vehicle decrease between 60,000 miles and 80,000 miles?

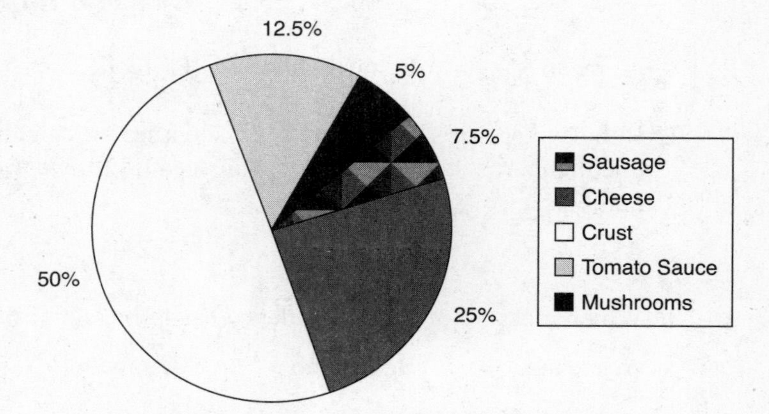

Percent of weight of each ingredient in a pizza.

1. What percent of the pizza's weight is the tomato sauce?

2. If the total weight of the pizza is 3 pounds, approximately how many pounds does the cheese used for the pizza weigh?

3. If the weight of the mushrooms used is 6 ounces, how many pounds does the pizza weigh?

Open-Ended Question: Over the past several years, you have recorded the number of automobiles that a used car dealer in your town has sold in different price ranges. First, decide whether a bar graph or a line graph is more appropriate for this situation.

Next, using the given data, construct several graphs of this situation using different intervals for the price ranges (vary the increment size, e.g., \$2000 or \$5000). If you were the car dealer, what increment size would you choose to advertise car sales? Why?

Price Range ($)	Number of Cars Sold
0–1000	3
1000–1999	5
2000–2999	12
3000–3999	25
4000–4999	40
5000–5999	75
6000–6999	52
7000–7999	35
8000–8999	15
9000–10000	9

MIXED PRACTICE, CLUSTER III, MACRO C

1. Based on the data given below, which measure of central tendency would remain unchanged if the 120 and 124 were dropped from the data?

100, 95, 98, 86, 95, 120, 105, 124, 90

A. mean **B.** median **C.** mode **D.** range

2. If the average of six numbers is 25 and after we calculate this average we realize that one number is supposed to be 38 instead of 26, what is the actual average of the six numbers?

A. 25 **B.** 37 **C.** 23 **D.** 27

3. Which measure of central tendency has to be one of the numbers in the given data?

A. mean **B.** median **C.** mode **D.** range

4. The appraised values of several houses on a certain street are as follows.

House No.	Appraised Value ($)
150	170,000
153	155,000
156	160,000
159	350,000
162	170,000
165	210,000
168	190,000

(a) Calculate the mean and the median for the appraised prices.
(b) Which gives a better representation of the actual appraised prices, the mean or the median? Why?
(c) Which statistical measure would a realtor trying to convince you to move to this neighborhood use and why?

5. Given five scores of 75, 80, 82, 86, and 94, answer the following questions.

(a) If the 86 were changed to a ______ the mode would be 80 and the median would be ______.
(b) To increase the average by two points, the 80 would have to become ______.
(c) To keep the median at 82, the 80 could be changed to ______.
(d) If the 94 is changed to a 90, the 75 would have to be changed to a ______ to keep the mean the same.
(e) To decrease the average by one point, the 75 would have to become a ______.

6. Use the graph below to answer the questions that follow.

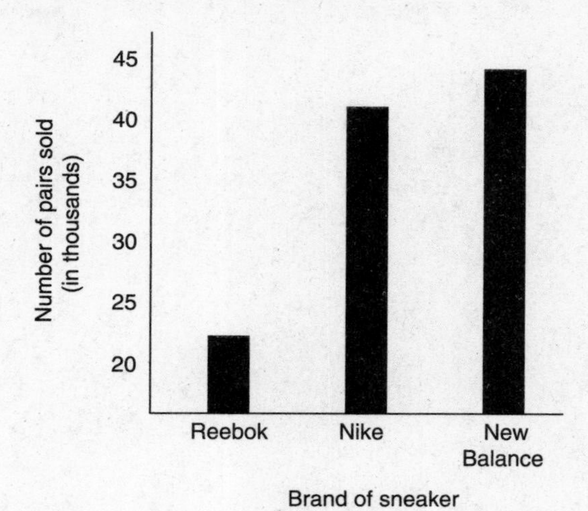

Teens Favorite Sneakers.

(a) From the graph, it *looks* like only about one-third as many teens like ______ as ______.

(b) Is your statement in (**a**) actually true? Why or why not?

(c) Who might use this graph and why?

MACRO D

COUNTING AND SORTING TECHNIQUES

One of the main topics in the area of discrete mathematics is counting techniques. In many instances, you will be asked to determine how many of a certain "thing" can be formed while following given restrictions or limitations. These types of problems involve ideas such as the Fundamental Counting Principle, factorials, permutations, tree diagrams, and combinations.

We begin by focusing on the **Fundamental Counting Principle**. Simply put, this states that if we have to fill a certain number of spots from a given set or sets, we multiply the number of possibilities for each spot to calculate the total number of possible sets.

Examples

A. Determine how many possible outfits are possible given 5 different colors of pants, 3 different colors of shirts, and 2 different pairs of shoes.

You simply multiply the number of choices for each component of the outfit: $5 \times 3 \times 2 = 30$ outfits.

The use of the Fundamental Counting Principle can become more involved when taking into account restrictions such as whether or not to allow repetition and mandating that certain criteria be met.

B. How many different three-digit numbers can be formed using the digits 1 through 3 and not allowing any repetition?

We are being asked to fill three slots. Keep in mind that because of the Fundamental Counting Principle, we will then multiply the numbers we determine represent the number of choices there are for each slot.

$$\_\_\_\_ \quad \_\_\_\_ \quad \_\_\_\_$$

Ask yourself how many different numbers can go in the first slot, then the second, and so on. Since we are not allowing repetition, this leads to

$$\_3\_ \times \_2\_ \times \_1\_$$

So, there are 6 three-digit numbers that can be made using the digits 1 through 3.

The above example leads to a sometimes useful notation called **factorial**. Since the number of slots matched the number of choices and we did not allow repetition, the product $3 \times 2 \times 1$ resulted. This can be abbreviated as 3!, which is read "three factorial." In general, $n!$ means we multiply $n \times (n - 1) \times (n - 2) \cdots \times 1$. Thus, as another example, $5! = 5 \times 4 \times 3 \times 2 \times 1$.

C. How many different three digit numbers could be formed using the digits 1 through 3 and allowing repetition?

Again, we start with 3 slots:

_____ _____ _____

Since we can repeat digits, each slot can be filled with any of the 3 possible digits.

So we have

_3_ · _3_ · _3_

which, as we know, means there are 3^3 or 27 possible numbers if we allow repetition.

D. How many three-digit numbers can be formed using the digits 1, 2, 5, and 9 without allowing repetition?

We again begin by looking at 3 slots and how these slots can be filled:

_4_ · _3_ · _2_

Since there are 4 numbers from which to choose, the first digit can be any of the 4 numbers; then, the second digit has to be 1 of the remaining 3 digits; and so on. Notice how this result is slightly different than a factorial since we don't multiply all the way down to 1. This is because we are not using all the available digits. When we have to choose a number of items out of a set without using up the entire set, we refer to this as a **permutation**. The problem we just did is referred to as a "permutation of 4 things taken 3 at a time." This can be written using a special notation as ${}_4P_3$. (*Note:* ${}_4P_3 = 4 \times 3 \times 2 = 24$.)

Thus, ${}_6P_4$ means we are using 4 out of 6 of some number, color, and so on. For instance, we could be creating four-digit numbers using the numbers 1, 4, 5, 6, 8, and 9. There would be $6 \times 5 \times 4 = 120$ such four-digit numbers assuming we do not allow repetition.

Another way to count the number of possible arrangements in a given situation is a visual representation called a **tree diagram**. In the example where we were asked to create three-digit numbers using the digits 1, 2, 5, and 9, a tree diagram could be constructed as shown below.

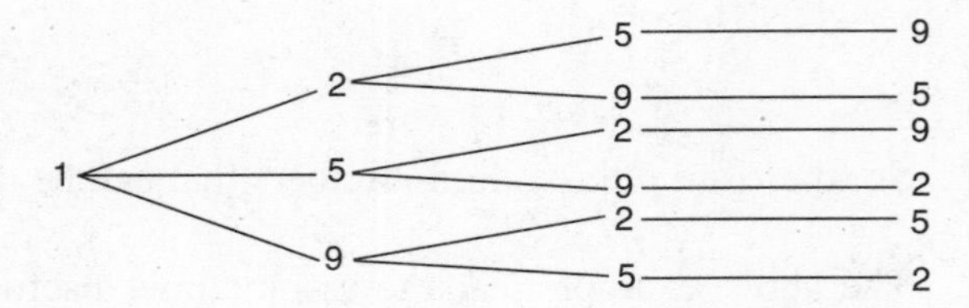

A similar setup could be used for 2, 5, and 9 as the first digit. As you can see, there are 6 different three-digit numbers formed using a 1 first and following a "path" to finish the number. Similarly, there could be 6 three-digit numbers using each of the remaining 3 numbers as the first digit. So, there would be a total of 24 numbers as we determined earlier using the permutation.

Sometimes you have to work with more involved restrictions other than simply allowing or not allowing repetition.

Examples

A. How many three-digit numbers can be formed using the digits 2, 5, 6, 7, and 8 if the number must be even and no digit can be repeated?

When you are given a restriction, always start by filling the slots that are affected by the restriction. In this problem, since the number has to be 3 digits and even, we know that the last digit has to be even (2, 6, or 8), which means there are three possibilities. So, initially, we have

$$\_\_ \cdot \_\_ \cdot \underline{\ 3\ }$$

Now, since we are going to use 2, 6, or 8 in the last slot, we are left with 4 of the original 5 numbers to fill the remaining two slots. Since we are not allowed to repeat, the remaining two slots must be filled by 1 of 4 possible numbers and then 1 of 3 possible numbers. Thus:

$$\underline{\ 4\ } \cdot \underline{\ 3\ } \cdot \underline{\ 3\ }$$

So, there are $4 \cdot 3 \cdot 3 = 36$ different three-digit numbers that meet the stated criteria.

B. Suppose you are taking a multiple-choice test consisting of 5 questions each with choices A through D and you know that the answer to question 3 is C and the answer to question 5 is either A or B. How many different sets of answers are possible?

To determine how many different sets of answers are possible, take a look at filling 5 slots with the given restrictions:

$$\underline{\ 4\ } \cdot \underline{\ 4\ } \cdot \underline{\ 1\ } \cdot \underline{\ 4\ } \cdot \underline{\ 2\ }$$

The three questions for which there are no given restrictions can be answered with A, B, C, or D (4 possibilities), while the answer to the third question must be C and the answer to the last question can be 1 of 2 possibilities (A or B). Thus, there are 128 possible answers on this test.

C. Use the letters in the word MATHEMATICS to form a four-letter combination whose second letter is a vowel, with no letter repeating.

We begin with the following since there are 3 vowels (A, E, I).

$$\_\_ \cdot \underline{\ 3\ } \cdot \_\_ \cdot \_\_$$

Next, since some letters repeat, after we choose one of the 3 vowels, we are left with only M, T, H, C, S, and 2 vowels (whichever two were not chosen), for a total of 7 letters. So, we continue:

$$\underline{\ 7\ } \cdot \underline{\ 3\ } \cdot \underline{\ 6\ } \cdot \underline{\ 5\ }$$

Therefore, 630 four-letter combinations can be made from the word MATHEMATICS with the given restrictions.

PRACTICE SET

1. A bank is setting up four-digit pin numbers to be used for ATM machines. If the pin numbers are to be made using the digits 1, 2, 4, 7, and 9, with no repetition of digits permitted, how many possible pin numbers are there?

2. Construct a partial tree diagram for Problem 1.

3. If you have a choice of 2 entrees, 3 side dishes, and 2 desserts, how many different meals can be chosen if you choose one from each category?

4. Evaluate each of the following.

A. 6! **B.** 3! **C.** 8!
D. ${}_5P_2$ **E.** ${}_7P_4$ **F.** ${}_{10}P_3$

> **Calculator Tip:** Permutations and combinations can be done on your calculator. If your calculator does not have these buttons, hit the PRB key. As always, consult your calculator's manual for help before the test.

5. How many different license plates can be assigned assuming that the format is 2 different letters followed by 3 different digits from 0 through 9?

6. Using the letters A, B, E, I, L, N, and R, how many three-letter "words" (they don't have to be actual words) can be formed if the first letter must be a vowel and no letter can be used more than once?

7. Using the letters in the word CALCULATOR and not repeating any letter and using a consonant as the first letter, how many different three-letter arrangements can be made?

8. If a license plate number consists of 3 letters followed by 2 digits from 0 through 9 and another letter, which expression shows the number of possible plate numbers that can be issued if there are no special restrictions?

A. $26 \cdot 25 \cdot 24 \cdot 10 \cdot 9 \cdot 26$ **C.** $26 \cdot 26 \cdot 26 \cdot 10 \cdot 10 \cdot 26$
B. $26 \cdot 25 \cdot 24 \cdot 9 \cdot 8 \cdot 23$ **D.** $26 \cdot 25 \cdot 24 \cdot 10 \cdot 9 \cdot 23$

9. In how many different ways can 5 students sit in a row that consists of 5 seats?

10. If a factory produces motorcycles in 3 models, 10 colors, and 2 engine types, how many different motorcycles are made?

11. In how many different ways can the letters D through K be arranged?

12. What does ${}_8P_3$ equal? Then, compute 8! and 5!. Do you notice any relationship between the three quantities? If so, state the relationship that you see. Does this work for other permutations? Provide examples to justify your conclusion.

13. Models of towers can be built using 4 gray and/or blue squares as shown below.

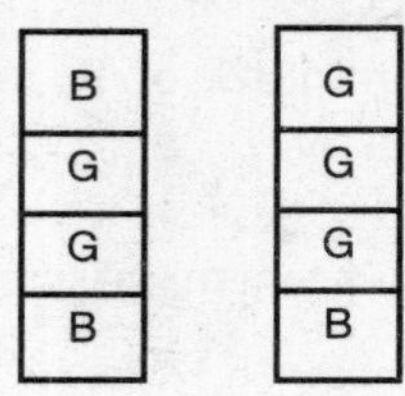

(a) Draw and label 2 additional towers.

(b) How many total towers can be built? (*Hint*: Keep in mind how many colors each square can be.)

COMBINATIONS

A **combination** is slightly different than a permutation in that it does not depend on order. For example, 321 is considered the same combination as 312. Frequently, combinations are used when forming groups of people.

Examples

A. Suppose that a committee of 3 students needs to be formed from a group of 5 students (Al, Bob, John, Kim, and Mark). It does not matter in what order we choose the 3 students. That is, Kim, Mark, John is the same as Mark, Kim, John. How many groups can be formed?

These calculations are called combinations. The notation for a combination is, for instance, ${}_5C_3$, which means we are forming a group of 3 from 5 where the order of selection is insignificant. To calculate this combination, we evaluate ${}_5P_3$ and then divide this result by 3!. Thus, there are $5 \cdot 4 \cdot 3 = 60/(3 \cdot 2 \cdot 1) = 10$ committees.

B. Suppose that instead of making committees of 3, we are asked to make committees of 2.

This is written as ${}_5C_2$ and calculated as ${}_5P_2 \ / \ 2! = (5 \cdot 4)/(2 \cdot 1) = 10$. Notice that we arrive at the same number of committees. This is because we are making use of the students left out after choosing the groups of 3. This is a powerful conclusion. Anytime we have combinations that are complements of one another such as ${}_{10}C_4$ and ${}_{10}C_6$ since 4 + 6 equals 10, these two combinations are equal.

PRACTICE SET

1. Compute each of the following.

(a) ${}_8C_5$ **(b)** ${}_9C_4$ **(c)** ${}_{10}C_7$ **(d)** ${}_{12}C_4$

2. If a school principal needs to choose a committee of 10 students out of a class of 28 students, write the combination(s) representing this situation and conclude how many committees can be formed.

3. How many groups of 6 can be formed from 12 people?

4. How many pairs of colors can be put together if there are 6 colors to choose from?

5. If a student committee of 12 is formed from a student body of 152, how many committees are possible?

Occasionally, a problem arises in which one choice is more prevalent than the others.

Examples

A. Suppose you are given 3 blue beads, 2 red beads, 2 yellow beads, and 1 bead of each of the colors green, orange, and purple. Determine how many different bracelets can be made.

Determining how many different arrangements of beads is a little different here. To solve such a problem, divide the factorial of the total number of items by the factorial of each multiple item. That is, here we take 10! (since there is a total of 10 beads) and divide by $3! \cdot 2! \cdot 2!$ (since there are 3 blue beads, 2 red beads, and 2 yellow beads). Thus, there are $10!/(3! \cdot 2! \cdot 2!) = 151{,}200$ possible bracelets.

B. In how many different ways can the letters in the word CALCULUS be rearranged?

Again, take the factorial for the total number of letters (in this case 8!). This number is to be divided by 2! (since 2 C's repeat) times 2! (since 2 L's repeat) times 2! (since 2 U's repeat). Thus, there are $8!/(2! \cdot 2! \cdot 2!) = 5040$ arrangements of these 8 letters.

PRACTICE SET

1. In how many different ways can the letters in COOL be rearranged?

2. In how many different ways can the following letters be rearranged?

A, B, C, C, D, D, D, E, F, F, F, F

3. If you have 5 blue marbles, 3 green marbles, 2 red marbles, and 1 each of white, black, and gray, in how many different ways can you line up these marbles?

Sometimes, a problem involving permutations or combinations involves multiple selections.

Examples

A. In how many ways is it possible to choose 3 cards from a standard deck with at least 2 of the cards being aces?

This means that either 2 are aces or 3 are aces. First, let's look at the case where there are 2 aces. This means we have a combination of 4 things (aces) taken 2 at a time since it doesn't matter if we choose the ace of spades and then the ace of hearts, or vice versa. Thus, ${}_4C_2 = 4 \cdot 3/2 \cdot 1 = 6$. So, there are 6 ways to choose 2 aces. The third card can be one of 50. Thus, there are 300 ways to choose 2 aces. If there are 3 aces, it is simply ${}_4C_3 = 4 \cdot 3 \cdot 2/3 \cdot 2 \cdot 1 = 4$ ways. Notice that this is the same as choosing 1 ace to leave out (which would be 4 ways). Thus, in total, there are 300 + 4 = 304 ways to choose 3 cards with at least 2 aces.

B. Suppose you are making a sandwich and can add lettuce, tomato, mayo, and/or pickles. If the choices for the sandwich are ham and turkey, how many sandwiches are possible with 0, 1, 2, 3, or 4 "extras"? First, realize that the same set of answers works for both ham and turkey, so we can just find the solution for a ham sandwich and double it. Zero extras is easy—there is only 1 way to make a sandwich with no extras. One extra would be any one of the 4 choices. For 2 extras, it's ${}_4C_2$, which equals 6. For 3 extras, it's ${}_4C_3$, which equals 4. For all 4 extras, there's obviously only one way. So, for a ham sandwich, there are $1 + 4 + 6 + 4 + 1 = 16$ ways to add 0 to 4 extras. Similarly, there are 16 ways to add 0 to 4 extras to a turkey sandwich, for a total of 32 possible sandwiches.

When solving problems, be sure to take into consideration whether or not the order in which you are choosing things makes a difference. If order matters, you should use a permutation. If the items taken in different order are considered the same, you can use a combination.

PRACTICE SET

1. For each situation described, decide whether you need to use a permutation or a combination.

(a) choosing a committee of 3 students from a group of 8
(b) choosing a president, vice-president, and secretary from a group of 10
(c) choosing a 3-digit number made up of 7 digits with no repetition
(d) choosing 3 colors out of 5.

2. You are making a quilt using squares that are red, blue, green, yellow, and violet. How many different quilts are possible using 1, 2, 3, 4, or all 5 color squares?

3. An ice cream parlor offers 2 flavors of ice cream (vanilla and chocolate) and 3 toppings (hot fudge, caramel, and nuts). How many different sundaes could be ordered if you are going to put 0, 1, 2, or 3 toppings on the sundae?

4. A school principal is forming a committee of 3 teachers and 2 students. If there are 20 teachers to choose from and 35 students to choose from, how many different committees are possible?

5. If you are asked to choose 5 cards from a standard deck, in how many ways is it possible to end up with a group consisting of 3 hearts and 2 spades?

6. You are taking a test that consists of 5 true-false questions and 3 multiple-choice questions (A, B, C, or D). In how many different ways is it possible to complete this test?

7. Look back at the example in the text involving making a sandwich and choosing 0, 1, 2, 3, or 4 extras. Compare the numbers we found for each number of extras to Pascal's Triangle shown below.

Explain what you notice about the numbers we found and Pascal's Triangle. Extend Pascal's Triangle another row. Then how many sandwiches can be made that have 3 extras out of 5 choices?

In the last exercise, we made reference to Pascal's Triangle. Pascal's Triangle is very useful in solving a particular type of problem in discrete mathematics called a *path problem*. In a path problem, you are asked to find the number of ways of getting between 2 points under certain constraints. It is tedious, if not impossible, to trace through all the ways without losing track or forgetting one. Therefore, a method of organization is used to count the paths.

Examples

A. In the diagram below, suppose you are asked to trace out the word HSPA by moving from one letter to another letter diagonally right or left of it in the row below. We label each letter with the number of ways to get there:

H_1

S_1 S_1

P_1 P_2 P_1

A_1 A_3 A_3 A_1

In the third row, since you can get to the middle P from either of the S's, there are 2 ways to get to the middle P. In the last row, since you can get to the second A from either of the two P's above it and there's 1 way to get to the first P and 2 ways to get to the second P, there is a total of 3 ways to get to the A. So, there is a total of 8 ways to spell HSPA. (*Note*: Add up the four numbers.)

B. Suppose we have to spell out the word SMART again by moving diagonally right or left between letters in the diagram below.

S

M M

A A A

R R

T

Again, a good approach is to label each letter with a number representing the number of ways to get to that spot. This yields

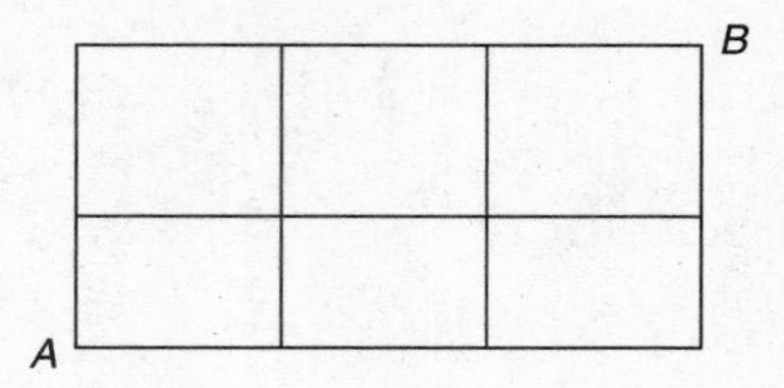

As one can see from the diagram after labeling, there are 6 ways to spell out the word SMART in this diamond-shaped diagram. Again, we see Pascal's Triangle coming into play. The letters bear the labels corresponding to the numbers in Pascal's Triangle.

The diagram we are asked to work with need not always be a triangle or diamond-shaped. Consider the following map and find the number of ways to get from point A to point B by moving either north or east.

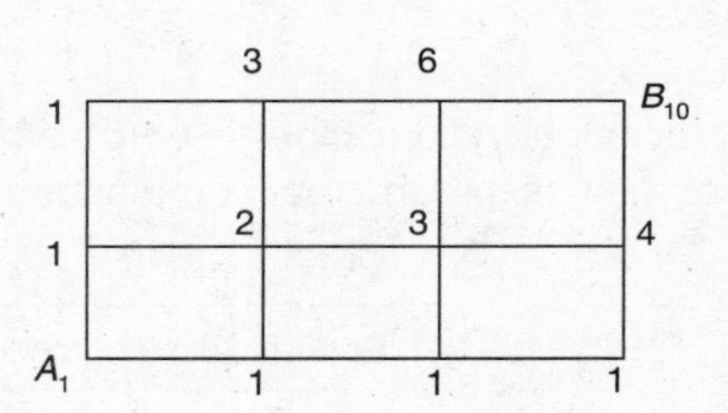

In this type of problem, we label each "corner" with a number representing the number of ways of getting there. Thus, we have

3 6
1 B_{10}
2 3
1 4
A_1 1 1 1

From the labeling and diagram that results, we can conclude that there are 10 ways to move from A to B moving only north or east each time.

PRACTICE SET

1. Determine how many ways you can spell out the word SCHOOLS in the diagram below if you must move diagonally right or left between letters.

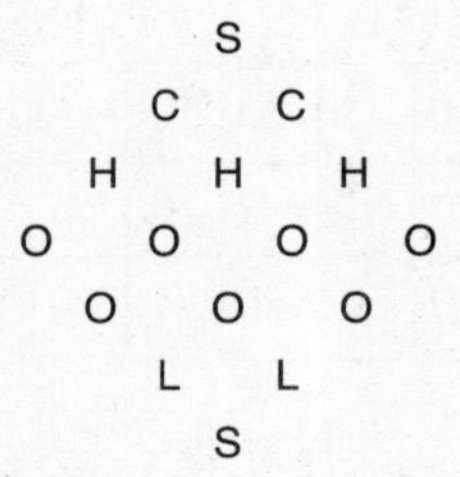

2. Determine how many ways there are to spell out the word ALGEBRA in the diagram that follows. Again, you can move only diagonally right or left to the letters in the next row.

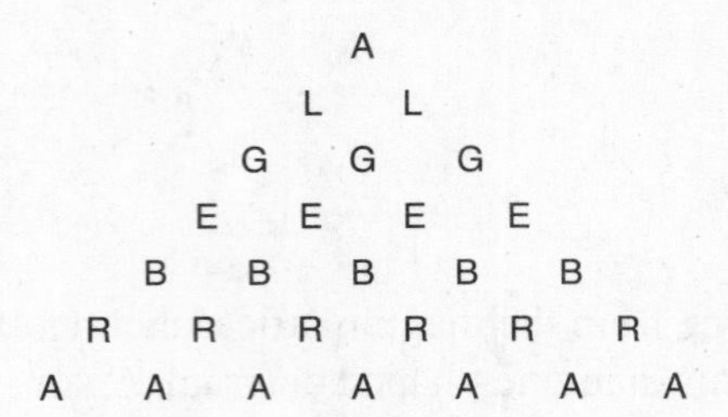

3. Determine how many ways there are to move from point A to point B moving only south or east.

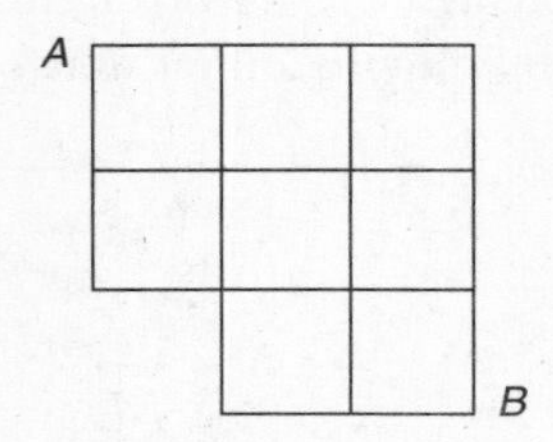

In addition to path problems, there are other types of problems that involve counting. Sometimes the best approach to these problems is to organize the information in a table.

Examples

A. A carpenter is making stools; some have 3 legs and some have 4 legs. If he has 60 legs to use, how many different combinations of stools can he make?

No. with 3 legs	20	16	12	8	4	0
No. with 4 legs	0	3	6	9	12	15

Frequently, in a table you will notice patterns. For example, here the number of 3-legged tables decreases by 4 (which means we use 12 fewer legs, which means we can make three 4-legged tables). So, there are 6 different combinations of stools he can make.

B. How many different ways are there of making 30 cents change using pennies, nickels, and dimes?

No. of pennies	30	25	20	20	15	15	10	10	10	5	5	5	0	0	0	0
No. of nickels	0	1	2	0	3	1	4	2	0	5	3	1	6	4	2	0
No. of dimes	0	0	0	1	0	1	0	1	2	0	1	2	0	1	2	3

As seen in the chart, there are 16 ways to make 30 cents change using pennies, nickels, and/or dimes. Again, you can observe some patterns.

PRACTICE SET

1. Use a table to find out how many different ways there are to give 50 cents change using only nickels and/or quarters.

2. If a company is manufacturing bicycles and tricycles and has 100 wheels, how many different combinations of the 2 types of cycles can be made using the 100 wheels?

3. A puppetmaker has 30 legs and is going to make 2 types of puppets—dogs and birds. How many different combinations of puppets can be made using the 30 legs?

   Describe any pattern(s) you see in the table you created to solve this problem.

   Suppose the profit on a dog puppet is $10 while the profit on a bird puppet is $7. Give the profit for two of the combinations you found above.

GAME STRATEGIES

There will be a few instances on the HSPA Exam where you might be asked to determine which of two players has a better chance of winning a particular game or you might be asked to analyze a particular strategy.

Examples

A. One of the most common strategy games is Nim. In this game, there are several piles or groups of stones and each player on his or her turn picks up one or more stones as long as they are from the same pile. The player who is forced to pick up the last stone loses.

Suppose that a game of Nim starts with piles of 1, 3, and 7 stones. Mary picks up 6 stones from the pile of 7 on her first turn. Now the piles have 1, 3, and 1 stones. On your turn, should you pick up a pile of 1 or pick 2 from the pile of 3 to guarantee a win?

Most of the time, to answer these types of questions, you actually have to act out playing the game. So, in this case, let's try the first suggestion:

Picking up a pile of 1 leaves a pile of 3 and a pile of 1. This means Mary can pick up the pile of 3 and force you to pick up the remaining pile of 1 and you would lose.

The second choice of picking 2 stones from the pile of 3 leaves 1, 1, 1, and Mary would have to pick up a pile of 1. This would leave 1, 1, and you would pick up a pile of 1. The last pile of 1 would have to be picked up by Mary, and you would win.

B. In a strategy game for two called Bachet's Game, players use between 15 and 25 counters. Each player must pick up 1, 2, or 3 counters on his or her turn. The winner is the player who picks up the last counter or counters.

If you are playing with 20 counters and play begins as follows, what should your next move be to guarantee a win?

Sue: Picks up 3, leaving 17
You: Pick up 2, leaving 15
Sue: Picks up 1, leaving 14
You: Pick up 3, leaving 11
Sue: Picks up 1, leaving 10
You: ?

Another common strategy when working with games or certain other types of problems is working backward. If you want to win, then you must leave 4 counters so that regardless of whether Sue picks up 1, 2, or 3, the amount she leaves (3, 2, or 1) can be picked up by you for the win. Prior to 4, in order to guarantee that you can leave 4, you must leave her with 8 (thus, if she picks up 1, 2, or 3, she will have to leave either 7, 6, or 5, all of which you can decrease to 4). This would continue: You could try to leave 4, 8, 12, 16, and so on.

So, you should pick up 2 on your next turn, leaving 8. No matter what Sue picks up (1, 2, or 3) and leaves (7, 6, or 5), you can pick up the appropriate number to leave 4, and again no matter what she picks up and leaves, you can win.

PRACTICE SET

1. In a game of Nim as described in Example A above, if you start with piles of 2, 8, and 10 and play begins as follows, should you pick up 1 from the 2 pile or pick up a pile of 1?

Tom: Picks up 5 from the 10 pile, leaving 2, 8, and 5
You: Pick up 1 from the 2 pile, leaving 1, 8, and 5
Tom: Picks up 3 from the 5 pile, leaving 1, 8, and 2
You: Pick up 3 from the 8 pile, leaving 1, 5, and 2
Tom: Picks up 4 from the 5 pile, leaving 1, 1, and 2
You: ?

2. In a particular game, you and your opponent keep a running total. On each person's turn, you must add 1, 2, 3, 4, or 5 to the accumulating total. The person who causes the total to reach 30 wins. What can you add on your turn in the game below to guarantee that you win?

Total: 0
Richard adds 5; total: 5
You add 3; total: 8
Richard adds 5; total: 13
You add 2; total: 15
Richard adds 4; total: 19
You add ????

3. In a certain game, you roll two dice and then cover up a number or combination of numbers equal to or adding to the total rolled. If the only numbers uncovered are 1, 2, 4, 8, and 9, what is the fewest number of rolls necessary for you to possibly win? Is this very likely? Why or why not?

NETWORKS

A **graph** or **network** is a figure consisting of points (called **vertices**) and lines or **edges** connecting them. Below are several examples of networks.

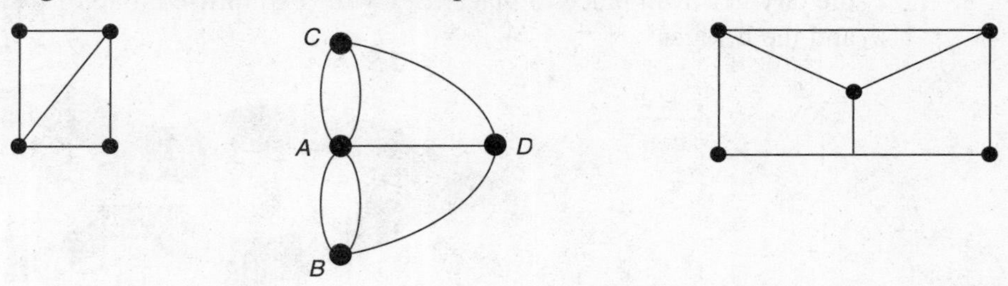

A route along edges and starting and ending at a vertex is called a **path**. Here are examples of paths:

If a path starts and ends on the same vertex, it is then called a **circuit**. The network below shows a path that is a circuit:

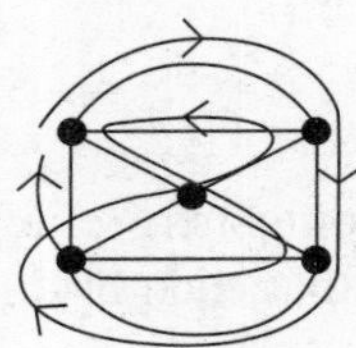

If a path uses every edge only one time, it is called a **Euler Path**. If a path is both a Euler Path as well as a circuit, then it qualifies as a **Euler Circuit**. The previous graph is also an example of a Euler Path and a Euler Circuit.

A network is said to be **traversable** if it can be traced without redrawing any edges already covered and without lifting one's pencil while drawing it. Whether or not a graph or network is traversable depends on the number of odd and even vertices. A vertex is **odd** if an odd number of edges meet at that point. Similiarly, a vertex is **even** if an even number of edges meet at that point. If a network has *exactly* zero or two odd vertices, then it is traversable (start at one odd vertex and end at the other); if a network does not meet this condition, it is not traversable.

Applications of Graphs (Networks)

Frequently, a real-life situation can be modeled using a network. Typically, vertices represent different towns, people, or groups, and those that are connected share something. For example, in the graph below, the children who are friends are joined with an edge.

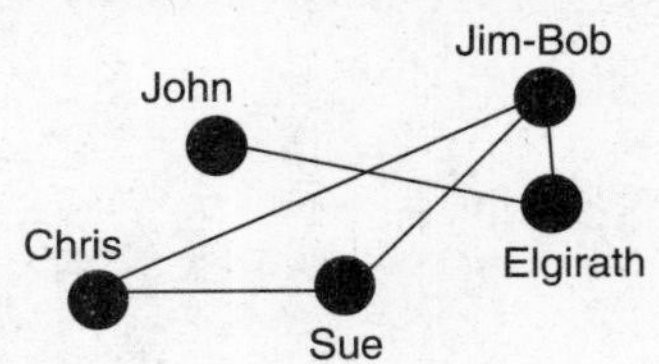

We can see from this graph, for instance, that John is friends only with Elgirath and that Jim-Bob has the most friends (three).

A well-known problem has to do with graphs—the Konigsberg Bridge problem. In this problem, based on Konigsberg, Germany, a river runs through the city, creating a center island, and then forks off into two parts. Seven bridges were built to help people from the city get from place to place. Below is a simplified map of the town showing the river and the bridges:

The people of Konigsberg wondered if it would be possible to travel through the city in a way that would involve crossing each bridge exactly once. Mathematicians over the years found that it was impossible. As this example shows, it is possible to come close, but you will never be able to cross all seven bridges once.

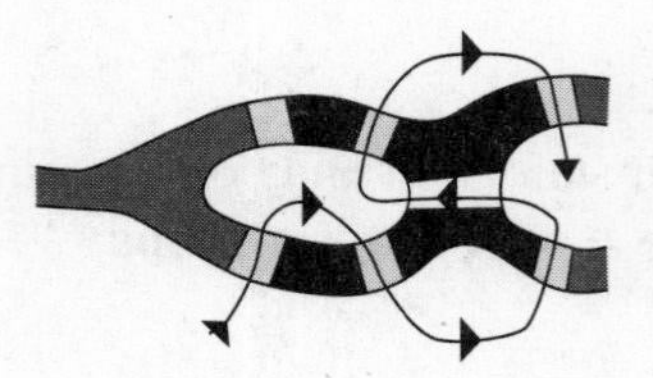

If the people of Konigsberg eliminated one bridge, as shown below, then the problem could be solved so that each bridge is crossed only once.

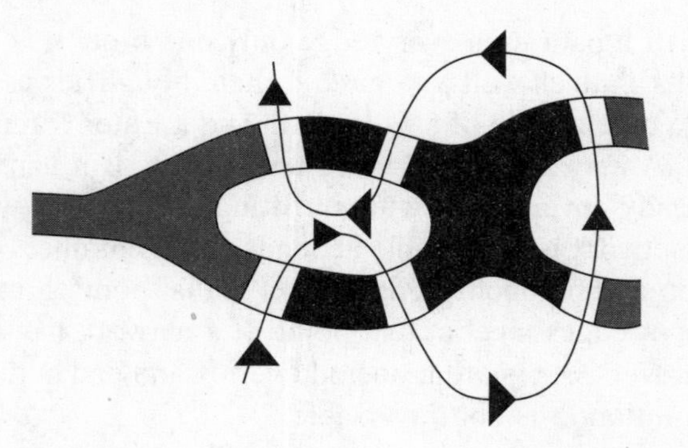

PRACTICE SET

1. Which of the networks below is *not* traversable?

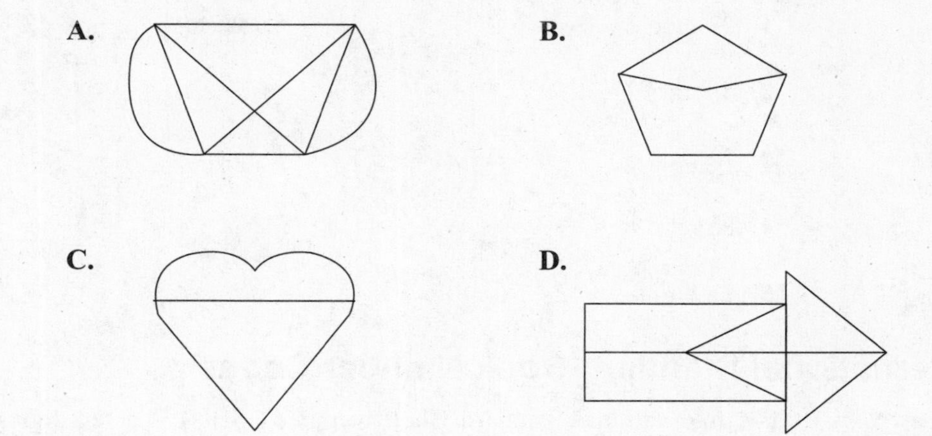

2. Sketch a network of five buildings and the roads connecting these buildings so that the network is not traversable.

3. How many odd vertices does network D in Problem 1 have?

4. How many even vertices does network B in Problem 1 contain?

5. A network is referred to as **complete** if there is at least one edge between every pair of vertices. Which of the following graphs (if any) are complete networks? There may be more than one answer.

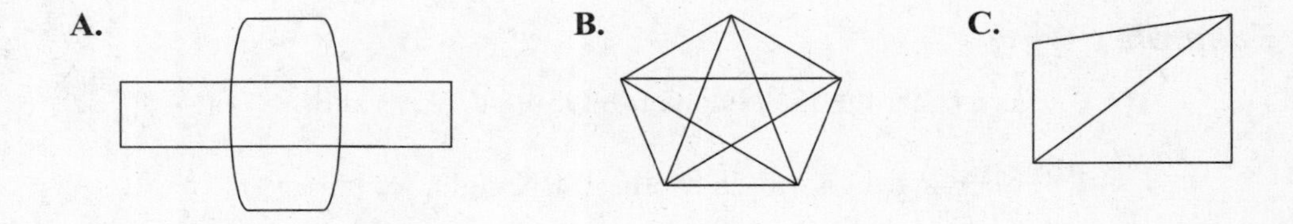

CODES

Another area of discrete mathematics that is part of the HSPA Exam is coding theory, which primarily has to do with how different organizations or agencies create codes that can be checked for security reasons.

Many codes rely on what is known as modular arithmetic, which centers around remainders. For example, in mod 3, $12 = 0$ since the remainder is zero when 12 is divided by 3. In mod 3, $14 = 2$ since $14 \div 3 = 4$ with a remainder of 2. Similarly, 57 (mod 4) is 1 since $14 \times 4 = 56$ and the remainder is therefore 1 when 57 is divided by 4.

PRACTICE SET

Find the value of each:

1. 23 mod 2

2. 45 mod 3

3. 73 mod 5

4. 85 mod 8

5. 129 mod 9

International Standard Book Number Codes

The International Standard Book Number (ISBN) is a ten-digit code used to uniquely identify books.

The ten digits are separated into four groups which are often, although not always, separated by hyphens. The first group consists of a single digit representing the country or language in which a publisher is incorporated. The second group of numbers identifies the publisher using two to seven digits. The third group of numbers are specific to the individual book—this group can consist of anywhere from one to six digits. The last digit is referred to as a "check digit" and is a number from 0 through 9 or an X (for the Roman numeral 10). The check digit is computed from the following equation.

$$10d_1 + 9d_2 + 8d_3 + \cdots + d_{10} = 0 \pmod{11}$$

where d_n represents the nth digit.

Example

Verify that the ISBN code 0-8493-9640-9 is valid.

Following the formula for the ISBN code, we have

$$10\times0+9\times8+8\times4+7\times9+6\times3+5\times9+4\times6+3\times4+2\times0+9$$
$$=0+72+32+63+18+45+24+12+0+9=275$$
$$275\div11=25\text{ r }0$$

Therefore, this is a valid ISBN number.

Universal Product Codes

The universal product code (UPC) is a 12-digit number along with a machine-readable bar code that is used to identify particular products. The UPC encodes the product but not the price. The UPC is maintained by the Uniform Code Council in Dayton, Ohio.

The first and last digits of a UPC appear in smaller type and are separated from the other ten digits. The first six digits identify the product's manufacturer, while the next five digits identify the specific product under that manufacturer. Again, the last digit is a check digit.

The check digit is calculated as follows.

$$d_{12} = 10 - \left[\left(3\sum_{\substack{i=1\\ \text{odd}}}^{11} d_i + \sum_{\substack{i=2\\ \text{even}}}^{10} d_i\right) \text{mod } 10\right] \text{mod } 10$$

Example

The UPC from Tradewinds Lemon-Lime tea is 0 88130 44019 4. Verify that the check digit is 4.

$$\begin{aligned} d_{12} &= 10 - \{[3\times(0+8+3+4+0+9)+(8+1+0+4+1)] \text{ mod } 10\} \text{ mod } 10 \\ &= 10 - \{[3\times 24+14] \text{ mod } 10\} \text{ mod } 10 \\ &= 10 - [(86) \text{ mod } 10] \text{ mod } 10 \\ &= 10 - [6] \text{ mod } 10 \\ &= 4 \end{aligned}$$

PRACTICE SET

1. Verify that the following ISBN is valid.

1-56765-544-0

2. Find the necessary check digit assuming that the rest of this ISBN is valid.

0-9653529-5-?

3. Verify that the check digit is correct in the UPC for Rold Gold pretzels below.

0 28400 02186 9

4. Find the necessary check digit assuming that the rest of this UPC is valid.

0 72000 15057 ?

MIXED PRACTICE, CLUSTER III, MACRO D

1. How many arrangements are possible using all the letters of the word GEOMETRY if each arrangement must begin and end with an E?

A. 6! **B.** 7! **C.** $\frac{8!}{2}$ **D.** 8! − 2

2. Starting at the letter V and moving downward on a diagonal to the left or the right, how many different paths spell the word VOLUMES?

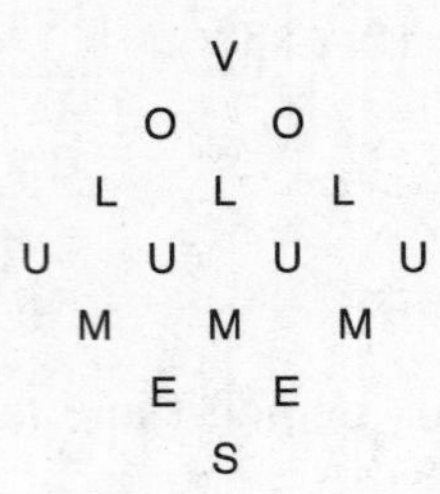

3. Using the digits 6, 7, 8, and 9, how many 4-digit numbers can be formed if you are allowed to repeat digits but can't use all 4 digits the same?

A. $4^4 - 4$ **B.** ${}_4P_4$ **C.** $({}_4P_4) - 4$ **D.** ${}_4C_4$

4. Given 5 points A, B, C, D, and E, none of which are collinear, how many unique triangles can be determined by these points?

(a) Is this a permutation or a combination? Why?
(b) What is the probability that a randomly chosen triangle will have A as a vertex?

5. Many of the workers in a certain department are on more than one subcommittee. In the chart below, an X means that the two committees have at least one common member. If at most two meetings can be held per afternoon, what is the minimum number of days meetings will have to be held?

	Advertising	Policies	Events	Schedule	Payroll	Social
Advertising						X
Policies			X		X	
Events		X		X	X	X
Schedule			X			X
Payroll		X	X			X
Social	X		X	X	X	

MACRO E

ITERATION AND RECURSION

Iteration or recursion refers to a process in which each new result depends on the preceding results. Iterations involve either numbers or geometric figures and contain an initial value or geometric figure (referred to as the seed), a rule, and output.

Examples

A. Below is an example of a numerical iteration.

Begin with 8.

Rule: Multiply the previous value by 2.
Result: 8, 16, 32, 64, …

B. Below is an example of a geometric iteration.

Begin with a rectangle with length 15 inches and perimeter 38 inches.

Rule: Decrease the length by 3 inches.
Result:

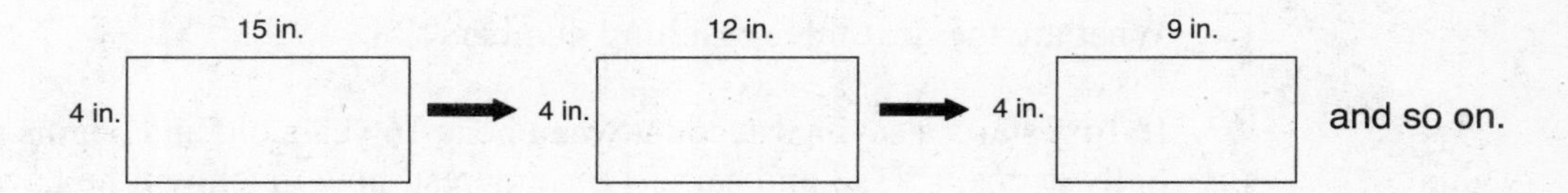

C. If Mary's parents open a savings account for her when she is 8 years old and start with \$500 and earn 3.5% interest every year, show the amount in Mary's account for the first 5 years.

$$\$500, \$500 \times 1.35 = \$675, \$675 \times 1.35 = \$911.25, \$911.25 \times 1.35 = \$1230.19, \$1230.19 \times 1.35 = \$1660.76$$

Note: This is a recursive or iterative process since each year's interest is based on 3.5% of the previous year's balance.

D. Find the next three terms based on the given recursive definition.

Start: 5, 7

Rule: Beginning with the third term, add the previous 2 terms to get the next term.
Result: 5, 7, 5 + 7 = 12, 7 + 12 = 19, 12 + 19 = 31, and so on

Sometimes when working with iterative or recursive processes, we use arrow notation to indicate the rule to be used. For instance, $x \rightarrow x$^2 means we square the previous term to get the next term.

Examples

A. Seed: 150.
Rule: $x \rightarrow x - 8$.
Result: 150, $150 - 8 = 142$, $142 - 8 = 134$, $134 - 8 = 126$, $126 - 8 = 118$, $118 - 8 = 110$, and so on.

B. Seed: 16.
Rule: $x \rightarrow \sqrt{x}$.
Result: 16, $\sqrt{16} = 4$, $\sqrt{4} = 2$, $\sqrt{2}$, and so on.

PRACTICE SET

1. A rectangle begins with length 7 inches and width 4 inches. If the length increases by 3 inches at a time, sketch and label the first four rectangles.

2. Begin with 12.

Rule: Add 5 to the previous term to get the next term.
Result: 12, 17, 22, ...

What are the next three resulting numbers?

3. If Tom starts a savings account when he is 16 years old and begins working, if he initially deposits $350 and earns 4.5% interest, how much will he have after 4 years?

4. Seed: –8.
Rule: $x \rightarrow x + 3$.
Result: –8, –5, ...

Find the next three results.

5. Find the next three terms based on the recursive definition.

Start: 3, 7.
Rule: Add the previous two terms to get the next term.
Result: 3, 7, 10, 17, ...

FRACTALS

A **fractal** is a figure that is obtained through iteration. In a fractal, you can see self-similarity, meaning that if you look at a smaller section of the fractal, it looks like the previous stages or the whole fractal.

A famous example of a fractal is the Sierpinski Triangle. For this fractal, you begin with an equilateral triangle and at each stage follow this rule: Connect the midpoints of the sides of each unshaded triangle and shade the resulting smaller triangle.

The results after three stages are

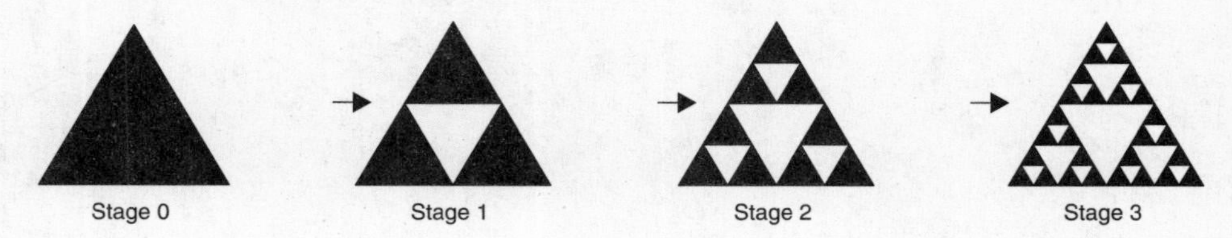

The following is another fairly well-known fractal, a fractal that appears in the book *Jurassic Park.*

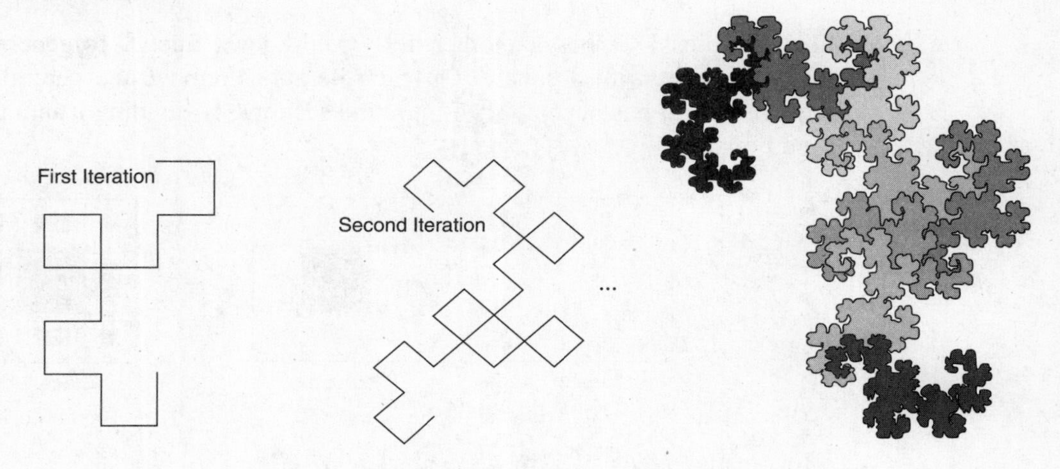

Last, let's take a look at another very famous fractal, the Koch Snowflake. To get the Koch Snowflake, you begin with an equilateral triangle. On each side, remove the middle third and replace it with a "triangular bump" with sides as long as the length removed. At each stage, do this to all triangle-shaped points. The resulting fractal is shown here.

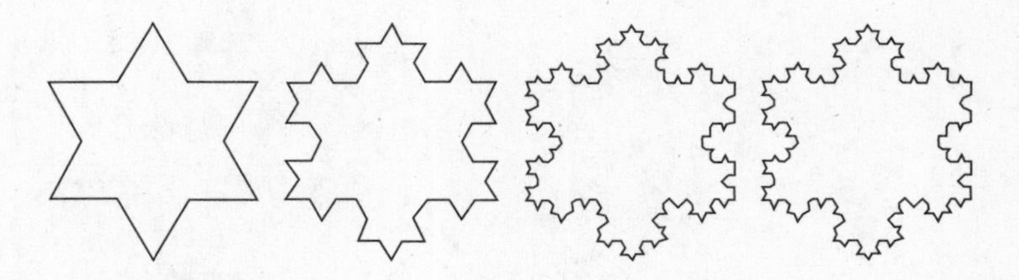

PRACTICE SET

1. A *fractal tree* is created by drawing two smaller branches approximately one-third of the way down each existing branch. Below are stages 0, 1, and 2. Draw the next stage.

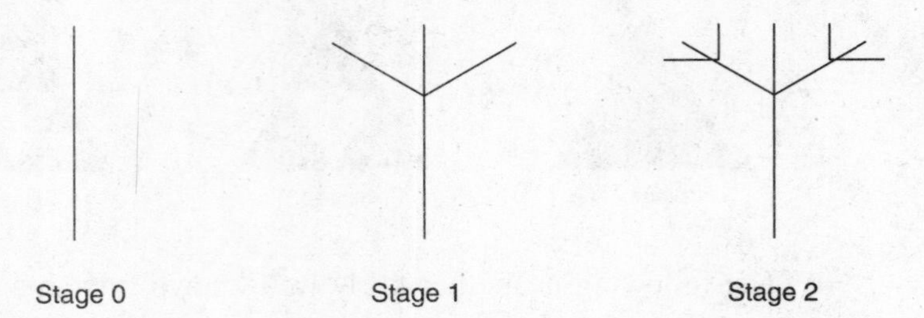

2. The Sierpinski Carpet is another fairly well-known fractal. To generate a Sierpinski Carpet, begin with a square. On each iteration, remove the central section of the square once it has been divided into nine sections by dividing it into thirds vertically and horizontally.

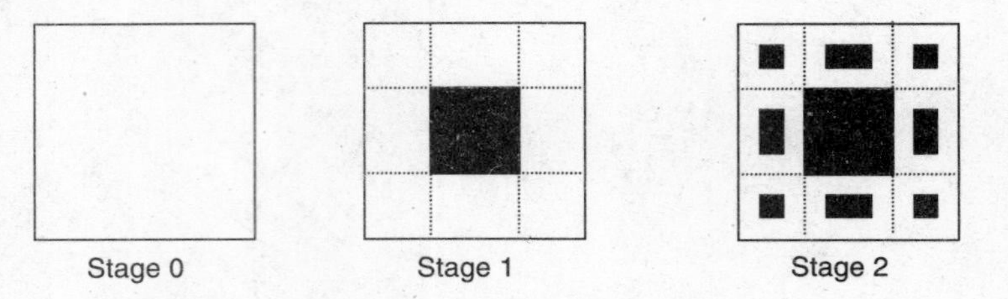

How many "holes" will there be at stage 3?

3. If the perimeter of the initial equilateral triangle for a Koch Snowflake is 9 units, find the perimeters of the next two stages. The following diagram will assist you with these calculations.

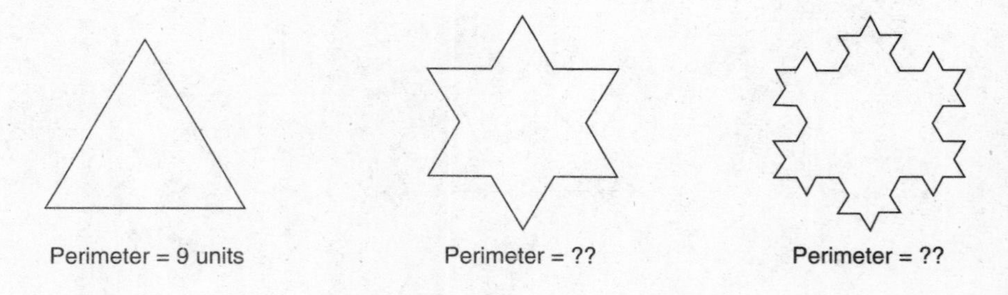

Do you notice a pattern in these perimeters? Explain.

ALGORITHMS AND FLOWCHARTS

An **algorithm** is a list of steps to follow to complete a procedure or accomplish a task. Some familiar applications of algorithms include following a recipe, putting a list of words in alphabetical order, making a factor tree for a given number, and performing long division.

Algorithms are often given in terms of some sort of graphical representation. These include **iteration diagrams**, which show the seed and the rule for an iterative process. Below is an example.

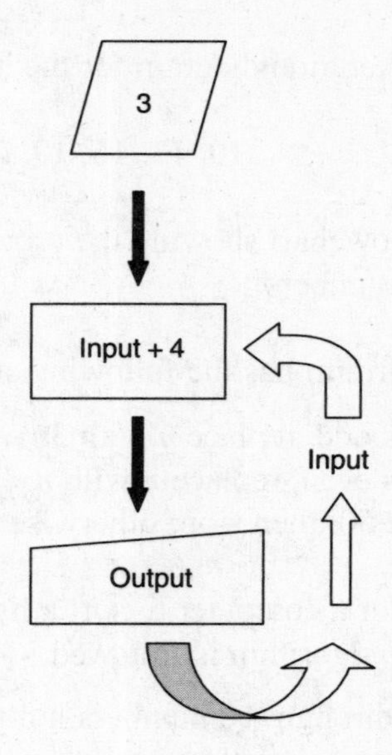

Another type of visual aid used to show an algorithm is a **flowchart**. Flowcharts used to be an essential part of computer programming and are still used today to visually show processes that are complicated and often involve repetition or decision making.

Below is a flowchart showing the algorithm one would follow to determine if a given number is a factor of 48.

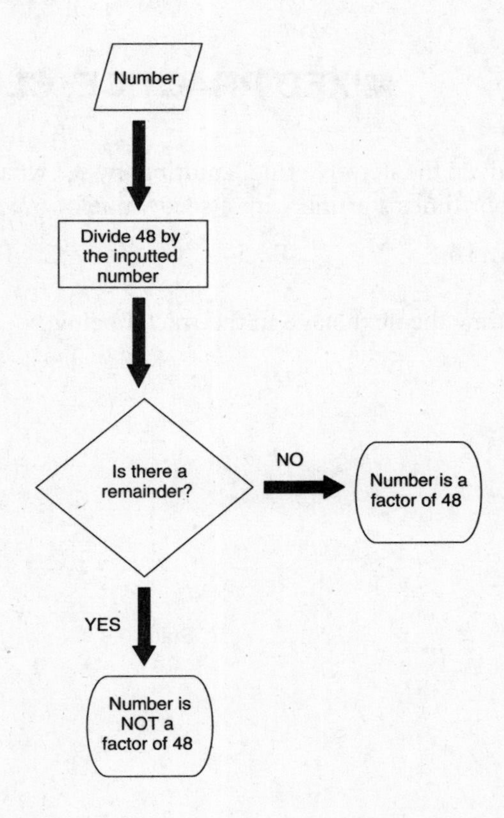

PRACTICE SET

1. Write an algorithm showing how to perform long division.

2. Draw an iteration diagram for the iterative process that would generate the following.

   10, 13, 16, 19, 22, 25, 28, . . .

3. Make a flowchart showing the process of determining the Least Common Multiple of a pair of numbers.

4. If an algorithm has the following steps, give the results produced if $n = 30$.

   (a) If n is odd, replace n with $3n - 7$.
   (b) If n is even, replace n with $n \div 2$.
   (c) If $n \geq 50$, then stop; otherwise go back to step a.

5. In order for a computer to sort a list of numbers and put them in ascending order, the following algorithm is followed.

   (a) Pass through the numbers left to right.
   (b) If a number to the right is smaller than the number to its left, switch them.
   (c) Repeat.

   Show the steps that take place inside a computer to sort the following list into increasing order:

   1, 4, 3, 8, 6, 10, 5

MIXED PRACTICE, CLUSTER III, MACRO E

1. Given the iterative rule "multiply by 3," what will the value be after the rule is applied four times starting with a seed value of 6?

   A. 18 **B.** 54 **C.** 162 **D.** 486

2. Draw the next stage in the fractal below:

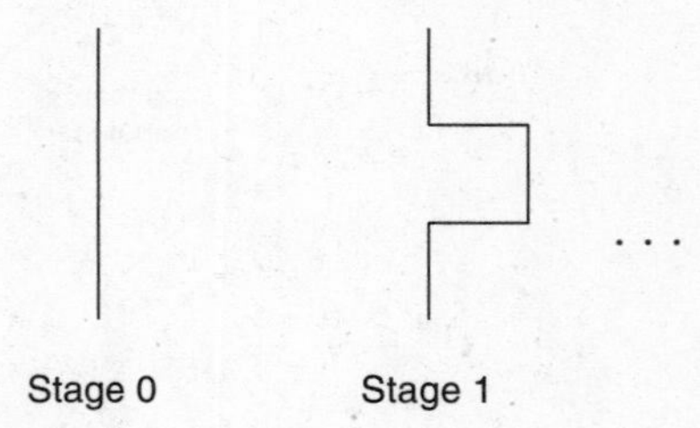

CLUSTER IV

PATTERNS, FUNCTIONS, AND ALGEBRA

MACRO A

PATTERNS

Observing patterns is a helpful way to approach many problems; patterns are usually either visual or numerical.

Numerical patterns are typically determined by the number of complete cycles in a repeated group of numbers and using the remainder, or by performing a similar operation repeatedly.

Examples

A. Determine the next three numbers in the pattern below.

$$7, 10, 13, 16, \ldots$$

In this pattern, we add 3 to each term to get the next term. So, the next three terms are 19, 22, 25.

B. Determine the next three numbers in the following pattern.

$$4, 6, 9, 13, 18, \ldots$$

Here, we first add 2, then add 3, then add 4, and so on. Therefore, we add 6 to 18 to get 24, then we add 7 to that to get 31, and finally we add 8 so the third number is 39.

C. What number is in the 50th decimal place of .474747 . . . ?

The digits obviously alternate between 4 and 7, with 4 being in the 1st, 3rd, 5th, . . . positions (that is, 4 is in the odd positions) while 7 is in the 2nd, 4th, 6th, . . . positions (that is, 7 is in the even positions). Since 50 is an even number, the digit in the 50th place is 7.

D. What number appears in the units digit of 3^{35}?

This question is similar to Example C with one marked difference. In Example C, there was a cycle of two numbers repeating; in this problem the cycle is longer, and thus it is a bit more complicated to figure out the required digit. *Remember*: The units digit refers to the number all the way to the right. For example, in 326, we have 3 hundreds plus 2 tens (20) plus 6 ones or units.

On examination, the powers of 3 begin as follows.

$3^1 = \mathbf{3}$	$3^5 = 24\mathbf{3}$
$3^2 = \mathbf{9}$	$3^6 = 72\mathbf{9}$
$3^3 = 2\mathbf{7}$	$3^7 = 218\mathbf{7}$
$3^4 = 8\mathbf{1}$	$3^8 = 656\mathbf{1}$

The units digits are in boldface type, and we can see that there is a cycle of four numbers that repeats (3, 9, 7, 1, 3, 9, 7, 1, . . .). Our job is find out how many complete groups of four there are in 35 (since that is the power) and what the remainder is. So, we divide: $35 \div 4 = 8$ with remainder 3 (since $4 \times 8 = 32$ and $35 - 32 = 3$).

Thus, the pattern 3, 9, 7, 1, 3, 9, 7, 1 . . . contains eight full cycles and the 35th is the third number (since 3 is the remainder). So, 3^{35} ends in a units digit of 7.

Calculator Tip: You can always use your calculator to get the quotient and remainder for a division problem. For example, if you divide 30 by 4 and get 7.5, don't mistakenly think the remainder is 5. Rather, multiply 4 by 7 to get 28 and then take 28 from 30 to get the remainder, 2.

Note: Calculators are not extremely helpful in solving these types of problems. If you use a calculator to divide, you can't use the decimal answer that you get ($35 \div 4 = 8.75$). If you do use a calculator, the result (8.75) will tell you that 4 "goes into" 35 eight times. Then you can multiply 4 by 8 to get 32 and subtract this from 35 to interpret the remainder.

Visual patterns obviously rely a lot more on what you see. Images can be rotated little by little or increase in size. In some cases, the image has several different parts and there is cyclic behavior much like that in the last example.

Examples

A. What are the next five letters in the pattern below?

BDDFFF . . .

In examining the pattern, you should notice two things: It's the second, fourth, sixth, and so on, letters of the alphabet (skips a letter in between), and each time there appears to be one more of the letter (that is, the pattern begins with one B, then two D's, and then three F's).

So, the next letter will be H (we skip G), and there should be four G's. Then, the next letter (the fifth continuing the pattern) will be a J (we skip I).

B. What symbol will be in the 107th position in the pattern below?

]-- --|-- --[

Since it is a cycle of three symbols, we divide 107 by 3 to get 35 remainder 2. The 35 means there are 35 complete groups of the three symbols, while the remainder of 2 means there are two extras. Thus, the 107th position is the second symbol, --|--.

C. How many shaded squares will be in the fifth picture?

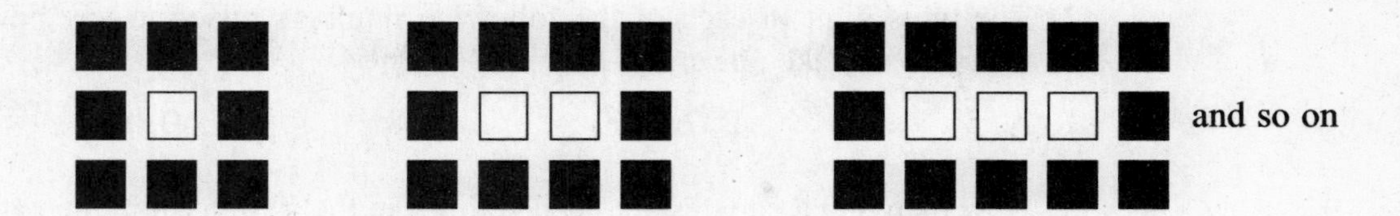

By observing the pattern and counting the shaded squares, we see that there are 8 shaded squares in the first picture, 10 in the second, and 12 in the third. Thus, 8, 10, 12, 14, 16, . . . , and so there will be 16 shaded squares in the fifth picture.

D. If this pattern continues, how many total triangles will be needed to represent all of the first five terms?

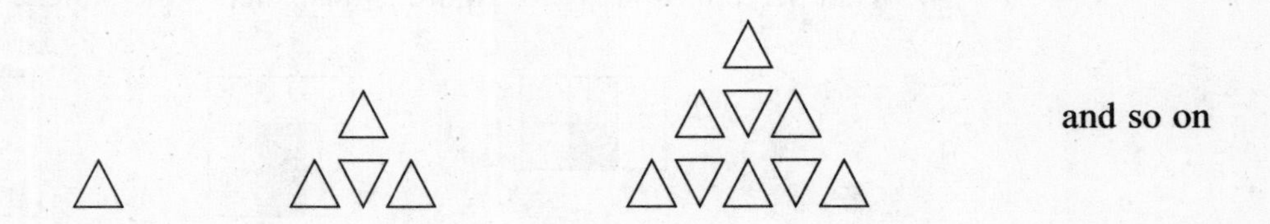

Again, by observing the pattern and simply counting, we see that there is 1 triangle in the first diagram, 4 triangles in the second diagram, and 9 triangles in the third diagram. This pattern may not be as easy for you to recognize, but it appears that the numbers are perfect squares (1^2, 2^2, 3^2). So the next two figures will have 4^2 or 16 triangles, and the fifth will have 5^2 or 25 triangles.

Thus, totaling the number in each figure gives $1 + 4 + 9 + 16 + 25 = 55$.

E. If the pattern below continues, what percent of the sixth diagram will be shaded?

The sixth figure will have 7 squares on each side. (Since the first has 2 and the second has 3, we can see that the length of each figure is one more square than the position number.) So, there will be 49 total squares, 13 of which will be shaded. Thus, the percent is 13/49 or $13 \div 49 \approx .2653 \approx 27\%$.

Practice Set

1. What digit is in the 56th decimal place in the decimal value of the fraction $\frac{8}{11}$?

A. 1 **B.** 2 **C.** 5 **D.** 7

2. The units digit in each of the following numbers raised to any power is the same, except if we take powers of

A. 5 **B.** 6 **C.** 8 **D.** 10

3. How many total squares are needed to make the sixth term in the pattern below?

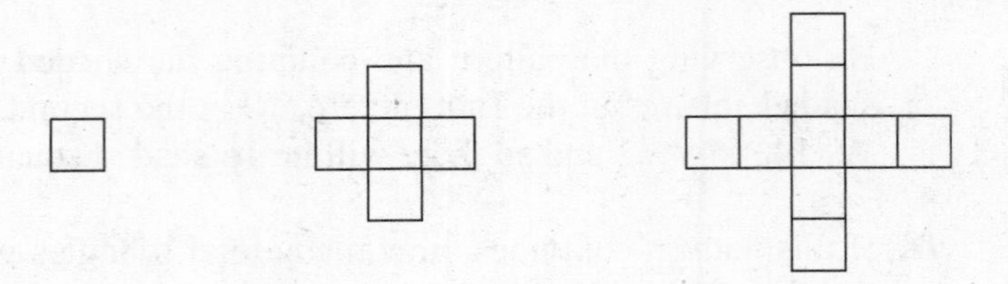

4. What percent of the tenth figure is unshaded if the pattern below continues?

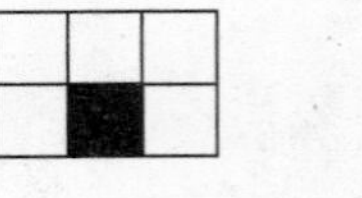

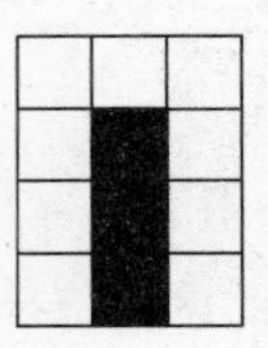

and so on

5. In each stage of the pattern below, the length of each side of an equilateral triangle becomes 120% of what it was previously. To the nearest tenth, what is the perimeter of the triangle at the fifth stage?

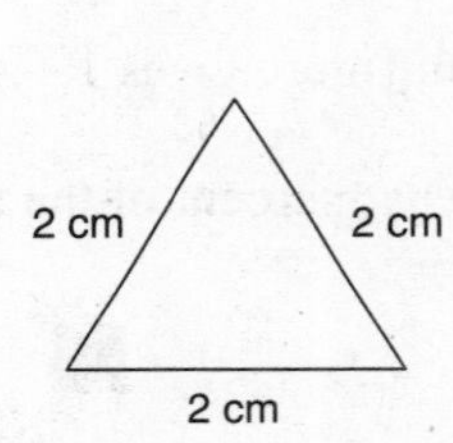

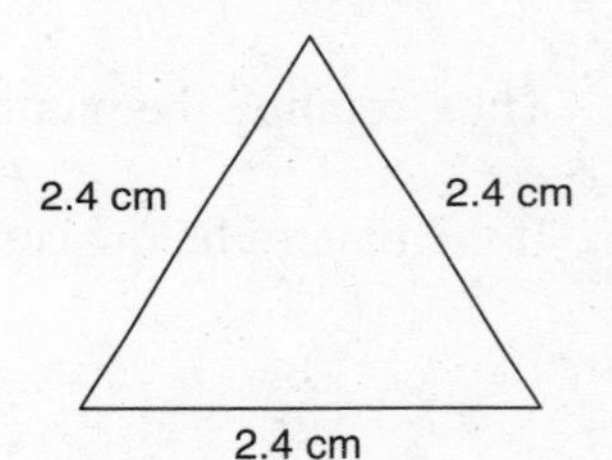

and so on

SEQUENCES

A **sequence** is a list of numbers that are in a particular order based on a pattern. There are two main types of sequences: arithmetic and geometric.

Arithmetic Sequences

In an **arithmetic sequence**, we add or subtract the same number to get from one term to the next. This is sometimes referred to as the fact that an arithmetic sequence has a **common difference**.

Examples

A. Determine if the sequence below is arithmetic. If so, identify the common difference and find the next three terms:

7, 11, 15, 19, 23, . . .

Since we add 4 to the previous term to get the next term or because 11 – 7 and 15 – 11 and 19 – 15, and so on, each equal 4, the sequence is arithmetic with a common difference of 4.

Thus, the next three terms are

$$23 + 4 = 27 \qquad 27 + 4 = 31 \qquad 31 + 4 = 35$$

B. Give the first five terms in an arithmetic sequence whose first term is 28 and whose common difference is –6.

A common difference that is negative signifies that the terms of the sequence are decreasing. Thus, we subtract 6 to get the next term.

So, the terms are 28, 22, 16, 10, 4, ...

Finding Specified Terms in an Arithmetic Sequence

If you are asked to find a certain term in an arithmetic sequence, you can use a formula to get that term. It's based on the following.

$$3, 8, 13, 18, 23, \ldots$$

To get to 8 (the second term), we add 5 to 3 (the first term). To get to 13 (the third term), we add 5 to 8 (the second term) or we add 2×5 or 10 to 3 (the first term). To get to 18 (fourth term), we add 5 to 13 (third term) or we add 3×5 or 15 to 3 (first term).

In general, to find the nth term in an arithmetic sequence, we add $(n - 1) \times d$ to the first term.

Examples

A. Find the 12th term in the arithmetic sequence 10, 13, 16, 19, ...

To find the 12th term, realize that we have added the common difference 11 times. The common difference is 3, so we have added 11×3, or 33, to 10 (the first term) and we have 43.

B. Find the 105th term in the arithmetic sequence 9, 13, 17, 21, 25, ...

To find the 105th term, realize that we have added the common difference 104 times. The common difference is 4, so we have added 104×4 or 416, to 9 (the first term) and we have 425.

C. Suppose we know that an arithmetic sequence begins with 12 and that the 25th term is 84. What is the common difference?

To answer this question, we need to realize that we are adding 72 (84 – 12) to 12 in 24 increments (since 84 is in the 25th position). Thus, $72 \div 24 = 3$. Thus, 3 is the difference.

PRACTICE SET

1. State the common difference for the arithmetic sequence below and then give the next three terms:

   $$19, 24, 29, 34, 39, \ldots$$

2. Find the first five terms of an arithmetic sequence that begins with 8 and has a common difference of 6.

3. Give the first five terms of an arithmetic sequence that begins with 135 and has a common difference of -12.

4. Find the 50th term in the arithmetic sequence below.

   $$4, 7, 10, 13, \ldots$$

5. Find the 95th term in the arithmetic sequence below.

   $$6, 11, 16, 21, \ldots$$

6. If an arithmetic sequence's 1st term is 9 and the 35th term is 213, find the common difference.

7. An arithmetic sequence's 3rd term is 14, and its 30th term is 176. Find the first term and the common difference.

Geometric Sequences

A **geometric sequence** is very similar to an arithmetic sequence with one main difference: To find a term, you multiply the previous term by a constant number called the **common ratio.**

Examples

A. The following is a geometric sequence with common ratio 3 (since we multiply every term by 3 to get the next term):

$$2, 6, 18, 54, 162, \ldots$$

Note: You can also find the common ratio by looking at the ratio of any term to the previous term, which is why it is called a common ratio. For instance, if you take $54 \div 18$ or $18 \div 6$, both results are equal to 3.

B. Find the first five terms in a geometric sequence that has an initial term of 4 and a common ratio of 5.

Simply start with 4 and multiply by 5 to get the second term, 20. Then multiply by 5 again to get the third term, 100. Multiply by 5 again, and the fourth term is 500, and then by 5 once more to get the fifth term, which is 2500.

Thus, the first five terms are 4, 20, 100, 500, 2500.

C. Give the first five terms in a geometric series that begins with 120 and has a common ratio of $\frac{1}{2}$.

Again, we simply begin with 120 and multiply by $\frac{1}{2}$ to get 60; then we multiply by $\frac{1}{2}$ again to get 30, multiply by $\frac{1}{2}$ again to get 15, and finally multiply by $\frac{1}{2}$ again to get 7.5.

So, the first five terms are 120, 60, 30, 15, 7.5.

To find a certain term in a geometric series, we use a formula based on the realization that, for example, to get to the eighth term in the geometric sequence below, we multiply by 3 seven times:

$$4, 12, 36, \ldots$$

So, to get the eighth term, we multiply 4 by 3 seven times, or 4×3^7, and this gives us 8748, which is the eighth term.

In general, to find the nth term in a geometric sequence with a common ratio of r, we multiply the first term by r raised to the $(n - 1)$*st power.*

Examples

A. Find the 11th term in the geometric sequence

$$5, 15, 45, \ldots$$

To get the 11th term, realize that we have multiplied 5 by 3 (the common ratio) 10 times. So, the 11th term is given by 5×3^{10}, which equals 295,245.

B. Find the 20th term in the geometric sequence

$$240, 120, 60, 30, \ldots$$

Here we are multiplying by $\frac{1}{2}$, so we must realize that to get to the 20th term, we must multiply 240 by $\frac{1}{2}$ 19 times. So, the 20th term is $240 \times \left(\frac{1}{2}\right)^{19}$, which equals approximately .000458.

C. Find the common ratio in the geometric sequence

243, 162, 148, . . .

To find the common ratio, divide any term by the term preceding it. So, taking 162/243, we get 2/3.

PRACTICE SET

1. Find the common ratio for the following geometric sequence.

3, 6, 12, 24, 48, . . .

2. What is the common ratio for the following geometric sequence?

176, 132, 99, . . .

3. Find the 12th term in the geometric sequence

2, 8, 32, 128, . . .

4. If the 5th term in the geometric sequence below is 75, find the common ratio.

1200, . . .

5. A machine's value depreciates 30% per year. If the machine originally cost $15,000, approximately how much is it worth after 5 years?

A Special Sequence

There is a special type of sequence called a **Fibonacci-like sequence.** In this type of sequence, rather than the term being based on performing some operation on the preceding term, the two previous terms are added together.

Examples

A. The following is an example of a Fibonacci-like sequence.

3, 5, 8, 13, 21, 34, . . .

Notice that to find the third term, 8, we add 3 + 5. To get the fourth term (13), we add the two previous terms: 5 + 8.

B. Find the first five terms of a Fibonacci-like sequence that begins with the terms 4 and 6.

Of course the first two terms are 4 and 6. To find the third term, we add 4 + 6, and so the third term is 10. To find the fourth term, we add 6 + 10 to get 16.

So, the first five terms of the Fibonacci-like sequence described are 4, 6, 10, 16, 26.

PRACTICE SET

1. Find the first five terms of the Fibonacci-like sequence that begins with 6 and 8.

2. Find the first five terms of the Fibonacci-like sequence whose first two terms are –2 and 5.

3. Which of the sequences below is a Fibonacci-like sequence?

   A. 2, 6, 10, 14, . . .　　**B.** 2, 6, 18, 54, . . .
   C. 2, 4, 8, 16, . . .　　**D.** 2, 4, 6, 10, 16, . . .

4. A special variation of a Fibonacci-like sequence is formed by taking twice the sum of the previous two terms to get a term. If a sequence of this sort begins with 3, 4, 14, 36, . . . , find the sixth term.

5. Write your own Fibonacci-like sequence and give the first five terms.

SERIES

A **series** is the sum of the numbers (terms) of a squence. It differs from a sequence in that the terms are combined by addition. All of its other properties are the same.

Examples

A. Below is an example of an arithmetic series.

$$3 + 7 + 11 + 15 + 19 + 23 + 27 + 31 + 35$$

Note: The common difference is 4 since we are increasing each term by 4 over its preceding value. However, instead of simply listing the terms separated by commas, we add the terms in a series.

B. Below is an example of a geometric series.

$$20 + 10 + 5 + \frac{5}{2} + \frac{5}{4} + \frac{5}{8}$$

Note: The common ratio is $\frac{1}{2}$ since we are multiplying each term by $\frac{1}{2}$ to find the next term. Again, instead of listing the terms and separating them by commas we add them to get a series.

C. This is an example of a Fibonacci-like series:

$$2 + 3 + 5 + 8 + 13 + 21 + 34$$

Note: We still add the previous two terms to get a given term. However, since it is a series, we join the terms with addition rather than simply listing them.

Sums of Series

For a Fibonacci-like series, the only way to find the sum is to find all the terms and add them. However, with arithmetic and geometric series, you can follow a formula to obtain the sum of the first n terms.

For an arithmetic series, to find the sum of the first n terms follow the formula below, where a_1 refers to the first term and a_n refers to the nth term.

$$S_n = n\left(\frac{a_1 + a_n}{2}\right)$$

For a geometric series, to find the sum of the first n terms follow the formula below, where again a_1 refers to the first term and r is the common ratio.

$$S_n = \frac{a_1(r^n - 1)}{r - 1}$$

Infinite Geometric Series

It is possible to have a geometric series that does not end; such a series is known as an **infinite geometric series.**

If $|r| > 1$, an infinite geometric series *diverges*, which means there is no limiting value for the sum. We are referring to the idea that as you add terms and continue generating the sum of a series, for certain series the sum approaches a number and never goes beyond that number. This happens when the value of r is such that $|r| < 1$. In this case, the sum is being increased by smaller and smaller values, and thus it eventually limits out.

For an infinite geometric series with $|r| < 1$, the sum of the series is

$$S = \frac{a_1}{1 - r}$$

Examples

A. Find the sum of the first eight terms of the following arithmetic series.

$$7 + 12 + 17 + \cdots$$

Following the formula

$$S_8 = 8\left(\frac{7 + a_8}{2}\right)$$

Notice that we now have to find a_8 by a previous formula, which is

$$a_8 = 7 + 5(7) = 42$$

Now,

$$S_8 = 8\left(\frac{7+42}{2}\right) = 8\left(\frac{49}{2}\right) = 4(49) = 196$$

B. Find the sum of the first ten terms of the following geometric series.

$$2 + 6 + 18 + \cdots$$

Following the formula,

$$S_{10} = \frac{2(3^{10} - 1)}{3 - 1} = 59{,}048$$

C. Find the sum of the first ten terms of the Fibonacci-like series below.

$$4 + 10 + 14 + \cdots$$

Unfortunately, for a Fibonacci-like series, there is no "magic formula" to follow to get the sum of the first n terms. So, instead, you have to list the terms and then add them using either a calculator or pencil and paper.

$$4 + 10 + 14 + 24 + 38 + 62 + 100 + 162 + 262 + 424 = 1100$$

D. Find the sum of the infinite geometric series if it exists:

$$100 + 50 + 25 + \frac{25}{2} + \frac{25}{4} + \cdots$$

Since $r = \frac{1}{2}$ for this series, its absolute value is less than 1, and so the series converges. Following the formula,

$$S = \frac{100}{1 - \frac{1}{2}} = 200$$

And if you actually use a calculator to add 100 + 50 + 25 + . . . and continue for at least 15 or 20 terms, you will see that the value gets closer and closer to 200.

E. Find the sum of the infinite geometric series if it exists:

$$6 + 24 + 96 + 384 + \cdots$$

Here $r = 4$, so this series has no sum and is said to diverge.

In fact, if you simply think about what is happening to the terms, they are increasing quite quickly since we are multiplying by 4. Therefore, the sum is increasing quickly as well and is not limited to a certain value.

Practice Set

1. Find the sum of this infinite geometric series if it exists:

$$45+30+20+\frac{40}{3}+\cdots$$

2. Find the sum of the first 12 terms of the following arithmetic series.

$$-5+3+11+19+\cdots$$

3. Find the sum of the first eight terms of the following geometric series.

$$3+12+48+\cdots$$

4. Find the sum of the first ten terms of the following Fibonacci-like series.

$$6+8+14+\cdots$$

VISUAL REPRESENTATIONS OF RELATIONSHIPS AND PATTERNS

When we place values from a table or pattern on a coordinate plane, we often notice a visual pattern (often a line).

Examples

A. Make a table of values for the perimeter of an equilateral triangle if each side is 1 in., 2 in., 3 in., 4 in., 5 in. and then plot the ordered pairs. Finally, give an equation.

Side (in.)	Perimeter (in.)
1	3
2	6
3	9
4	12
5	15

First, notice how the perimeters follow an arithmetic sequence with a common difference of 3 in.

When we plot the ordered pairs, we plot (side, perimeter); that is, (1, 3), (2, 6), and so on.

The resulting graph looks like the following.

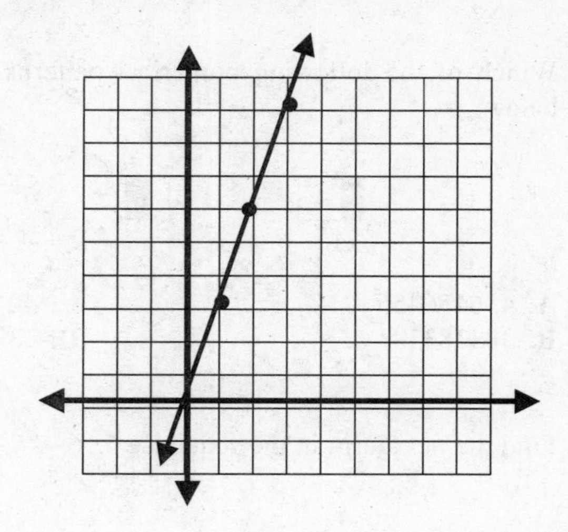

The equation for the perimeter of an equilateral triangle with side x is given by $3x$ (we multiply the length of one side by 3 since all three sides are equal). We could more appropriately give the equation as $P = 3s$, where P = perimeter of the equilateral triangle and s = length of one side.

B. If the rate a plumber charges is $40 plus $55 an hour, write an equation to show how much the plumber charges for a job that lasts h hours.

$C = 40 + 55h$ since the plumber charges $40 outright and then adds $55 per hour, h, that she works.

Practice Set

1. Based on the pattern shown in the table of values below, what is the missing value for y?

x	2	4	6	8
y	5	11	?	23

2. A telephone call lasting m minutes costs *what* if the first minute costs .48 and every minute after the first minute costs .17? Make a table of values for a telephone call lasting 1 minute, 2 minutes, 3 minutes, . . . , 6 minutes. Then, write an equation for the cost of a phone call lasting m minutes.

MIXED PRACTICE, CLUSTER IV, MACRO A

1. Which of the following numerical patterns is similar to the visual pattern shown below?

A. 456456456 . . .
B. 383383383 . . .
C. 4242424242 . . .
D. 3333333333. . .

2. Find the next term in the sequence 5, $6\frac{1}{4}$, $7\frac{1}{2}$, $8\frac{3}{4}$,....

A. $9\frac{1}{4}$ **B.** $9\frac{1}{2}$ **C.** $9\frac{3}{4}$ **D.** 10

3. If the pattern below is continued, what letter will be in the 154th position?

HSPAHSPAHSPAHSPAHSPA . . .

A. H **B.** S **C.** P **D.** A

4. Find the sum of the infinite geometric series below.

$$\frac{7}{10}+\frac{7}{100}+\frac{7}{1000}+\cdots$$

5. If the pattern below is continued, how many shaded squares will there be in the 20th diagram?

A. 380 **B.** 360 **C.** 60 **D.** 20

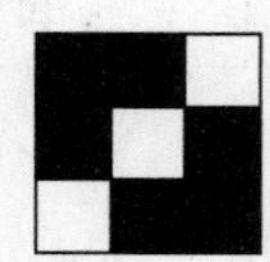

6. Which of the following is an example of a geometric sequence?

A. 3, 5, 7, 9, 11, . . .
B. 3, 6, 12, 24, . . .
C. 3, 6, 12, 48, 384, . . .
D. 3, 5, 8, 13, 21, . . .

7. Find the 25th term in the arithmetic sequence below.

5, 11, 17, 23, 29, . . .

A. 155 **B.** 149 **C.** 125 **D.** 131

8. Which of the following cannot be a term of the sequence below?

$$\frac{3}{4}, 1\frac{1}{2}, 3, 6, 12, \ldots$$

A. 24 **B.** 96 **C.** 200 **D.** 384

9. In an arithmetic sequence, the 10th term is 48 and the 12th term is 62. Find the common difference and the 9th term.

10. If 5, 10, 20, 40, . . . are the first four terms of a geometric sequence, what is the 20th term?

11. What is the pattern in the units digit of $3^1, 3^2, 3^3, \ldots$?

A. 3, 1, 9, 7 **B.** 3, 9, 7, 1 **C.** 3, 1, 7, 9 **D.** 3, 9, 1, 7

12. If the following pattern is continued, how many hearts will be needed to represent the first six terms?

    and so on

13. If the first term of an arithmetic series is 8 and the common difference is 4, how many terms are in the series for the sum to be 140?

A. 6 **B.** 7 **C.** 8 **D.** 12

MACRO B

RELATIONS AND FUNCTIONS

Any equation can be presented in terms of a table and ordered pairs. This list of ordered pairs is called a **relation**. In a relation, the variable that depends on the other variable's value is called the **dependent variable**, and the other variable is called the **independent variable**. For example, if we look at the area of a circle, the formula is $A = \pi r^2$ and we can create a table of values for various radii. Since the area of a circle depends on its radius, area is the dependent variable and radius is the independent variable.

If a set of ordered pairs (a relation) is such that every x-value for the independent variable pairs up with exactly one y-value for the dependent variable, the relation is a **function**. For a function, we typically refer to all the permissible x-values as the **domain** and all the output or resulting values for y as the **range**.

Examples

A. Given the following table of values, decide if the relation is a function.

x	4	7	−2	4	6	−4
y	6	−3	2	−1	−5	9

Remember, for a relation to be a function, every x must be paired up with only one y. So, since 4 is paired up with 6 and with −3, this relation is *not* a function.

B. Given the table of values below, determine if the relation is a function.

x	−5	−2	0	3	4	7
y	−7	−1	3	9	11	17

Just as in the previous example, we recall that a relation is a function if each x-value corresponds to just one y-value. So, if we look at our x-values here, we see that all of them are unique, and so there are no x-values paired up with more than one y-value. Thus, this is an example of a relation that is a function.

Visual Test

It is possible to tell by looking at a graph of a relation whether the relation is a function. If when you plot the ordered pairs, *no* vertical line passes through more than one point in the relation, then it is a function. This is called the **vertical line test**.

Examples

A. The following is a graph of a relation that is a function.

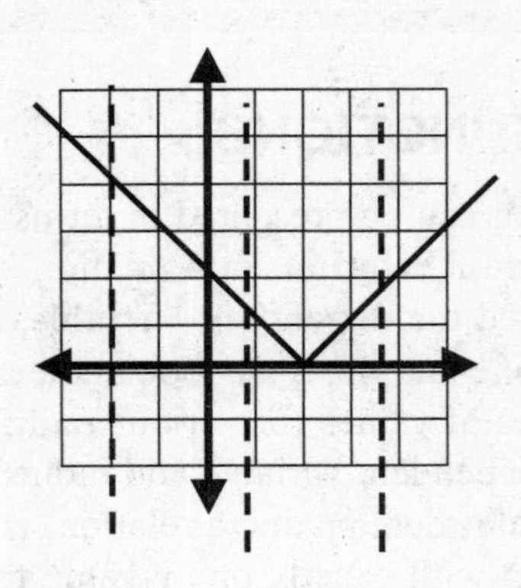

This is a function since every vertical line (such as the few shown) passes through the graph of the relation.

B. The graph of a relation that is *not* a function is shown below.

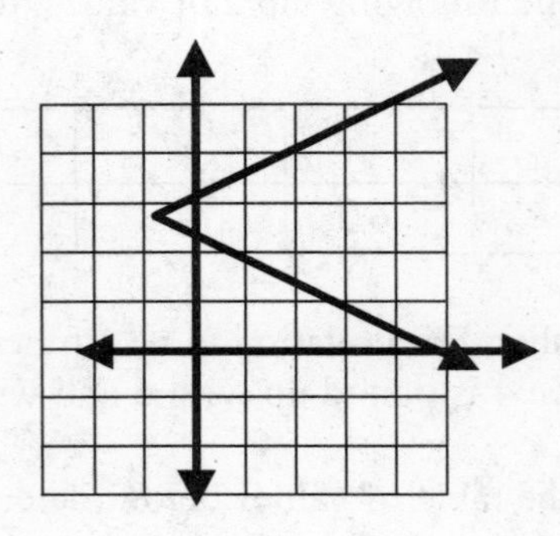

PRACTICE SET

1. Do the given ordered pairs represent a function? Why or why not?

(−2, 5) (−1, 4) (0, 3) (2, 4) (3, 7)

2. Do the given ordered pairs represent a function? Why or why not?

(3, 4) (−1, 7) (2, 6) (−1, −2) (−4, 6)

SPECIAL FUNCTIONS

There are certain special functions given by equations that you should be familiar with before taking the HSPA exam. They are described on the following pages.

Constant Function

Format: $y = c$, where c is a constant real number

Example: $y = -2$

Graph: A horizontal line going through the point $(0, c)$; in this case, going through $(0, -2)$

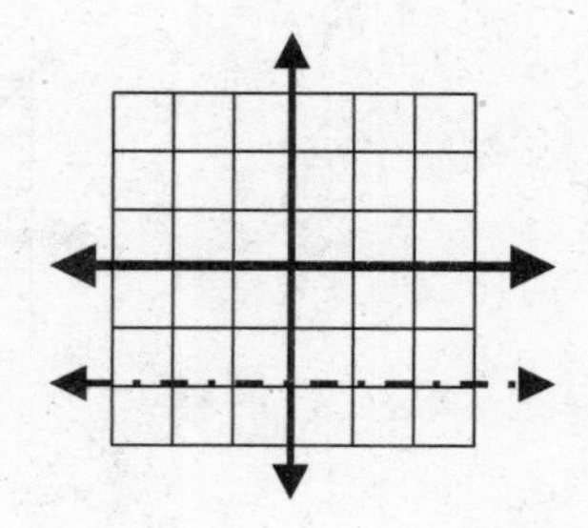

Linear Function

Format: $y = mx + b$, where m is the slope and b is the y-intercept (which you will learn more about shortly)

Example: $y = 3x + 2$

Graph: A slanted line going through $(0, 2)$ with slope 3

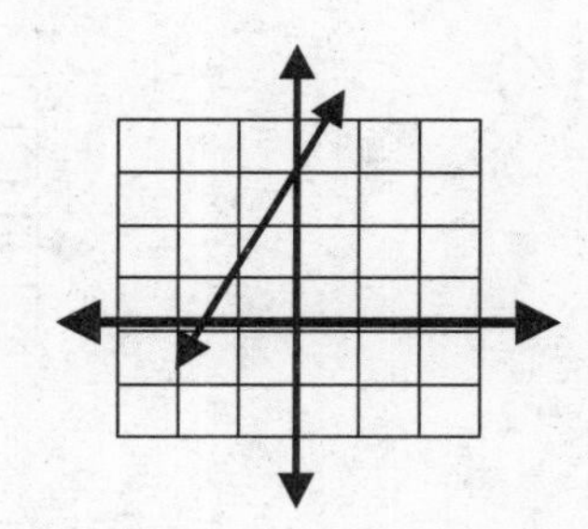

Quadratic Function

Format: $y = ax^2$, where a is a constant number

Example: $y = x^2$

Graph: A U-shaped graph with a vertex at the origin

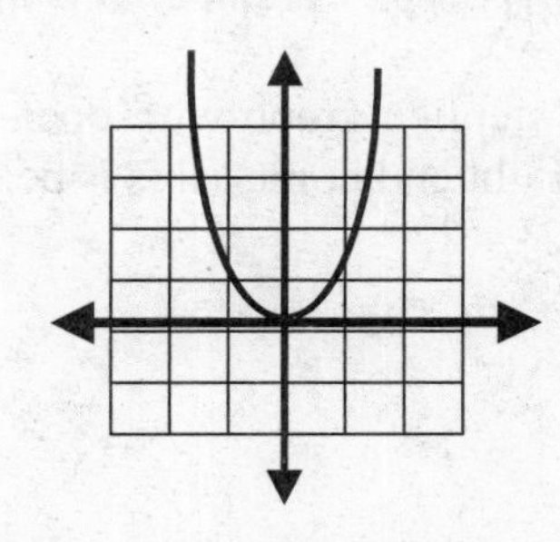

Absolute Value Function

Format: $y = |x|$

Example: $y = |x|$

Graph: A V-shaped graph with a vertex at the origin

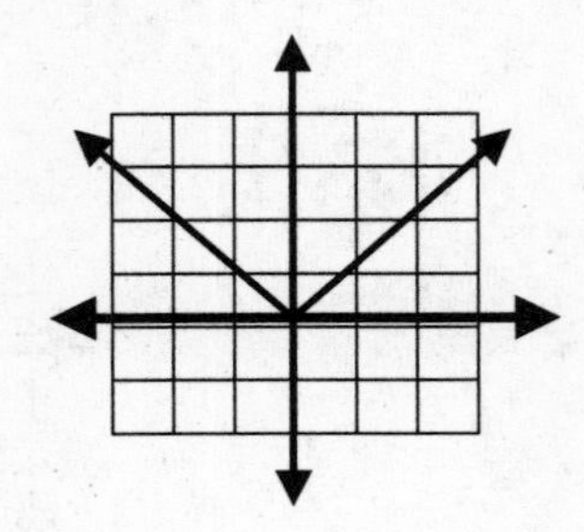

Exponential Function

Format: $y = b^x$, where b is a constant integral base greater than 1

Example: $y = 3^x$

Graph: A graph that at first increases slowly and then increases more quickly; the graph of a basic exponential always passes through (0, 1) since any base to the zero power is 1.

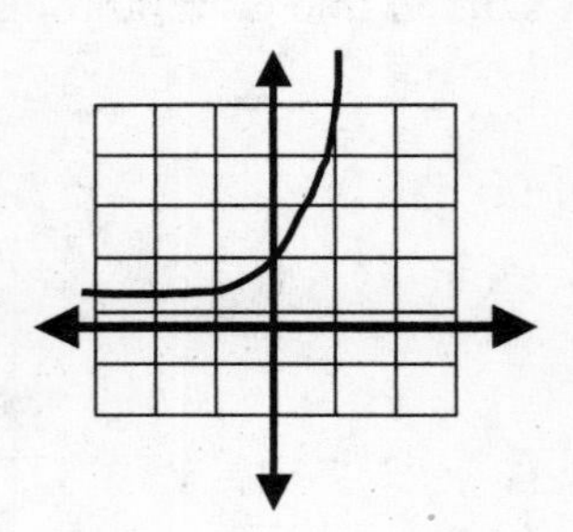

Greatest Integer Function

Format: $y = [x]$

This function equals the greatest integer less than or equal to the inputted value.

Example: $[3.1] = 3$ since 3 is the largest integer less than or equal to 3.1.
$[2.999] = 2$ since 2 is the largest integer less than or equal to 2.999.
$[-2.4] = -3$ since -3 is the largest integer less than or equal to -2.4.

Graph: A steplike graph with open end points on the left end points and closed end points on the right end points (this is because, for example, $[1] = 1$ while $[1.9999999]$ also equals 1 yet $[2] = 2$.

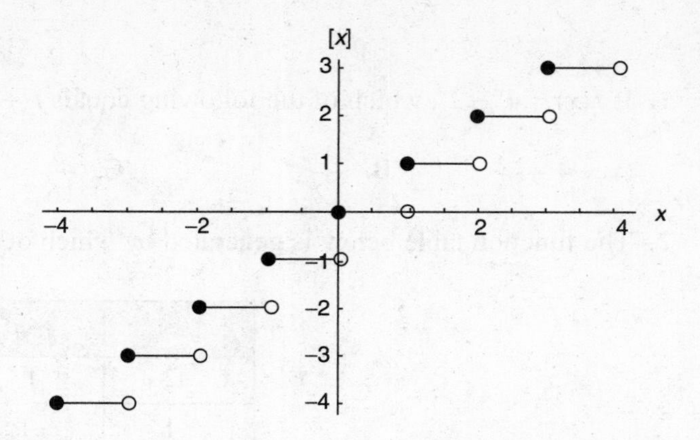

The greatest integer or step function is most commonly seen in real-life situations where there is a cost associated with a time or weight interval and the cost jumps to the next rate once a limit is exceeded. For example, when mailing a package, the postage is typically based on the weight; there is a certain cost for packages weighing certain amounts.

Most of these functions (with the exception of the greatest integer function, which is very specialized) can be graphed by simply choosing several values for x and plugging them in to evaluate the expression stated for y in the equation.

A function can also be labeled using *function notation*: $y = 3x - 1$ can be written as $f(x) = 3x - 1$, which emphasizes the idea that the function is dependent on the value of x. We then say that $f(2)$ is the value of the function when $x = 2$.

Examples

A. Fill in the missing values in the table below for the function $y = 2x + 5$.

x	y
2	
–1	
–3	

To get the values, you simply plug in each value for x and see what $2x + 5$ equals.

For example, using 2, $2(2) + 5 = 4 + 5 = 9$. Using –1, $2(-1) + 5 = -2 + 5 = 3$. And using –3, $2(-3) + 5 = -6 + 5 = -1$.

B. Given $f(x) = x^2 - 4$, find $f(3)$ and $f(1)$.

Even though the notation is different, you are still asked to do something simple. You just need to plug the 3 and the 1 into the equation for x. So, $f(3) = (3)^2 - 4 = 9 - 4 = 5$ and $f(1) = (1)^2 - 4 = 1 - 4 = -3$.

PRACTICE SET

1. If $f(x) = 4^x - 2^x$, which of the following equals $f(-2)$?

   A. -4 **B.** $\frac{1}{12}$ **C.** $\frac{3}{16}$ **D.** $-\frac{3}{16}$

2. The function table below is generated by which of the following functions?

x	$f(x)$
−2	1
−1	−1
0	−3
2	−7

   A. $f(x) = x + 3$ **B.** $f(x) = -2x - 3$ **C.** $f(x) = -2x + 3$ **D.** $f(x) = x$

3. Which of the following is an example of a situation that could be modeled by a constant function?

   A. an athlete's pulse rate as he or she is practicing
   B. the number of people standing in a checkout line at the grocery store
   C. a person's salary over the course of their career
   D. the ratio of a circle's circumference to its diameter

4. Which of the following is *not* an example suitable for modeling with a greatest integer function?

   A. the cost of mailing a package
   B. the temperature over the course of a day
   C. the cost of parking a car in a parking garage for a certain number of hours
   D. taxicab fare for certain distances traveled

5. Two postal employees were looking at the list of zip codes shown below. Mark argued that the zip codes were not a function because of Deptford and Woodbury, while JoAnn argued that the zip codes were not a function because of Toms River. Who is correct and why?

Town	Zip Code
Deptford	08096
Woodbury	08096
Long Branch	07740
Delran	08075
Toms River	08753
Toms River	08755
Toms River	08757

SLOPE

We mentioned earlier that you would be learning more about slope soon. **Slope** is the constant ratio that determines a line's steepness and direction.

Notice in the table of values below that there is a steady change in y that accompanies a steady change in x. That is, as x goes up by 2's, y goes up by 3's.

x	y
–2	–7
0	–4
2	–1
4	2
6	5

Slope is calculated by the following ratios.

$$\text{Slope} = \frac{\text{rise}}{\text{run}} = \frac{\text{change in } y}{\text{change in } x} = \frac{y_2 - y_1}{x_2 - x_1}$$

For example, in the function above, since two of the coordinates are (6, 5) and (4, 2), to get the slope we subtract the y-values in the numerator and the x-values in the denominator:

$$\text{Slope} = \frac{5-2}{6-4} = \frac{3}{2}$$

Visually, we can determine the slope by plotting the ordered pairs on a graph and counting the *rise* and the *run*. A portion of the graph is shown below.

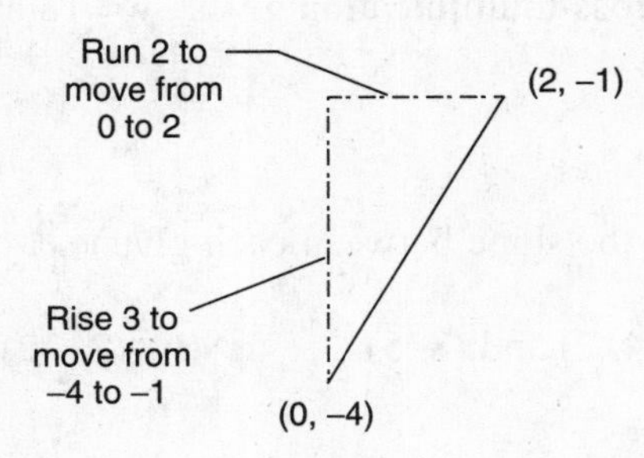

Types of Slope

Just by looking at a line, you will be expected to determine whether its slope is positive, negative, zero, or undefined. Following is a guide to help you determine what type of slope a line has.

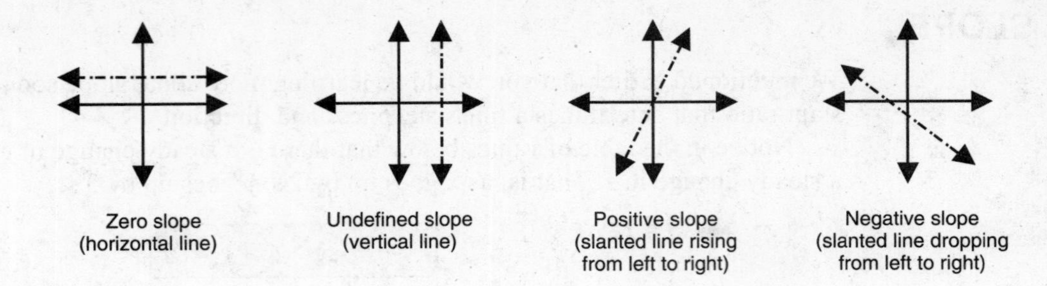

Examples

A. Find the slope between (–6, –2) and (4, –7).

To find the slope, subtract the first y-value from the second y-value and use this result as the numerator. Then subtract the second x-value from the first x-value and use this result as the denominator:

$$\text{Slope} = \frac{-7-(-2)}{4-(-6)} = \frac{-5}{10} = -\frac{1}{2}$$

B. If the line through (–6, 0) and (12, y) has slope $\frac{1}{3}$, find the value of y.

Begin with the setup for calculating slope:

$$\text{Slope} = \frac{y-0}{12-(-6)} = \frac{y}{18} = \frac{1}{3}$$

So cross-multiplication gives $3y = 18$ and $y = 6$.

PRACTICE SET

1. Find the slope between each given set of points:

(a) (4, 2) and (8, 5) **(b)** (2, –2) and (4, –7) **(c)** (–1, –9) and (2, 3)

2. Find the value of x so that the slope between the points (5, 7) and (x, –2) is $\frac{3}{5}$.

SLOPE-INTERCEPT FORMAT FOR AN EQUATION

Earlier we alluded to the format of a line being $y = mx + b$, where m is the slope and b is the y-intercept. This is called the **slope-intercept form** of an equation and is considered fairly standard because it is easy to use in obtaining a graph and/or using a graphing calculator.

In order to graph an equation in this format, you follow three steps:

1. Identify and plot the y-intercept.
2. Move from the y-intercept to a second point using the slope to count.
3. Connect the two points with a straight line.

Examples

A. Graph: $y = \frac{2}{3}x - 1$.

The y-intercept is −1, so we plot (0, 1).

The slope is $\frac{2}{3}$, so we rise 2 and run 3 from (0, 1).

Then we connect the points, resulting in the following graph.

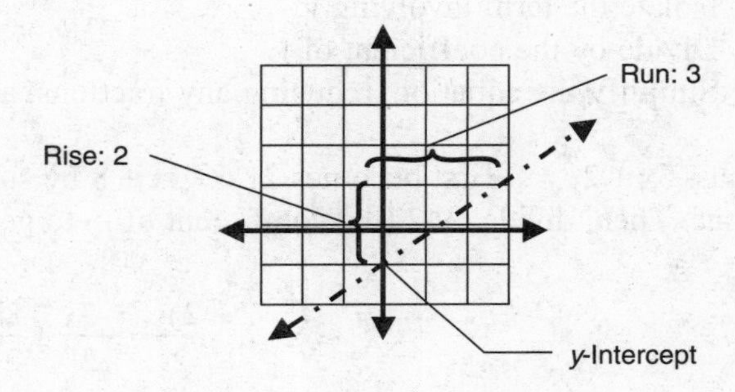

B. Graph: $y = -\frac{5}{2}x + 3$.

The y-intercept is +3, so we plot (0, 3).

The slope is $-\frac{5}{2}$, so we drop 5 and run 2 from (0, 3).

Then we connect the points and obtain the following graph.

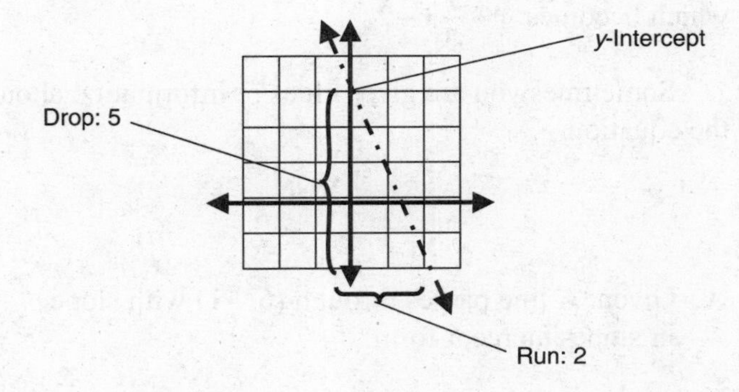

Note: Be careful in cases where the slope is an integer, such as $y = 2x - 5$. Here, you must realize that the slope should be thought of as $\frac{2\,2}{1\,1}$.

PRACTICE SET

1. What is the slope of the line $y = 3x + 4$?
2. What is the y-intercept of the equation $y = \frac{2}{3}x - 7$?
3. Graph: $y = 4x + 2$
4. Graph: $y = -\frac{3}{2}x + 5$

Rewriting and Writing Equations

On occasion you will be given a linear equation not already in slope-intercept form. For example, you may have an equation like $3x + 2y = 8$. To change this equation to the $y = mx + b$ form follow these steps:

1. Isolate the term involving y.
2. Divide by the coefficient of y.
3. Simplify the equation, reducing any fractions, and so on.

Thus, $3x + 2y = 8$ first becomes $2y = -3x + 8$ by subtracting $3x$ from both sides to get $2y$ alone. Then, divide by 2 (the coefficient of y) to get

$$\frac{2y}{2} = \frac{-3x}{2} + \frac{8}{2}$$

which becomes $y = -\frac{3}{2}x + 4$.

As another example, consider $6y - 4x = 12$. The first step is to add $4x$ to both sides to isolate the term involving y. This yields $6y = 4x + 12$. Then, divide by 6. So, we get

$$\frac{6y}{6} = \frac{4x}{6} + \frac{12}{6}$$

which becomes $y = \frac{2}{3}x + 2$.

Sometimes you are given clues or information about a line and asked to come up with the equation.

Examples

A. Given: A line passes through $(6, -1)$ with slope $\frac{2}{3}$. Write the equation for this line in slope-intercept form.

Set up the slope calculation, cross-multiply, and solve for y as shown above.

So, here

$$\text{Slope} = \frac{y-(-1)}{x-6} = \frac{2}{3}$$

Cross-multiplication then gives $3(y + 1) = 2(x - 6)$, and if we distribute, we get

$3y + 3 = 2x - 6$. Solving for y gives $3y = 2x - 9$ and finally $y = \frac{2}{3}x - 3$.

B. Given: A line passes through two points (2, –5) and (–3, 10). Write the equation for this line in slope-intercept form.

Find the slope and then use the slope and *either* point and proceed as above.

So, the slope is

$$\frac{10-(-5)}{-3-2} = \frac{15}{-5} = -3$$

Remember: This integral slope must be thought of as a fraction! Then, choose a point to use from the given two points. Suppose we use (2, –5). Thus,

$$\frac{y-(-5)}{x-2} = \frac{-3}{1}$$

and cross-multiplying gives $y + 5 = -3(x - 2)$, which results in $y + 5 = -3x + 6$ and finally $y = -3x + 1$.

PRACTICE SET

1. Write the equation $5x - 2y = 12$ in slope-intercept form.

2. Write an equation in slope-intercept form for the line with slope $\frac{2}{5}$ and passing through the point (10, –3).

3. Write the equation of the line passing through (4, 2) and (–8, 11). Use the slope-intercept form for the equation.

4. How many of the following lines have a slope of 3?

$y = 4x + 3$ $\quad y = 3x$ $\quad 3y = x$ $\quad 5x - 15y = 30$ $\quad y + 3x = 9$

A. 1 **B.** 2 **C.** 3 **D.** 4

5. If an electrician charges $42 simply to come to your home to perform a job and then charges $55 an hour, write an equation representing his total charge for a job taking *h* hours. Is this a linear function? Why or why not? Be sure to explain.

TRANSFORMATIONS PERFORMED ON FUNCTIONS

Similar to the various transformations that can be applied to a geometric figure as covered in Cluster II, we can perform various transformations or changes on functions.

There are three main transformations you should be familiar with:

1. Vertical shift
2. Horizontal shift
3. Reflection.

Given $f(x)$, a vertical shift up or down is achieved by $f(x) +/- c$, where c is some constant.

Given $f(x)$, a horizontal shift right or left is achieved by $f(x +/- c)$, where c is some constant.

Given $f(x)$, a reflection through the y-axis is achieved by $f(-x)$, and a reflection through the x-axis is accomplished by $-f(x)$.

Examples

A. Sketch $f(x+2)$ based on the graph of $f(x)$.

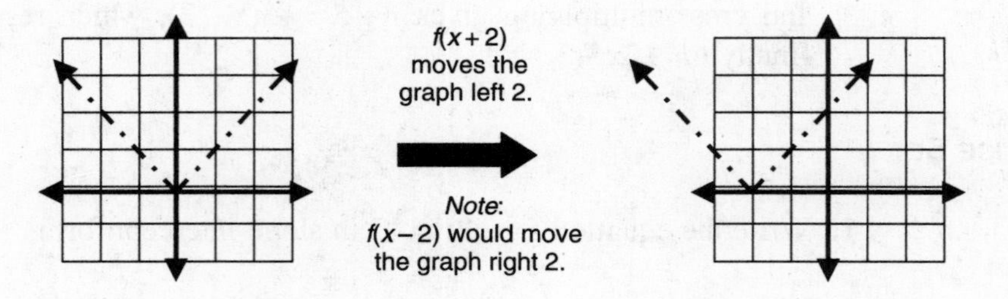

f(x + 2)
moves the
graph left 2.
Note:
f(x – 2) would move
the graph right 2.

B. Sketch $f(x) + 2$ given $f(x)$ as shown.

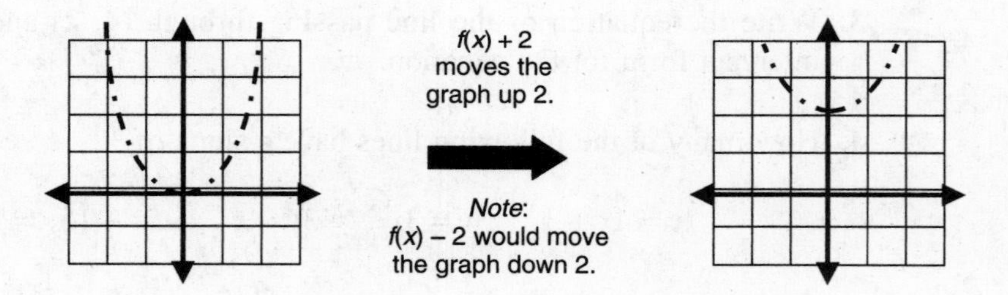

f(x) + 2
moves the
graph up 2.
Note:
f(x) – 2 would move
the graph down 2.

C. Given $f(x)$ sketch a graph of $f(-x)$.

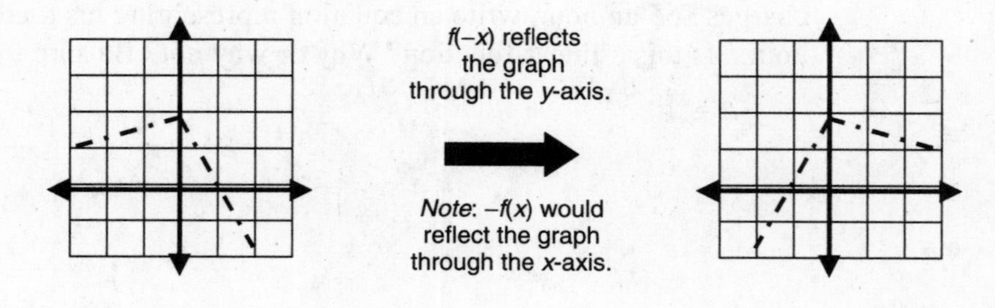

f(–x) reflects
the graph
through the y-axis.
Note: –f(x) would
reflect the graph
through the x-axis.

MIXED PRACTICE, CLUSTER IV, MACRO B

1. Which of the diagrams below suggest the definition of a function?

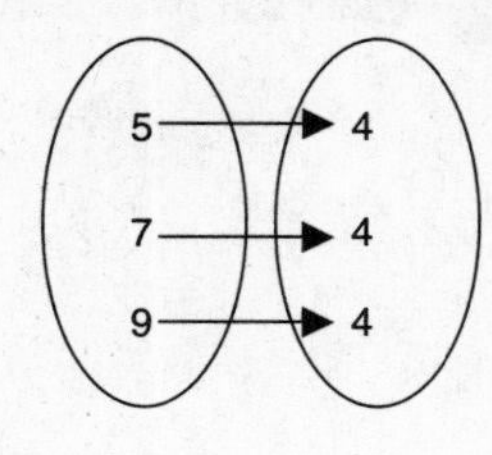

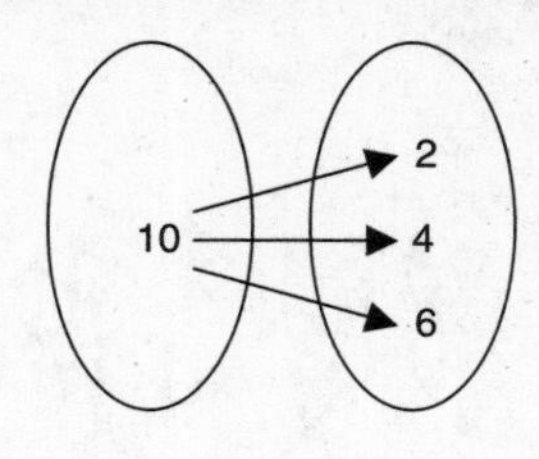

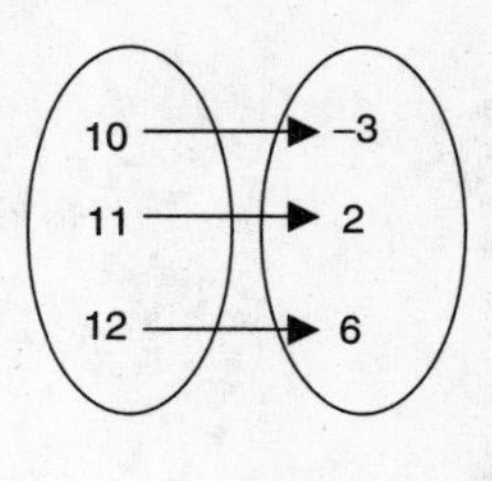

A. III only **B.** II only **C.** I and III **D.** II and III

2. Which of the functions below does *not* have the entire set of real numbers as its domain?

A. $f(x) = 2x - 9$ **B.** $f(x) = \frac{3}{x}$ **C.** $f(x) = 4^{-x}$ **D.** $f(x) = 3x^2$

3. Which of the graphs of functions shown below shows a function in which, for at least one pair of x values, $f(a) = f(b)$?

A.

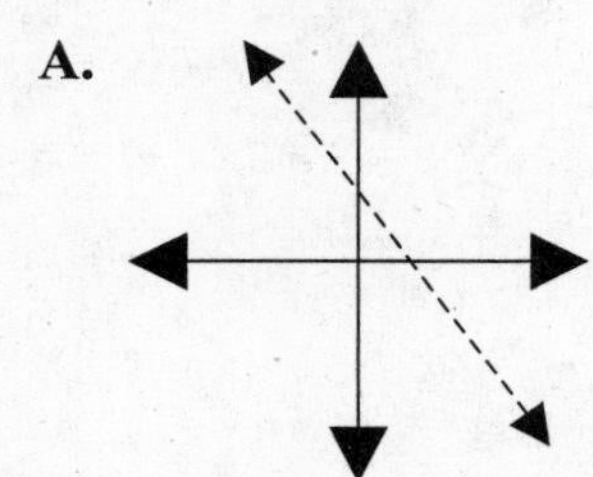

C.

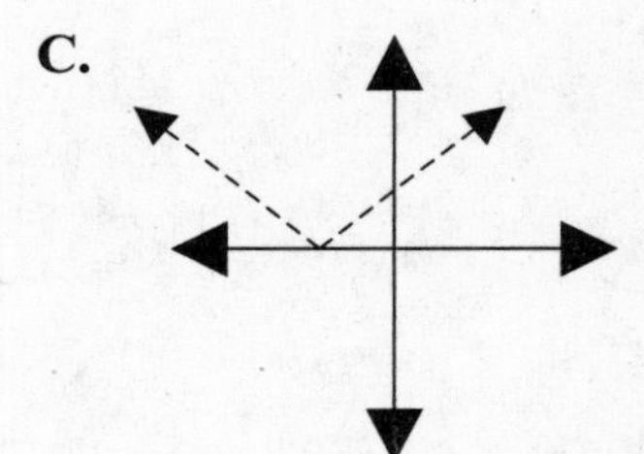

B.

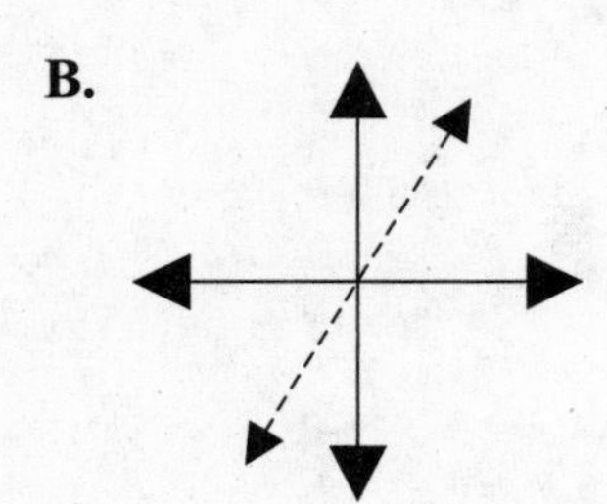

D.

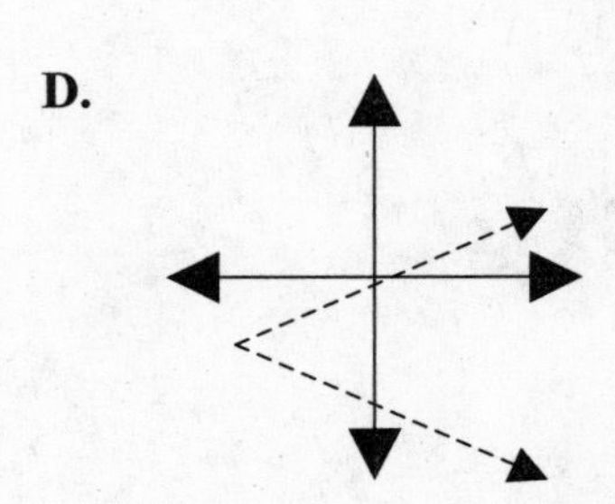

4. Write an equation in slope-intercept form for the line containing (6, –2) and parallel to the line through (–6, 0) and (8, 7).

5. If the point (2, 5) is on the graph of $f(x)$, where would this point be on the graph of $-f(x)$?

MACRO C

VARIABLE EXPRESSIONS AND OPEN SENTENCES

In algebra, a lowercase letter standing for a number is called a **variable**. It is called a variable because its value can vary or change. A number, a variable, or the product or quotient of a variable and a number is called a **term**. One or more terms make up a **variable expression**.

Examples

5, 8, π, 8.17 are constant terms.

$3x$, $2m^2$, $6x^3y + 99$ are variable expressions.

Note: In variable expressions, the number in front of the variable is called the **coefficient**. For $3x$, 3 is the coefficient of x.

Variable expressions can be evaluated once we know the value of the variable(s). In a variable expression, we can combine like terms. *Like terms* are two or more terms that contain exactly the same combination of variables and exponents.

Examples

$5m$ and $8m$ are like terms.

$10x$ and $4y$ are unlike terms.

$2x^4y$ and $-11x^4y$ are like terms.

$3x^4y$ and $7xy^4$ are unlike terms.

A. Simplify: $4x^3 + 6x^3 = 10x^3$

B. Simplify: $10m - 3n + m + 8n = 11m + 5n$ (*Note*: m means $1m$, and since it's $-3n$ we treat that as $-3 + 8$, hence $5n$.)

Sentences or descriptions given can be translated into algebraic language in the form of a variable expression.

Examples

Write each of the following sentences in algebra.

A. The sum of a number and 8.

$$x + 8$$

B. Double a number increased by 11.

$$2x + 11$$

C. Seven less than triple a number is 17.

$$3x - 7 = 17$$

D. Five times the difference between a number and 9.

$$5(x - 9)$$

E. The product of two consecutive even integers.

$$x(x + 2)$$

Note: Consecutive integers are 3, 4, 5, 6, ... In algebra, consecutive expressions are x, $x + 1$, $x + 2$, and so on.

PRACTICE SET

1. What expression should be added to $3x - 5$ to get a sum of $x + 1$?

2. Which of the following expressions means "if 4 is subtracted from 9 times some number, the result is 23"?

   A. $9(x - 4) = 23$ **B.** $9x - 4 = 23$ **C.** $4 - 9n = 23$ **D.** $9(4 - x) = 23$

3. If $x + 3$ represents an odd number, the next odd number is

   A. $x + 1$ **B.** $x + 5$ **C.** $2x + 6$ **D.** $5x + 3$

4. Write an open sentence for the following statement: Seven less than triple a number is nine more than the number.

5. If the length of a rectangle is $3x + 4$ and its width is $2x$, write a simplified expression that represents the rectangle's perimeter.

6. Evaluate $6 + x(x + 2)$ if $x = 4$.

   A. 60 **B.** 30 **C.** 16 **D.** 38

7. Write an equation that could be modeled by the following:

   □ □ □ = □ □ □ □ □

   \+ + + + + = − − − − − − −

EQUATIONS AND INEQUALITIES

To solve an equation in algebra means to find the value of the variable that makes the equation true. Solving an equation involves isolating the variable through the use of inverse operations. The basic steps for solving an equation are as follows.

1. Distribute if necessary (to clear any parentheses in the equation) on either or both sides.
2. Combine like terms on each side if possible.
3. Add or subtract the smaller variable from both sides if the variable is on both sides.

4. Add or subtract the number that is subtracted or added to the variable in the equation.
5. Multiply or divide by the coefficient of the variable.

Examples

A. Solve: $3x - 2 = 13$

To solve this equation, add 2 to both sides: $3x = 15$. Then divide by 3, so $x = 5$.

B. Solve: $8x - 11 = 2x + 13$

To solve this equation, subtract $2x$ from both sides to get $6x - 11 = 13$. Then add 11 to both sides to get $6x = 24$ and then divide by 6 to get $x = 4$.

C. Solve: $5(2x - 3) + x = 18$

To solve this equation, begin by distributing to get $10x - 15 + x = 18$. Then, combine like terms to yield $11x - 15 = 18$. Add 15 to both sides of the equation: $11x = 33$, so division by 11 gives $x = 3$.

When solving an inequality, follow the same steps *except* that if, in the process of solving, you divide or multiply by a negative number, you reverse the direction of the inequality; that is, if you divide or multiply by a negative number, $<$ becomes $>$, and vice versa.

Examples

A. Solve: $2x - 11 > 3$

To solve this inequality, add 11 to both sides to get $2x > 14$. Then divide by 2, so $x > 7$.

B. Solve: $14 - 3x \geq 35$

To solve this inequality, subtract 14 from both sides to get $-3x \geq 21$. Then, since we divide by -3 to complete solving, we must make the inequality $\leq$.

Thus, the solution is $x \leq -7$.

There is a special notation that you should familiarize yourself with for the HSPA Exam. This notation is for showing that a number is between two numbers. To show that x is between -2 and 7, we write $-2 < x < 7$.

PRACTICE SET

Solve each equation or inequality.

1. $2x - 11 = 17$

2. $3(4x - 9) + 10 = 7$

3. $5x - 9 = 2x + 15$

4. $3x + 11 < 5$

5. $8 - 5x \leq -7$

MIXED PRACTICE, CLUSTER IV, MACRO C

1. Within the set of integers, which of the following sets is the solution to $3x + 2 \geq -9$?

   A. $\{-3, -2, -1, 0, 1, 2, 3, \ldots\}$
   B. $\{-4, -3, -2, -1, 0, 1, 2, 3, 4, \ldots\}$
   C. $\{\ldots, -6, -5, -4\}$
   D. $\{\ldots, -6, -5, -4, -3\}$

2. Which of the following expression(s) has(have) been simplified correctly?

   I $2x + 3y - 5x + y = -3x + 3y$
   II $4rs + 3t - rs - 5t = 3rs - 2t$
   III $2x^3 + 5x^2 - 3x - 5x^2 + 8x = 2x^3 + 5x$

   A. I and II only **B.** I and III only **C.** II and III only **D.** all three

3. Evaluate the following expression using $x = 20$ and $y = -5$.

$$\frac{x}{y} + y$$

   A. -1 **B.** 1 **C.** -20 **D.** -9

4. Solve: $-5x < 30$

5. Which of the following is equivalent to $5(x - 2) - 3 > x - 10$?

   A. $x > -\frac{3}{4}$ **B.** $x > \frac{3}{4}$ **C.** $x > -\frac{17}{4}$ **D.** $x > 0$

6. Which set of steps are used to solve the following equation?

$$5(x - 3) = 60$$

A. Distribute the 5; then add 3 to both sides and finally divide by 5.

B. Distribute the 5; then add 15 to both sides and finally divide by $\frac{1}{5}$.

C. Distribute the 5; then divide both sides by 5 and finally add 15.

D. Distribute the 5; then add 15 to both sides and finally divide by 5.

7. If the sum of four consecutive odd integers has to be less than 200, what is the largest possible value any one of these four integers can have? Why? Explain!

8. Solve for x:

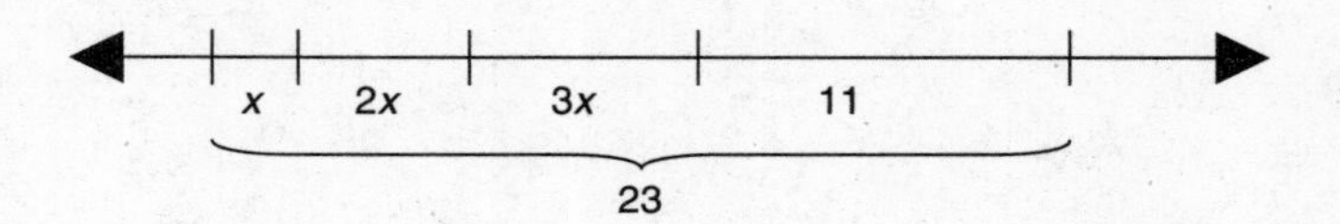

9. Write a simplified expression in terms of x for the area of the trapezoid shown below:

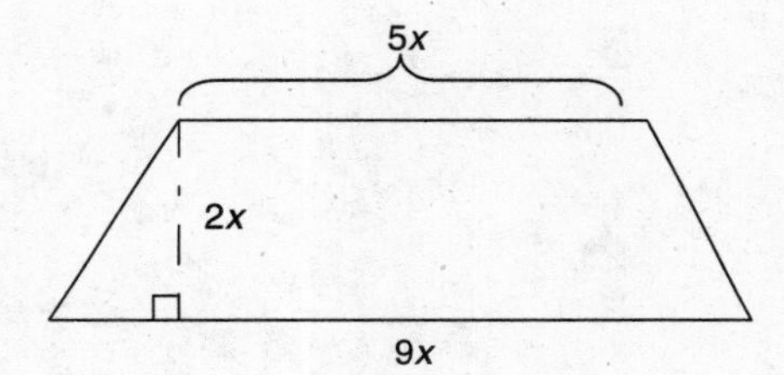

10. Which of the choices below represents the following scenario: Mr. Jones started an account with $1200. He withdrew the same amount of cash weekly for 5 weeks and then deposited $218. The balance was then $668.

A. $1200 - 5x = 668 + 218$

B. $1200 - 5x + 218 = 668$

C. $1200 + 5x - 218 = 668$

D. $1200 - 5x - 218 = 668$

ANSWERS AND EXPLANATIONS

Practice Set, p. 14

1. Since the whole numbers are the set {0, 1, 2, 3, . . . } and the natural numbers are the set {1, 2, 3, 4, . . . }, the only number that is an element of the whole numbers but not the natural numbers is **0**.

2. To show that a number is rational, you need to show that it can be written as a fraction. Since $\mathbf{-4} = \frac{-4}{1}$, for example, it is **rational**.

3. To find out what type of decimal a given fraction is, divide the numerator by the denominator. In this case, divide 7 by 9 and the result is .77777 or $.\overline{7}$ which is a **repeating decimal**.

4. As with #3, divide 3 by 8 and the result is .375 so this fraction equals a **terminating decimal**.

5. Again, divide: 14 ÷ 33 = .424242 or $.\overline{42}$, so this fraction is equivalent to a **repeating decimal**.

6. Remember the natural numbers is the set {1, 2, 3, 4, 5, . . . }. If you subtract any two natural numbers, the result is not necessarily another natural number. For example, 8 – 8 = 0, which is not a natural number, and 3 – 7 = –4, which is also not a natural number. Therefore, the set of natural numbers is **not** closed for subtraction.

Practice Set, p. 15

1. To round 11.34 to the nearest whole number, look at the tenths place (immediately to the right of the decimal place), which is a "3," and since this is less than 5, the number rounds to **11**.

2. To round 28.657 to the nearest tenth, look at the hundredths place (immediately to the right of the tenths place), which is a "5," and since this is 5 or more, the number rounds up to **28.7**.

3. To round 34.2341 to the nearest thousandth, look at the ten-thousandths place (immediately to the right of the thousandths place), which is a "1," and since this is less than 5, we keep the number at **34.234**.

4. To round 56.876 to the nearest hundredth, look at the thousandths place (immediately to the right of the hundredths place), which is a "6," and since this is more than 5, we round up to **56.88**.

Practice Set, p. 16

1. Since the signs are the same, add and keep the sign: **–8**.

2. Since the signs are different, subtract the absolute values (10 – 3 = 7) and take the sign of the larger absolute value (positive). **7**.

3. This problem changes to $12 + (-15)$ and again, since the signs are different, we subtract the absolute values $(15 - 12 = 3)$ and take the sign of the larger absolute value (negative). **-3.**

4. This problem changes to $-1 + (-6)$, and since the signs are the same, we add and keep the sign with a result of **-7.**

5. Opposites always add to **zero**.

6. The problem changes to $-2 + 9$ and since the signs are opposite, we subtract absolute values $(9 - 2 = 7)$ and take the sign of the larger absolute value (positive). **7.**

7. The problem changes to $4 + 3$ which is a basic addition problem. **7.**

8. The problem changes to $-5 + 3$ and since the signs are opposite, we subtract absolute values $(5 - 3 = 2)$ and keep the sign of the larger (negative). **-2.**

9. Since the signs are opposite, subtract absolute values $(17 - 9 = 8)$ and keep the sign of the larger absolute value (negative). **-8.**

10. The problem changes to $-8 + 13$ and we subtract absolute values $(13 - 8 = 5)$ and take the sign of the larger absolute value (positive). **5.**

Practice Set, p. 17

1. The signs are the same, so the result is **positive**.

2. The signs are different, so the result is **negative**.

3. The signs are different, so the result is **negative**.

4. The signs are different, so the result is **negative**.

5. The signs are the same, so the result is **positive**.

6. The signs are different, so the result is **negative**.

7. Since we are given that $a > 0$, this means ***a* is a positive number** and since $b < 0$, ***b* is a negative number**.

   (a) $a \div b$ is never greater than zero (**positive**), since a positive divided by a negative is **always negative**.
   (b) Depending on which absolute value is larger (a's or b's), $a + b$ could be positive or negative or even zero so $a + b$ is **sometimes less than zero (negative)**.
   (c) $a - b$ is always greater than zero (positive) since the problem is going to change into $a + (-b)$. Plus, since $b < 0$, $-b$ is a positive number, and the sum of two positive numbers is **guaranteed to be positive**.

Practice Set, p. 18

1. $2^5 = 2 \times 2 \times 2 \times 2 \times 2 = \mathbf{32}$

2. $(-4)^2 = -4 \times -4 = \mathbf{16}$

3. $(-2)^3 = -2 \times -2 \times -2 = \mathbf{-8}$

4. $3^8 = 3 \times 3 \times 3 \times 3 \times 3 \times 3 \times 3 \times 3 = \mathbf{6{,}561}$

5. By the first clue, the number could be 10 (since 10^2 is 100) up to 15 (since 15^2 is 225 and 16^2 is over 250). By the third clue, we can eliminate the odd numbers so the number is either 10, 12, or 14. The only one that is over 2,000 when cubed is **14**.

6. Since the result has to be between 4,000 and 5,000, the number to the fourth power has to be between (4,000 – 125) and (5,000 – 125) or between 3,875 and 4,875. By estimation and trial and error, you should arrive at a base of **8**.

Practice Set, p. 19

1. Since you are multiplying two like bases, **add** the exponents. 6^7

2. Since you are raising a power to another power, **multiply** the exponents. 7^{12}

3. Since you are dividing, **subtract** the exponents. 3^6

4. Since you are multiplying, **add** the exponents. 5^{10}

5. Since you are dividing, **subtract** the exponents. 4^6

Practice Set, p. 20

1. Both 9 and 27 can be written with a base of 3 as follows: $(3^2)^3 \times (3^3)^4 = 3^6 \times 3^{12}$, since we multiply exponents when we are raising a power to another power. Finally, since we are multiplying like bases, we can add the exponents to get a final answer of $\mathbf{3^{18}}$.

2. Both 25 and 125 can be written as powers of 5, so the problem becomes:

$$\frac{(5^2)^6}{(5^3)^3} = \frac{5^{12}}{5^9} = 5^3.$$

Practice Set, p. 21

1. 7.8×10^8.

2. 3.1×10^{-5}.

3. 9.4×10^6.

4. 4.4×10^{-3}.

Practice Set, p. 22

1. When asked to give the range of values for $(6.738 \times 10^3) \times (8.\_\,\_\,\_ \times 10^4)$, you need to focus on the smallest and largest possible numbers for the blank. The second quantity can be between 8.000 and 8.999, and these results yield:

   53.904×10^7 (multiply 6.738 by 8 and add the exponents on 10)

   60.635262×10^7 (multiply 6.738 by 8.999 and add the exponents on 10)

   Since neither of these is written in scientific notation, we must adjust the numbers and exponents so 53.904 becomes $\mathbf{5.3904 \times 10^8}$ and similarly, 60.635262 becomes $\mathbf{6.0635262 \times 10^8}$.

2. Similar to #1 above, multiply 5.261 by 3 and by 3.999 and add the exponents to get:

   15.738×10^8 and 21.038739×10^8; these become $\mathbf{1.5738 \times 10^9}$ and $\mathbf{2.1038739 \times 10^9}$.

Practice Set, p. 23

1. $\sqrt{81} = \mathbf{9}$ since $9 \times 9 = 81$.

2. $\sqrt{30} \approx \mathbf{5.477}$ (use a calculator).

3. $\sqrt{119} \approx \mathbf{10.909}$ (use a calculator).

4. $\sqrt{36} \approx \mathbf{6}$ since $6 \times 6 = 36$.

5. $\sqrt{200} \approx \mathbf{14.142}$ (use a calculator).

Practice Set, p. 24

Given $x = 3$, $y = 4$, and $z = -5$, evaluate each of the following:

1. $2x^2 = 2 \cdot 3 \cdot 3 = \mathbf{18}$.

2. $(2y)^2 = (2 \cdot 4)^2 = \mathbf{8^2 = 64}$.

3. $2z^3 = 2 \cdot (-5) \cdot (-5) \cdot (-5) = \mathbf{-250}$.

4. $3y^2 = 3 \cdot 4 \cdot 4 = \mathbf{48}$.

5. $(2x)^2 = (2 \cdot 3)^2 = 6^2 = \mathbf{36}$.

Practice Set, p. 25

1. Begin with multiplication and division from left to right, since there are neither parentheses nor exponents. So, $8 \times 2 - 9 + 4$ and then $16 - 9 + 4$. Finish with addition and subtraction from left to right to get $7 + 4 = \mathbf{11}$.

2. Begin with parentheses to get $108 \div 9 - 2^3$, and then do the exponents to get $108 \div 9 - 8$. Next is division, which gives you $12 - 8 =$ **4**.

3. Begin in the parentheses inside the brackets. This gives you $92 - [37 + (5)^2 \times 2]$, and then we do the exponents to get $92 - [37 + 25 \times 2]$, followed by multiplication, $92 - [37 + 50]$, and then $92 - 87 =$ **5**.

Practice Set, p. 27

1. The operation # as defined is **commutative**, since $3 \# 4 = 3^2 + 4^2 = 4 \# 3 = 4^2 + 3^2$. Basically, the new operation is commutative since addition is commutative. The new operation # is not associative, however, since $2 \# (3 \# 4) = 2 \# (3^2 + 4^2) = 2 \# (25) = 2^2 + 25^2$ while $(2 \# 3) \# 4$ equals $(2^2 + 3^2) \# 4 = 13 \# 4$.

2. The relationship "is perpendicular to" is **not transitive** since two lines perpendicular to the same line are most likely parallel to each other.

3. To make the problem easier to do in your head, re-arrange the order of the numbers (using commutativity) to become: $12 + 28 + 45 + 25 = 40 + 70 = 110$. Notice that the associative property is also used to help group the numbers two at a time.

4. No, John is **not correct**. The relationship "is sitting next to" is not transitive because, for instance, Mary could be "sitting next to" John and so could Susan. They could be sitting on either side of John, yet not sitting next to each other.

Practice Set, p. 27

1. The easiest system for listing all the factors is in pairs. So, for 40, begin with 1 and 40. Continue with 2 and 20. Then, 4 and 10. Then, 5 and 8.

2. The factors of 64 are: 1 and 64, 2 and 32, 4 and 16, 8 or, in order, **1, 2, 4, 8, 16, 32, 64.**

3. The factors of 45 are: 1 and 45, 3 and 15, 5 and 9 or, in order, **1, 3, 5, 9, 15, and 45.**

Practice Set, p. 28

1. In order to get the sum of the first five prime numbers, we must first identify the first five primes and then add them. The first five primes are: 2, 3, 5, 7, and 11. Remember, 1 is NOT prime. So, the sum is $2 + 3 + 5 + 7 + 11 =$ **28**.

2. Since 57 can be divided by 3, 57 is **composite**.

3. $40 = 2 \times 20 = 2 \times 4 \times 5 = \mathbf{2 \times 2 \times 2 \times 5}$.

   $95 = \mathbf{5 \times 19}$.

   $120 = 2 \times 60 = 2 \times 2 \times 30 = 2 \times 2 \times 2 \times 15 = \mathbf{2 \times 2 \times 2 \times 3 \times 5}$.

Practice Set, p. 29

1. The first five multiples of 8 are: $8 \times 1 = \mathbf{8}$, $8 \times 2 = \mathbf{16}$, $8 \times 3 = \mathbf{24}$, $8 \times 4 = \mathbf{32}$, $8 \times 5 = \mathbf{40}$.

2. If a number is a multiple of 6 and also of 8, then the smallest number it could be is 24. However, the number could also be a multiple of 24 such as 48 or 72, etc.

Practice Set, p. 30

1. To get the GCF, list the factors of each number and find the largest common factor:

   18: 1, 2, 3, 6, 9, 18
   45: 1, 3, 5, 9, 15, 45

   So, the GCF is **9**.

2. As above:

   35: 1, 5, 7, 35
   56: 1, 2, 7, 8, 28, 56
   84: 1, 2, 3, 4, 7, 12, 21, 28, 42, 84

   So, the GCF is **7**.

Practice Set, p. 30

1. To get the LCM, list the multiples and find the first one that is common

   10: 10, 20, 30, 40
   8: 8, 16, 24, 32, 40

   Thus, the LCM is **40**.

2. As above:

   18: 18, 36, 54, 72, 90
   15: 15, 30, 45, 60, 75, 90
   6: 6, 12, 18, 24, 30, 36, 42, 48, 54, 60, 66, 72, 78, 84, 90

   Thus, the LCM is **90**.

Practice Set, p. 31

1. To find the number of times, find the number of common multiples within the first 3 hours or 180 minutes:

Steve: 10, 20, 30, **40**, 50, 60, 70, **80**, 90, 100, 110, **120**, 130, 140, 150, **160**, 170, 180
Mark: 8, 16, 24, 32, **40**, 48, 56, 64, 72, **80**, 88, 96, 104, 112, **120**, 128, 136, 144, 152, **160**, 168, 176

The common multiples are in bold—there are **4**.

2. Similar to #1 above, list the multiples up to 250 and count the common ones:

18: 18, **36**, 54, **72**, 90, **108**, 126, **144**, 162, **180**, 198, **216**, 234
12: 12, 24, **36**, 48, 60, **72**, 84, 96, **108**, 120, 132, **144**, 156, 168, **180,** 192, 204, **216,** 228, 240

The common multiples are in bold—there are **6**.

Mixed Practice, Cluster I, Macro A, pp. 31–33

1. **B** 3.8 is about 4 and .42 is about .5 so half of 4 is 2 and the answer closest is 1.6.
2. **B** The laws of exponents say to add exponents when multiplying like bases so statement B is incorrect.
3. **D** Since the square root of 27 is a non-terminating, non-repeating decimal, it is irrational.
4. To find the height after 3 seconds, simply plug in $t = 3$ to get $30(3) - 6(3)(3) = 90 - 54 =$ **36 m**.
5. **B** Since $4^3 = 64$ and $3^4 = 81$, the best choice is that 4^3 is about 2/3 as much as 3^4.
6. Since the number is rounded to the nearest hundred, the digit that determines if it rounds up or remains the same is the tens. Thus, the largest value would be 1,449 and the smallest value would be 1,350, both of which would round to **1,400**.
7. **C** $3^6 \cdot 9^6 = 3^6 \cdot (3^2)^6 = 3^{18} = (3^2)^9 = (27^2)^3 = [(3^3)^2]^3$ so C is not equivalent.
8. **C** Estimation gives: $.03 \times .06 = .0018$, so .002 is closest.
9. **B** In looking at the powers of 7, the units digits are 7, 9, 3, 1, 7, 9, 3, . . . so $83 \div 4 = 20$ r 3 so the units digit is a **3**.
10. $4^2 + 4^5 = 16 + 1{,}024 = 1{,}040$ and $4^7 = 16{,}384$ so the two are **not equal**.
11. **A** Since the square root of 50 is a non-terminating, non-repeating decimal, it is **irrational**.
12. **D** Since it is non-terminating and non-repeating, 0.0200200020002 . . . is an irrational number.
13. **C** If a number is a multiple of 4 and 9, it is a multiple of 36, so it must divide by 36. The only choice that divides by 36 is 3,744.
14. Since the first arrowhead is on 5, the problem begins with 5. Then we go to the left 12 units so it is $5 + (-12)$. Then we go to the right 3 units so the problem finishes as $5 + (-12)$, or **+3**.
15. The number could be between 1.300 and 1.399, so multiply 3.821 by 1.3 *and* by 1.399, and then add the exponents to get 10^{10}. So, the range of values is $\mathbf{4.9673 \times 10^{10}}$ **to** $\mathbf{5.345579 \times 10^{10}}$.
16. Since the multiples of 9 is a large list as is the list of odd numbers, we read through the list to the last two clues, which give us a range of values. The first multiple of 9 that is between 480 and 550 is 486 but this is not odd. The next multiple of 9 is

495, which is odd. So, this is the first such number. The next multiple of 9 is 504 but this is even. It will turn out that we skip every other multiple of 9 (or multiples of 18). Thus, the list of numbers satisfying all four conditions is: **495, 513, 531, 549**.

17. **B** If there are 25 students, put them in 8 groups of 3 and there will be one student left over. If there are 32 students, put them in 4 groups of 7 and 1 group of 3 and there will be 1 left. If there are 26 students, put them in 6 groups of 3 and 1 group of 7 and there will be 1 left.
So, the only number in the choices that doesn't work is 24.

18. $2 \ \& \ 5 = 2^2 + 3(2)(-5) = 4 + (-30) = \mathbf{-26}$.

19. **D** Start with 119, which is closest to 120 but still smaller. 119, however, divides by 7, so is not prime. The next largest, 117, divides by 3. 115 divides by 5, which leaves **113**.

20. A perfect square is a number resulting from squaring an integer such as 1, 4, 9, 16, 25, 36, 49, 64, or 81. These are all the perfect squares less than 100 (resulting from squaring 1 through 9). Looking at 4 more than each of these gives: 5, 8, 13, 20, 29, 40, 53, 68, 85. We can get rid of the even ones, leaving 5, 13, 29, 53, 85. Of the remaining list, the sum of the digits are: 5, 4, 11, 8, 13 and we need the ones whose digits' sum is prime; thus, the list of numbers meeting all four conditions is: **5, 29, 85**.

21. **B** Two pairs are relatively prime (7 and 9 and 43 and 100). Note that 21 and 24 can each divide by 3, and 54 and 93 can each divide by 3 as well.

22. No, it is not possible for a multiple of 13 other than 13 itself to be prime, since any multiple of 13 can be written as the product of 13 and 2 or 13 and 3, and so on.

23. The factors of each number are as follows:
72: 1, 2, 3, 4, 6, 8, 9, 12, 18, 24, 36, 72
63: 1, 3, 7, 9, 21, 63
Since both 72 and 63 share a factor of 9, we can create **9 baskets with 8 pencils and 7 pens.**

24. If we list the multiples of each number up to 195:
12: 12, 24, 36, 48, **60**, 72, 84, 96, 108, **120**, 132, 144, 156, 168, **180**, 192
10: 10, 20, 30, 40, 50, **60**, 70, 80, 90, 100, 110, **120**, 130, 140, 150, 160, 170, **180**, 190

(a) The store gave away **16 sundaes** and **19 ice cream cones**.

(b) Tom was the 60th customer, so there were **59** people ahead of him. **Two** others got both items free.

Practice Set, p. 36

1. To write a mixed number as an improper fraction, multiply the denominator by the whole number. In this case, $4 \times 5 = 20$. Then, add the numerator; in this case, $20 + 2 = 22$. The result goes over the original denominator. So, $4\frac{2}{5} = \mathbf{\frac{22}{5}}$.

2. To change $\frac{14}{3}$ to a mixed number, divide 14 by 3. The result is 4 with a remainder of 2. The remainder is written as a fraction $\frac{2}{3}$ so the mixed number is $\mathbf{4\frac{2}{3}}$.

3. To get equivalent fractions, you multiply both the numerator and denominator of a given fraction by the same number. Thus, $\frac{5}{8} = \frac{5 \times 2}{8 \times 2} = \frac{10}{16}$ or $\frac{5}{8} = \frac{5 \times 3}{8 \times 3} = \frac{15}{24}$ and so on.

Practice Set, p. 37

1. $\frac{18 \div 6}{24 \div 6} = \frac{3}{4}$ **2.** $\frac{12 \div 6}{30 \div 6} = \frac{2}{5}$ **3.** $\frac{32 \div 32}{96 \div 32} = \frac{1}{3}$ **4.** $\frac{36 \div 12}{60 \div 12} = \frac{3}{5}$

Practice Set, p. 39

1. Since the numerator is very close to the denominator, $\mathbf{\frac{15}{19}}$ is closest to 1.

2. Since half of 13 is 6.5 and the numerator, 7, is barely more than 6.5, the fraction $\mathbf{\frac{7}{13}}$ is closest to $\frac{1}{2}$.

3. Since the numerator is significantly smaller than the denominator, the fraction $\mathbf{\frac{5}{34}}$ is closest to 0.

4. $\mathbf{\frac{16}{25}}$ is more than one half, since half of 25 is 12.5, and the numerator, 16, is more than 12.5.

5. Insert < or > between each pair of fractions to make a true statement.

(a) $\frac{6}{11} > \frac{8}{17}$ since $\frac{6}{11}$ is more than half, while $\frac{8}{17}$ is less than half.

(b) $\frac{7}{17} < \frac{7}{10}$ since $7 \times 10 < 17 \times 7$.

(c) $\frac{9}{16} > \frac{5}{16}$ since $9 > 5$, and the denominators are the same.

(d) $\frac{7}{12} > \frac{8}{15}$ since $7 \times 15 > 12 \times 8$.

Practice Set, p. 41

1. Since there are 4 odd numbers and 8 even numbers, the ratio of odd to even is 4:8 or **1:2**.

2. Since there are 6 squares and 10 hearts, the ratio of squares to hearts is 6:10 or **3:5**.

3. Since there are 5 arrows and 6 squares, the ratio of arrows to squares is **5:6**.

Practice Set, p. 41

1. To find the unit rate per hour, divide \$630 by 35 hours to get **\$18/hour**.

2. To figure the unit rate, divide \$6.25 by 5 pounds to get **\$1.25/pound**.

3. To figure the unit rate, divide 45 miles by 5 hours to get **9 mph**.

Practice Set, p. 43

1. Cross multiplication yields $3x = 24$, and division gives $\mathbf{x = 8}$.

2. Cross multiplication gives $5x = 80$, and division yields $\mathbf{x = 16}$.

3. Cross multiplying gives $2x = 42$, and dividing by 2 means $\mathbf{x = 21}$.

4. Cross product is $4x = 60$, and dividing both sides by 4 means $\mathbf{x = 15}$.

5. When we cross multiply, we get $4x = 120$ and then by dividing, $\mathbf{x = 30}$.

Practice Set, p. 44

1. Set up the following proportion: $\frac{MAP}{ACTUAL} = \frac{2}{25} = \frac{x}{650}$. Cross multiplication yields $25x = 1300$; and dividing gives $\mathbf{x = 52}$ **cm**.

2. Set up the following proportion: $\frac{Recommend}{Asked} = \frac{3}{8} = \frac{75}{x}$ and cross multiplying gives $3x = 600$ so that $x = 200$, which means **200** doctors were surveyed.

3. Set up the following proportion: $\frac{Cashews}{Pecans} = \frac{3}{4} = \frac{x}{68}$. Cross products equal $4x = 204$ and $x = 51$, which means the mixture should contain 51 pounds of cashews, and the entire mixture would weigh $51 + 68 =$ **119 pounds**.

4. Set up the following proportion: $\frac{Blueprint}{Actual} = \frac{1}{5} = \frac{2.25}{x}$. Cross multiplication yields $x =$ **11.25 ft.**, the actual measurement.

5. Set up the following proportion: $\frac{Rainy}{Total} = \frac{4}{15} = \frac{x}{60}$. Cross multiplication yields $15x = 240$, and $x = 16$, meaning the family should expect about **16 rainy days**.

Practice Set, p. 48

The completed chart:

Fraction	Decimal	Percent
$\frac{13}{20}$	Since $\frac{13\times5}{20\times5}=\frac{65}{100}$, the decimal is .65	When you move the decimal point over two places, 65%
The proper way to read .86 is "86 hundredths" so the fraction is $\frac{86}{100}$ which reduces to $\frac{43}{50}$	.86	When you move the decimal point over two places, 86%.
The proper way to read .24 is "24 hundredths" so the fraction is $\frac{24}{100}$ which reduces to $\frac{6}{25}$	Moving the decimal two places to the left yields .24	24%
$\frac{5}{8}$	To get the decimal, use a calculator to divide 5 by 8 and the result is .625	To get the percent, move the decimal two places to the right, which gives 62.5%
.325 is read "325 thousandths" so the fraction is $\frac{325}{1000}$ which reduces to $\frac{13}{40}$	.325	Moving the decimal two places to the right so the percent is 32.5%
The correct way to read .78 is "78 hundredths" which means the fraction is $\frac{78}{100}$ which reduces to $\frac{39}{50}$	Moving the decimal two places to the left means the decimal is .78	78%

Practice Set, p. 50

1. 75% is $\frac{3}{4}$, so $92 \div 4 = 23$, and $23 \times 3 = \mathbf{69}$.

2. $10.5 = .15x$, so $x = \mathbf{70}$.

3. $45\% = \frac{45}{100} = \frac{9}{20}$ and $\frac{9}{20} = \frac{x}{60}$ so $\boldsymbol{x} = \mathbf{27}$.

4. $\frac{36}{180}=\frac{1}{5}$ or **20%**.

5. $112 = .64x$, so $\mathbf{x = 175}$.

6. $\frac{98}{140}=\frac{7}{10}$ or **70%**.

Practice Set, p. 53

1. $149 × 1.07 = **$159.43**.

2. 3,400 × .35 = $1,190 and $3,400 − *$1,190 = **$2,210**.

3. $432 = .64x$, so $x = 675$, which represents the total number of people surveyed. Therefore, 675 − 432 = **243 people like summer**.

4. $1{,}400 = .4x$, so $\mathbf{x = 3{,}500}$.

5. 80 × .80 = 64. Then, 64 × .9 = **$57.60**.

Mixed Practice, Cluster I, Macro B, pp. 53–56

1. Since it takes six $\frac{1}{4}$ inches to make $1\frac{1}{2}$ inches, and each $\frac{1}{4}$ inch equals 50 miles, multiply 50 × 6. **300 miles**.

2. Since 75% of the participants prefer country music over jazz (this number is 450), and 75% is $\frac{3}{4}$, and 450 ÷ 3 = 150, each quarter is 150, which is the **25% who like jazz**.

3. There are a total of 2 + 4 + 6 = 12 "parts." So, dividing $216 by 12 gives $18. Therefore, the largest share is six times this or **$108**.

4. **C** $\$18{,}000 \times (.04)^5 \approx \$21{,}900$

5. **C** $\frac{\textit{Satisfactory}}{\textit{Manufactured}}=\frac{7}{12}=\frac{x}{450}$. Cross multiplication gives $12x = 3150$ and division gives 262.5 so the best approximation is 265.

6. **B** Since 3 × 34 < 8 × 13, the fraction $12\frac{13}{34}$ is the greater of the two.

7. **B** There are a total of 2 + 3 or 5 parts and since the entire length is 40″, each of the five parts is 8″ and the shorter piece is 2 times 8″ or **16″**.

8. **C** 6 out of 16 are white; this reduces to 3 out of 8, which is **37.5%**.

9. **B** Of the choices given, the ones that are larger than $\frac{9}{16}$ inch are B and D. Thus, we need to compare these two. Since $\frac{3}{4}=\frac{6}{8}, \frac{5}{8}''$ is the smallest of the two yet still larger than $\frac{9}{16}''$.

10. **B** Since the discount is $33\frac{1}{3}\%$, you are left to pay $66\frac{2}{3}\%$. So the question really becomes, \$108.66 is $66\frac{2}{3}\%$ of what price? Thus, $108.66 = .\bar{6}x$ and $x \approx$ **162.99**.

11. Set up the proportion: $\frac{\text{Miles}}{\text{Minutes}} = \frac{342}{36} = \frac{x}{132}$. *Note*: 2 hours = $60 \times 2 = 120$ minutes. Add 12 minutes to that and you get 132 minutes. Cross multiplying gives $36x = 45{,}144$ and division yields **1254 miles**.

12. First, add the total sales in systems alone. This is \$5599.96 and he earns 6.5% commission. So, multiply .065 times 5599.96, which is \$364. Then add the upgrades, which total \$1499.96. He earns 10% commission on these, so multiply 1499.96 by .1 which is \$150. Therefore, Mr. Dennis' total commission for November is \$364 + \$150 or **\$514**.

13. **C** The amount the price is reduced is 20,795 – 17,500 or 3295. Thus, to get the percent, divide 3295 by 20,795, which is approximately .158 or **15.8%**.

14. **B** To answer this question, divide each numerator by its denominator using a calculator:

    A $\frac{254}{99} = 2.\overline{56}$. Thus, this meets **all three conditions**.

    B $\frac{25}{11} = 2.\overline{27}$. Since the hundredths place in this result is NOT 6, we don't need to continue.

15. **D** $252 = 2.1x$; division gives you **120**.

16. Set up the proportion: $\frac{\text{Pages}}{\text{Minutes}} = \frac{6}{35} = \frac{30}{x}$. When you cross multiply, $6x = 1050$ and $x = 175$, which means it takes Pedro 175 minutes. Since there are 60 minutes in an hour, dividing 175 by 60 means it takes him 2 hours and 55 minutes, so he would finish at **11:55 A.M.**

17. Sneakers-R-Us: $\$120 \times .7 =$ **\$84** (30% off means you pay the remaining 70%).
    Sneaks Unlimited: $\$135 \times .65 =$ **\$87.75** (35% off leaves you with 65% to pay).
    Thus, the sneakers are cheaper at Sneakers-R-Us and you save 87.75 – 84 = **\$3.75**.

18. **D** $\$95{,}000 \times 1.55$ (100% plus 55% profit) = **\$147,250**.

19. **C** Divide the price by the number of ounces to get the price per ounce.
$2.25 \div 12 = .1875$
$2.95 \div 16 = .184375$
$4.25 \div 24 = .17708\overline{3}$
$5.75 \div 32 = .1796875$
Thus, the cheapest per ounce is the 24 oz. container.

20. $268 = 1.25x$ so division gives $x \approx$ **214 students** who took it last year.

21. Consider, for example, the fraction $\frac{5}{9}$. The numerator becomes 20 (5×4) and the denominator becomes 3 ($9 \div 3$). Since $4 \div \frac{1}{3} = 12$, the fraction becomes **twelve times as large.**

22. $\frac{Rotations}{Seconds} = \frac{40}{60} = \frac{2}{3}$, which means it takes **1.5 seconds** to rotate once.

23. **B** **A.** 18 to 20 is an increase of 2 out of 18, which is about **11%**.
B. 16 to 18 is an increase of 2 out of 16, which is about **12.5%**.
C. 95 to 100 is an increase of 5 out of 95, which is about **5.3%**.
D. 600 to 650 is an increase of 50 out of 600, which is about **8.3%**.

24. **C** $240 × .8 (since 20% off leaves 80% to pay) = $192. 10% off means ($192 × .9), which is **$172.80**.

25. **C** $160,000 = 1.25$x$ so by division x = $128,000.

26. **C** First, add up the prices of their dinners: $16.60 + $10.75 + $20.25 + $15.45 = $63.05. The tip is 15% of this, or .15 times 63.05 (approximately $9.46). Tax is 6% of $63.05 (.06 times 63.05 which is about $3.78). So, adding the amounts gives: $63.05 + $3.78 + $9.46 = **$76.29**.

27. **B** Of the 12,000 widgets produced per day, 3% are defective (12,000 times .03 = 360). Since 360 are defective, this leaves 12,000 – 360 or 11,640 that are not defective. With a .27 profit on every good widget, the earnings per day are 11,640 × .27, which is about **$3142.80**.

28. **A** $3,400 × .05 = $170 while $3,400 × .06 = $204 so she saves $204 – $170 = $34.

29. **D** Tyshawn's deductions total 25.7%, so he is left with 100% – 25.7% or 74.3%, so multiply $135.40. by .743 to get his net pay: **$100.60**.

30. **C** .50 ÷ $14.25 ≈ 3.5%, and .45 ÷ 8.95 is about 5%, so the difference is about **1.5%**.

Practice Set, p. 61

1. Obtuse.

2. Right.

3. Acute.

4. Acute.

5. Straight.

6. Obtuse.

Practice Set, p. 62

1. 17° and 73° are **complementary** since they add to 90°.
2. 126° and 54° are **supplementary** since they add to 180°.
3. 86° and 104° are **neither** because they add to 190°.
4. 158° and 22° are **supplementary** since they add to 180°.
5. 39° and 51° are **complementary** since they add to 90°.
6. 90° – 44° = **46°**.
7. 90° – 85° = **5°**.
8. 90° – 38° = **52°**.
9. 90° – 61° = **29°**.
10. 90° – 27° = **63°**.
11. 180° – 100° = **80°**.
12. 180° – 114° = **66°**.
13. 180° – 79° = **101°**.
14. 180° – 157° = **23°**.
15. 180° – 32° = **148°**.

Practice Set, p. 63

1. Acute.
2. Right.
3. Obtuse.
4. Since one angle is marked as a right angle and one is labeled 63°, add these two and subtract from the 180° that make up a triangle. Thus, the missing angle is **27°**.
5. Since the legs are marked as equal, take 180° – 118° = 62° and split this in half. Each missing angle is **31°**.

Practice Set, p. 64

1. Since $\angle 1$ and $\angle 2$ are supplements and $\angle 1 = 108°$, $m\angle 2 = 72°$. Then, angles 2, 3, and 4 should total 180° since they are the three angles of the triangle. $m\angle 2 + m\angle 4 = 72° + 66° = 138°$, so $m\angle 3 = 180° - 138° =$ **42°**.

2. Since $\angle 2$ and $\angle 4$ are supplements and $m\angle 4 = 141°$, subtracting this from 180° gives $m\angle 2$ as 39°. As you know, the three angles of the triangle ($\angle 1$, $\angle 2$, and $\angle 3$) add to 180°. Since $m\angle 2$ and $m\angle 3$ are known, we can add these (43° + 39°) and then subtract from 180° to get $m\angle 1 =$ **98°**.

Practice Set, p. 64

1. Since the sum of the two shorter lengths (3″ and 7″) exceeds the third length (7″), these measurements can form a triangle. It would be an **isosceles triangle** since two sides are both 7″.

2. Since $5 + 8 > 10$, these measurements can form a triangle. And, since the measurements are all different, the triangle would be **scalene**.

3. These can form a triangle since $8 + 8 > 8$. The triangle is **equilateral**.

4. $3 + 4 > 5$ so these measurements form a **scalene** triangle.

5. $6 + 6 > 11$ so these form an **isosceles** triangle.

Practice Set, p. 65

1. A hexagon has six sides so, by the formula, its angle sum is 180° times 4 or **720°**.

2. $8 \times 180° =$ **1440°**.

3. The angle sum in a quadrilateral is 360°, so add 60°, 140°, and 150°, and then subtract from 360°, which gives a missing angle of **10°**.

4. The angle sum of a pentagon is 180×3 or 540°, so add 80°, 95°, 95°, and 165° and then subtract from 540° to get the missing angle, which is **105°**.

5. To find each angle measure of a regular octagon, first find the angle sum (180 times 6 = 1080) and then divide by 8 (since all sides and angles are equal) to get each angle **135°**.

Practice Set, p. 67

Since we are given that $m\angle 6 = 73°$, begin with all the angles that are also equal to $\angle 6$ and thus are also 73°. $\angle 7$ is vertical to $\angle 6$ so, $m\angle 7 = 73°$. Also, $\angle 6$ corresponds with $\angle 2$ so $m\angle 2 = 73°$. Finally, $\angle 3$ is also equal to 73° since it is corresponding with $\angle 7$, alternate exterior angles with $\angle 6$ and vertical with $\angle 2$.

Then, we can get the supplement of $\angle 6$, which is 107°, and all of the remaining angles equal **107°**.

Practice Set, p. 68

1.

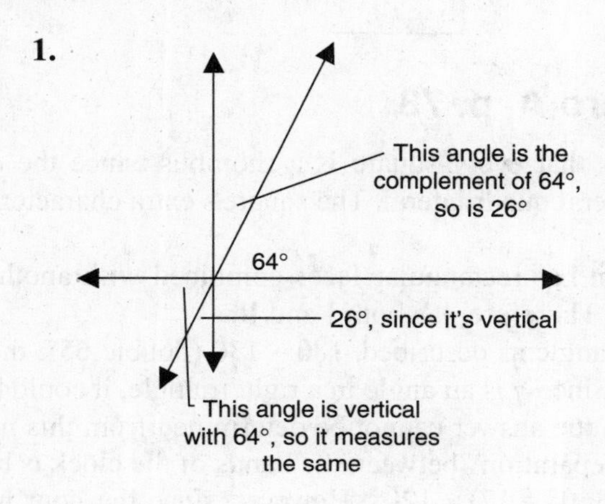

2.

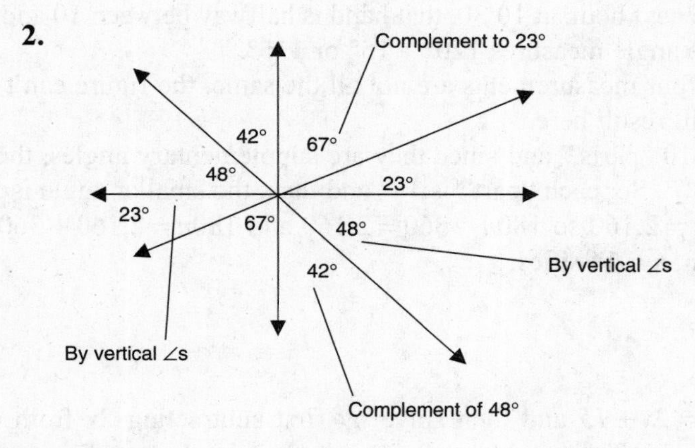

Practice Set, p. 69

1. Quadrilateral, parallelogram.
2. Quadrilateral, parallelogram, rhombus.
3. Quadrilateral, trapezoid.
4. Quadrilateral, parallelogram, rectangle.
5. Quadrilateral.
6. Quadrilateral, parallelogram, rectangle, rhombus, square.

Practice Set, p. 72

1. 16 cubes

2.

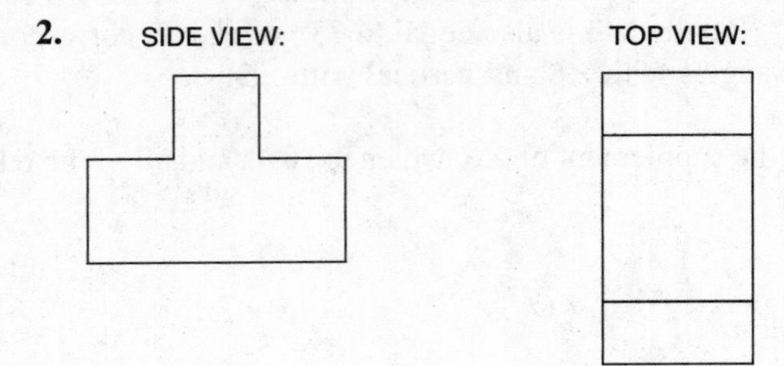

Mixed Practice, Cluster II, Macro A, p. 73

1. **B** The true statement is that every square is a rhombus, since the definition of a rhombus is an equilateral quadrilateral. The square's extra characteristic is its right angles.
2. **B** By definition, a prism has rectangular faces combined with another polygon (in this case, pentagons). Therefore, it's both I and II.
3. **D** Since p is the vertex angle as described, 180 – 130 (double 65°, the measure of a base angle), $p = 50°$. Since q is an angle in a right triangle, it could be 90° or some acute angle; therefore the answer cannot be determined from this information.
4. At 10:30, the basic "separation" between the hands of the clock is between 10 and 6, which is $4 \times 30°$ $(360° \div 12) = 120°$. However, since the hour hand is moving toward the next hour, at 10:30, that hand is halfway between 10 and 11, which adds 15°, so the angle measures 120° + 15° or **135°**.
5. **A** Since the four measurements are not all the same, the figure can't be a rhombus.
6. A **cone** will result here.
7. There are 10 "parts," and since they are supplementary angles, the total of the 10 parts is 180°. So, each "part" is 18° and thus, the smaller angle is $2 \times 18°$, or **36°**.
8. $180(n - 2) = 2{,}160$ so $180n - 360 = 2{,}160$ and $180n = 2{,}160 + 360 = 2{,}520$. Thus, by division, $n =$ **14 sides**.

Practice Set, p. 76

1. Make $5x + 9 = 2x + 15$ and then solve by first subtracting $2x$ from each side. This gives $3x + 9 = 15$. Then, subtract 9 from both sides to get $3x = 6$ and then finally divide and $x = 2$. Thus, $AB = 5(2) + 9 = 10 + 9 = 19$, as does BC, so $AC = 19 \times 2 =$ **38**.

2. $\angle AEB \cong \angle CED$ for example (any pair of vertical angles are congruent).

To find x, add the two adjacent angles, which are supplements; this gives $7x + 19 = 180$ and when solved, $x = 23$. $m\angle CED = 4(23) - 9 = 92 - 9 = 83°$. Then, $m\angle BED = 180° - 83° =$ **97°**.

Practice Set, p. 77

1. Since the triangles are similar, the angles are equal. So, m$\angle L = 60°$, since that's the measure of angle I. Also, m$\angle H = 75°$ since m$\angle K = 75°$. To find the missing lengths, we can set up the proportion:

$$\frac{36}{27} = \frac{GH}{18} = \frac{20}{KL}$$

This reduces to:

$$\frac{4}{3} = \frac{GH}{18} = \frac{27}{KL}$$

Solving gives $\boldsymbol{GH} = \mathbf{24}$ and $KL = \mathbf{15}$.

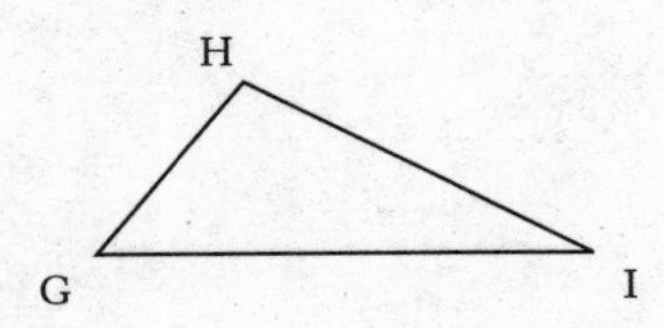

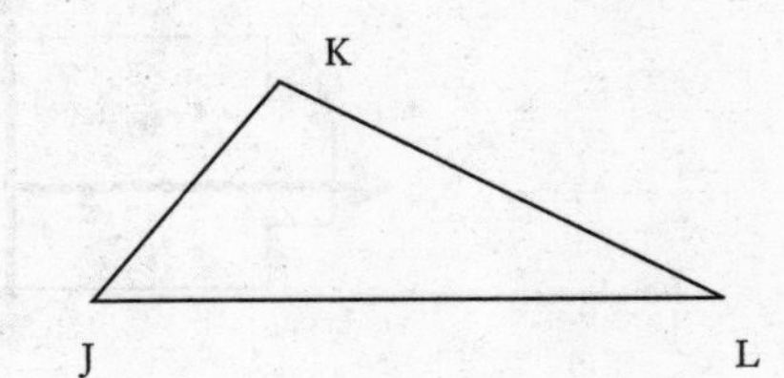

2. Set up the proportion below:

$$\frac{24}{30} = \frac{12}{QR} = \frac{8}{QT} = \frac{OP}{20}$$

When solved, $QR = \mathbf{15}$, $QT = \mathbf{10}$, and $OP = \mathbf{16}$.

Practice Set, p. 80

1.

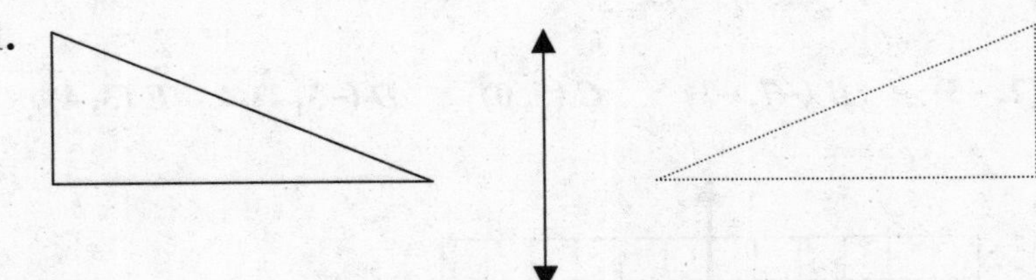

2.

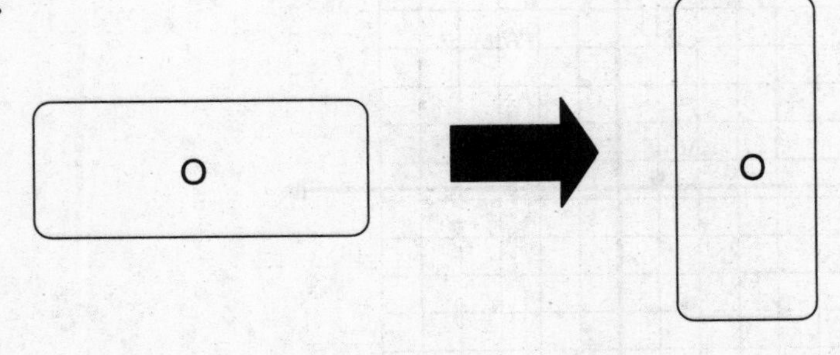

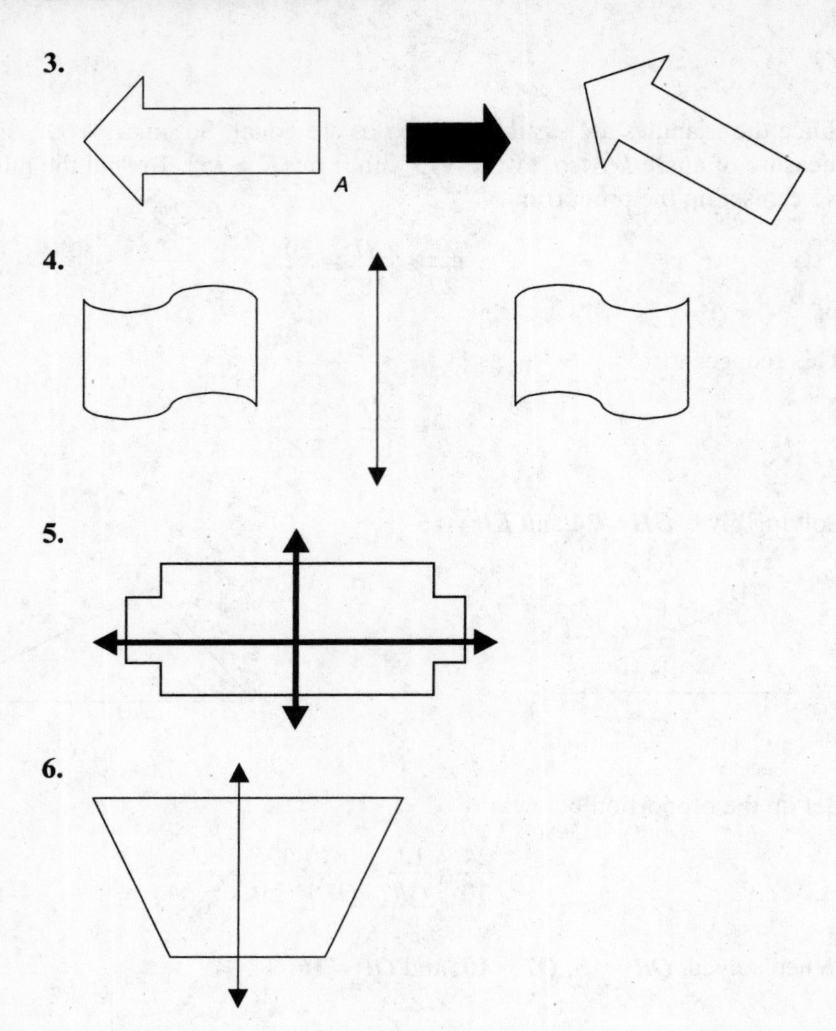

Practice Set, p. 83

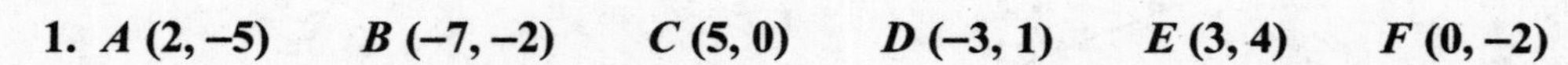

1. ***A*** **(2, –5)** ***B*** **(–7, –2)** ***C*** **(5, 0)** ***D*** **(–3, 1)** ***E*** **(3, 4)** ***F*** **(0, –2)**

2.

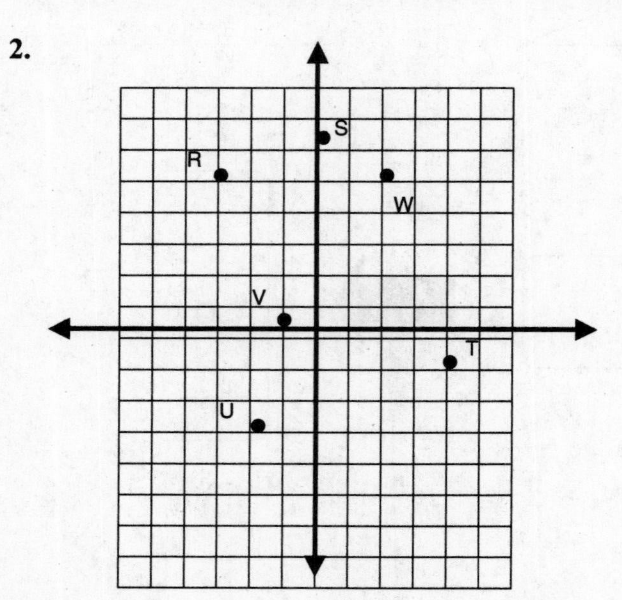

3. ***P*** **(6, –3)**

4. An **isosceles right triangle**

5. ***U*** **(0, 5)**

Practice Set, p. 85

1. ***P′*** **(2, –1)** ***Q′*** **(0, 2)** ***R′*** **(5, 3)** ***S′*** **(5, –5)**

2.

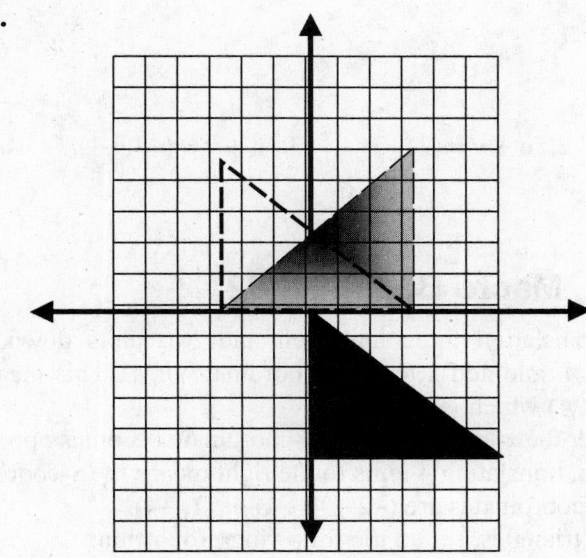

Original Triangle (Solid Black): (0, 0) (0, –5) (6, –5).
Final Triangle after translation and reflection (Gray): (3, 5) (3, 0) (–3, 0).
[*Note*: Dotted outline shows translated triangle, which is the intermediate step. This is just one example of many triangles that could have been drawn.]

3. *A′* (–1, –1) *B′* (1, –5) *C′* (5, –5) *D′* (5, –1)

Practice Set, p. 86

1. Translate **7** units right and **4** units down.

2. Translate **4** units up and **2** units left; then reflect through the *y*-axis.

Practice Set, p. 87

1. Translate up 1 and right 5.

2. Translate down 4 and left 2 and then reflect through the *x*-axis.

Practice Set, p. 88

1. $\begin{bmatrix} -3 & 6 \\ 5 & 7 \\ -5 & -5 \end{bmatrix}$ **2.** $\begin{bmatrix} 12 & -4 & 18 \\ -6 & 8 & 20 \end{bmatrix}$ **3.** $\begin{bmatrix} -6 & 3 \\ -9 & -12 \\ 3 & 18 \end{bmatrix}$ **4.** $\begin{bmatrix} 7 & 3 & -4 \\ -5 & 5 & 3 \end{bmatrix}$

5. $\begin{bmatrix} -22 & -29 \\ -5 & -8 \\ 15 & -10 \end{bmatrix}$

Practice Set, p. 91

1. $\mathbf{u} + \mathbf{v} = (4, 3)$ **2.** $\mathbf{u} + \mathbf{v} = (2, -5)$ **3.** $\mathbf{u} + \mathbf{v} = (-3, -1)$ **4.** $\mathbf{v} = (3, 4)$

5. $\mathbf{v} = (1, -5)$

Mixed Practice, Cluster II, Macro B, pp. 91–92

1. **A** To undo a translation three units left and two units down, add 3 to the x-coordinate of A' and add 2 to the y-coordinate of A'. This means that A $(-1 + 3, 2 + 2)$ or A (2, 4) which is choice A.
2. **C** Reflection over the x-axis means the y-coordinate becomes opposite so P becomes (–2, –4). Then, translating 4 units to the right means the x-coordinate increases by 4 so the final coordinates are $(-2 + 4, -4)$ or **(2, –4)**.
3. Using similar triangles, set up the following proportion:

$$\frac{24}{x} = \frac{60}{60-15} = \frac{60}{45} = \frac{4}{3}$$

So, when you cross-multiply, $4x = 72$ and $x = 18$. Thus, the support beam is **18′ long**.

4. **D** The last statement is not necessarily true, since the lengths of the legs of the two isosceles triangles could be different, so the triangles don't have to be congruent.
5. **C** The problem really becomes $(2 + ? = 7, 5 + ? = -2)$, so the missing vector has to be (5, –7).
6. **D** By definition, a reflection maintains the same size and shape; therefore, choices A, B, and C do not make sense. However, reflecting does change the orientation of the figure.
7. **B** In choice B, there are 4 units separating the y-coordinates (3 and 7) and only 3 units separating the x-coordinates (–1 and –4). These are not consecutive vertices of a square, since all sides have to be equal.
8. An **isosceles trapezoid** is formed.
9. The center is halfway between the end points of the diameter; so, find the average of the x- and y-coordinates of the end points to get $\left(\frac{-3+7}{2}, \frac{2+2}{2}\right)$ or **(2, 2)** as the center.
10. The distance between R and S as given is 5. Thus, the two widths are 5 units each, which is a total of 10 units. This leaves 16 if the perimeter is 26. The other coordi-

nates are each 8 units away from R and S (either up or down), so the third and fourth coordinates can be **(0, 8) and (5, 8)** or **(0, –8) and (5, –8)**.

Practice Set, p. 94

1. 30°.
2. 110°.
3. 150°.
4. 70°.

Practice Set, p. 95

m∠EAB = 74.9°
m∠HFG = 160.1°
m∠IJK = 115.0° (samples)

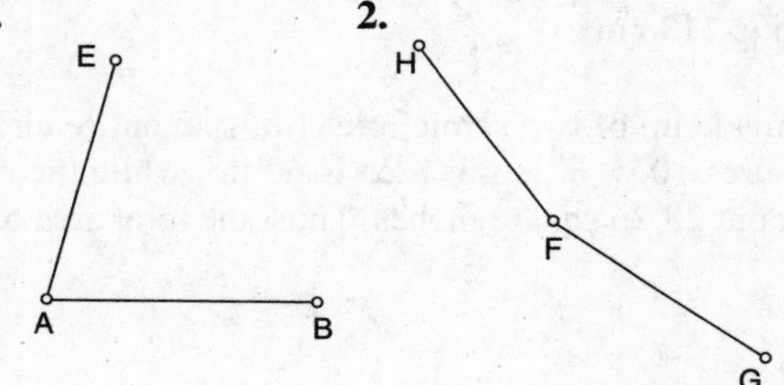

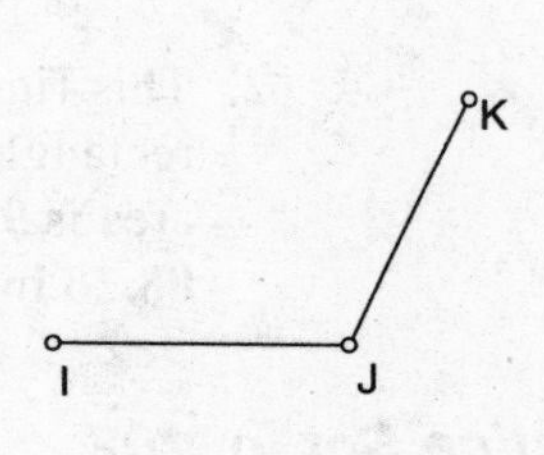

Practice Set, p. 96

1. Rectangle; perimeter = 2 × 5″ + 2 × 12″ = 10″ + 24″ = **34 in.**
2. Regular pentagon; perimeter = 5 × 6 cm = **30 cm**.
3. Scalene triangle; perimeter = 5″ + 8″ + 11″ = **24 in.**
4. Rhombus; perimeter = 4 × 5″ = **20 in.**
5. Quadrilateral; perimeter = 6 m + 9 m + 12 m + 16 m = **43 m**.

Practice Set, p. 97

1. Circumference = π × 6 (since the radius is 3) ≈ 3.14 × 6 ≈ **18.84**.
2. Circumference = π × 15 ≈ 3.14 × 15 ≈ **47.1**.

Practice Set, p. 100

1. Rectangle; area $= l \times w = 10$ cm $\times$ 4 cm = **40 cm²**.

2. Triangle; area $= \frac{1}{2}bh = \frac{1}{2}(12'')(5'') = \mathbf{30\ in^2}$.

3. Trapezoid; Area $= \left(\frac{\text{Base}_1 + \text{Base}_2}{2}\right) \times \text{Height} = \left(\frac{13'' + 7''}{2}\right) \times 5'' = \mathbf{50\ in^2}$.

4. Circle; area $= \pi r^2 = \pi\,(8\text{ cm})^2 = 64\pi \approx \mathbf{200.96\ cm^2}$.

5. Square; area $= (\text{Side})^2 = (9')^2 = \mathbf{81\ ft^2}$.

6. Triangle; area $= \frac{1}{2}bh = \frac{1}{2}(9'')(4'') = \mathbf{18\ in^2}$.

Practice Set, p. 101

1. This figure is made up of two rectangles—one is 15 cm × 7 cm so its area is 105 cm^2, and the other rectangle is 4 cm × 2 cm, so its area is 8 cm^2. Thus, the total composite area is **113 cm^2**.

2. This figure is made up of two semicircles (thus an entire circle) and a rectangle. The rectangle measures 10″ × 6″, so its area is 60 in^2, while the circle's radius is 3″ so its area is 9π or about 28.26 square inches. Thus, the total area of the composite figure is **88.26 in^2**.

Practice Set, p. 102

1. The front and back of the prism are 12″ × 7″, so their areas are each 84 square inches. The top and bottom of the prism are 12″ × 4″, so their areas are each 48 square inches. The sides are 4″ × 7″, so their areas are each 28 square inches.

   Thus, the combined surface area is: $(2 \times 84) + (2 \times 48) + (2 \times 28) =$ **320 square inches**.

2. The base of the pyramid is a square measuring 7 cm per side, so its area is 49 cm^2. Each of the four sides of the pyramid is a triangle with a base of 7 cm and a height of 10 cm, so the area of each face of the pyramid is $\frac{1}{2}$ (7 cm)(10 cm) or *35 square cm*. Since there are 4 faces, multiply 35 × 4 to get 140, and add 49 to this to get the total surface area of **189 cm^2**.

3. The cube has six identical faces—each one is a square 4″ by 4″ with an area 16 square inches. So, the total surface area is 6 × 16 or **96 square inches**.

Practice Set, p. 105

1. Cone; $V = \frac{1}{3}\pi r^2 h$; so, $V = \frac{1}{3}\pi(9)^2(10) = 270\pi \approx$ **847.8 cubic inches**.

2. Prism; $V = lwh$; so, $V = 12 \times 7 \times 4 =$ **336 cubic cm.**

3. Cylinder; $V = \pi r^2 h$; so, $V = \pi r^2 h = \pi(4)^2(8) = 128\pi \approx$ **401.92 cubic inches.**

4. Sphere; $V = \frac{4}{3}\pi r^3$; so, $V = \frac{4}{3}\pi(6)^3 = 288\pi \approx$ **904.32 cubic cm.**

Practice Set, p. 108

1. For a rectangle to have an area of 24 in^2, choose dimensions that multiply to 24, such as 6 in. × 4 in. or 8 in. × 3 in.

2. If a square has a perimeter of 20 cm, divide this by 4 to get the measurement of each side (5 cm). Thus, its area is 25 square cm. If this is cut from a rectangle measuring 7 cm × 12 cm (area of 84 square cm), just take it away: $84 - 25 =$ **59 square cm.**

3. If the length is increased by 5 inches, this adds 2×5 or 10 inches to the perimeter. However, since the width is decreased by 3 inches, this takes $2 \times 3''$ or $6''$ away from the perimeter. So, the net result is the perimeter **increases by 4 inches.**

4. Subtract the area of the smaller circle (unshaded) from the larger circle (shaded) to get the total shaded area: $(7)^2\pi - (3.5)^2\pi = 49\pi - 12.25\pi = 36.75\pi \approx$ **115.395 in^2.**

5. The composite area includes a rectangle and a semicircle. The rectangle is 13 cm × 6 cm, so its area is 78 sq cm. The semicircle has a radius of 3 cm, so its area is $\frac{1}{2}\pi(3)^2$ or 4.5π, which is about 14.13 sq cm. Thus, the total area is $78 + 14.13 =$ **92.13 cm^2.**

6. First, find the new measurements: 16×1.15, 20×1.15, and 24×1.15, which are 18.4, 23, and 27.6. Thus, the area is $\left(\frac{23+27.6}{2}\right) \times 18.4 =$ **465.52 square inches.**

Practice Set, p. 111

1. $(6)^2 + (8)^2 = x^2$ so $x^2 = 36 + 64 = 100$. Solve $x = \sqrt{100} =$ **10.**

2. $(8)^2 + x^2 = (4\sqrt{5})^2$, so $64 + x^2 = 80$, and $x^2 = 80 - 64 = 16$. Solve $x = \sqrt{16} =$ **4.**

3. $(3)^2 + (9)^2 = x^2$, so $x^2 = 9 + 81 = 90$, and $x = \sqrt{90} = 3\sqrt{10} \approx$ **9.5.**

4. $(4\sqrt{3})^2 + x^2 = (8)^2$, so $48 + x^2 = 64$, and $x^2 = 64 - 48 = 16$. Solve $x = \sqrt{16} =$ **4.**

Mixed Practice, Cluster II, Macro C, pp. 112–114

1. **D** Since the prism is two cubes high and two cubes deep, and each cube is 2.5″ per side, the height and width of the prism are each 5″. The prism is four cubes in length, which equals 10″. Thus, the volume is $10'' \times 5'' \times 5'' =$ **250 cubic inches.**
2. **D** To measure revolutions we need to use circumference. So, the circumference of the tire is $\pi \times$ diameter or $\pi \times 30$ inches, which is about 94.2 inches. Next, we have to convert 94.2 inches to feet, which we do by dividing by 12 (since there are 12″ in

a foot). Thus, the wheel's circumference is 94.2 ÷ 12, or **7.85 feet**. Lastly, to find out how many revolutions this is, divide 35 ft. by 7.85 ft., which is about **4.5**.

3. The volume of a box is 6 in. × 8 in. × 3 in. or 144 cubic inches. Thus, each box holds 144 one-inch cubes. Therefore, multiply 144 by 5 to find that five boxes hold **720 cubes**.

4. **A** To find the area of the shaded walkway, subtract the smaller circle's area from the larger circle's area. That gives: $(20)^2\pi - (12)^2\pi$, which equals $400\pi - 144\pi = 256\pi$, which is about **803.84 square feet**.

5. The perimeter can be found by multiplying the length of 8 cm × 12, doubling this, and then adding 8 cm to each side. Thus, the perimeter is 96 × 2 + 16 = **208 cm**.

6. The area of the 6 in. by 6 in. square is 36 sq. inches, and the area of the 12 in. by 12 in. square is 144 sq. inches, so the area is increased by 144 – 36 (or 108) and 108 divided by 36 is 3 or 300%; meaning the area **tripled**.

7. **B** The 50 foot ladder is the hypotenuse and the $2\frac{1}{2}$ foot base is the one leg.

   Using the Pythagorean Theorem, we get $(2.5)^2 + x^2 = (50)^2$, so $x^2 = 2500 - 6.25 = 2493.75$. Thus, $x = \sqrt{2493.75} \approx$ **49.9 feet**.

8. **B** Since the height of the second prism is 25% greater than the first prism's height, the height of the second prism can be calculated by 4.8 × 1.25, which equals 6 m. The volume of the first prism is $4.8 \times 8.5 \times 3.2 = 130.56$. The volume of the second prism has to be the same, so $V = (3)(6)(w) = 130.56$. The width is 130.56 ÷ 18, which is about **7.3 m**.

9. **A** To find the percentage that is shaded, first find the amount that is shaded. The shaded area is made up of a rectangle 5 × 7 (area is 35) and a trapezoid with bases 1 and $1\frac{1}{2}$ and height 5 so its area is 6.25 and thus the area that is shaded totals 41.25 out of 60 (area of whole rectangle) and this is about **68.75%**.

10. The area of a trapezoid is calculated by multiplying the height by the average of the bases.

    To get an area of 60 in^2, the height (6″) must be multiplied by an average height (10″). An average of 10″ can be obtained by any two bases that add up to 20″.

11. The perimeter of a rectangle which measures 5.5 cm × 8 cm is $2 \times 5.5 + 2 \times 8 = 11 + 16$, which is 27 cm. Then, if the equilateral triangle has the same perimeter, divide 27 cm by 3, since each side has to be the same. Thus the triangle measures **9 cm per side**.

12. **C** Estimate the area by counting the number of squares and half squares.

13. The missing length (12″) can be found by using the Pythagorean Theorem. Thus, the volume is 5″ × 12″ × 3″, or **180 cubic inches**. Then, find three other numbers that multiply to 180 but combine to have a smaller surface area. One such set of dimensions is 10″ × 6″ × 3″, as shown below.

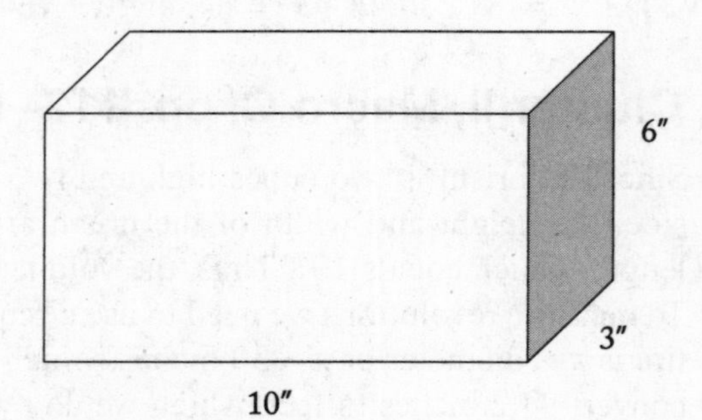

Practice Set, p. 118

1. Since 3 of the 12 letters are vowels (A, E, O), the probability that a randomly selected letter is a vowel is $\frac{3}{12}$ or $\mathbf{\frac{1}{4}}$.
2. Since 5 of the 6 numbers are odd, the probability of choosing an odd number is $\mathbf{\frac{5}{6}}$.
3. In the given list, the factors of 18 are: 1, 2, 3, 6, 9 so the probability of choosing a factor of 18 from the list is 5 out of 10 or $\mathbf{\frac{1}{2}}$.
4. In the word "MATHEMATICS," 7 of its 11 letters are consonants, so the probability of choosing a consonant is $\mathbf{\frac{7}{11}}$.
5. There is a total of 12 red marbles + 8 green marbles + 4 blue marbles = 24 marbles, and 4 of these are blue, so the probability of a randomly chosen marble being blue is $\frac{4}{24}$ or $\mathbf{\frac{1}{6}}$.

Practice Set, p. 121

1. The probability of choosing *red* × probability of choosing an odd number: $\frac{1}{5} \times \frac{4}{8} = \frac{1}{5} \times \frac{1}{2} = \mathbf{\frac{1}{10}}$.
2. $\frac{8}{12} \times \frac{7}{11} = \frac{2}{3} \times \frac{7}{11} = \mathbf{\frac{14}{33}}$.
3. The probability of choosing orange × probability of choosing aqua + probability of choosing aqua × probability of choosing orange $= \frac{3}{10} \times \frac{2}{9} + \frac{2}{10} \times \frac{3}{9} = \frac{1}{15} + \frac{1}{15} = \mathbf{\frac{2}{15}}$.
4. The following are factors of six that could occur from rolling a die: 1, 2, 3, 6. So, the probability of rolling a factor of 6 is $\frac{4}{6}$ *or* $\frac{2}{3}$. If we choose one of the 26 letters of the alphabet, there is an $\frac{8}{26}$ or $\frac{4}{13}$ chance it is one of the letters in the word "MATHEMATICS." Thus, the probability of rolling a number that is a factor of 6 and choosing a random letter from the alphabet that is also in the word "MATHEMATICS" is $\frac{2}{3} \times \frac{4}{13} = \mathbf{\frac{8}{39}}$.

5. A. The following are the two-digit numbers that are multiples of 12: 12, 24, 36, 48, 60, 72, 84, and 96. There are 99 – 10 = 89 + 1 or 90 two-digit numbers, so the probability of the two-digit number being a multiple of 12 is $\frac{8}{90}$ or $\mathbf{\frac{4}{45}}$.

B. In a standard 52-card deck, there are four aces, so the probability of choosing an ace from a standard 52-card deck is $\mathbf{\frac{4}{52}}$.

Since $\frac{4}{45} > \frac{4}{52}$, it is more likely that choice A will occur.

6. The numbers between 1 and 30 that contain at least one 2 are: 2, 12, 20, 21, 22, 23, 24, 25, 26, 27, 28, 29. So, if you randomly choose a number between 1 and 30, the probability that the number contains at least one 2 is $\frac{12}{30}$ or $\mathbf{\frac{2}{5}}$.

Practice Set, p. 122

1. $\frac{1}{4}$ **2. (a)** .620 **(b)** .063 **(c)** .144 **(d)** .75

Mixed Practice, Cluster III, Macro A, pp. 124–126

1. Since three sections of the eight on the spinner contain an odd number and all are equally sized, the probability of spinning an odd number is $\mathbf{\frac{3}{8}}$.

2. **D** From 1 to 40, the following numbers have at least one 3: 3, 13, 23, 30, 31, 32, 33, 34, 35, 36, 37, 38, 39. This means that the probability of choosing a number from 1 to 40 and having it contain at least one 3 is $\frac{13}{40}$, which means 13 ÷ 40, which is about **.325**.

3. **D** The angle containing the portion of the spinner labeled "2" measures $360 - (180 + 30) = 150$ so the probability of spinning a "2" is $\frac{150}{360}$ or $\mathbf{\frac{5}{12}}$.

4. **C** The area of the entire enclosure is $100 \times 40 = 4000$ square feet. If the tiger is within 5 feet of the long side of the enclosure, the area he is within measures 5 ft by 100 ft with an area of 500 sq ft. So, the probability of the tiger being in that section is $\frac{500}{4000} = \mathbf{\frac{1}{8}}$.

5. **D** Since 120 out of 180 reduces to 2 out of 3, the probability of Tyler getting a hit is 2 ÷ 3, which equals about **67%**.

6. **A** Each spinner except (a) has only one quarter of its spinner labeled "E" so therefore (a) has the best chance of spinning an "E," since it has more sections labeled "E."

7. If a bumblebee is buzzing around a room that is 12 ft × 24 ft × 10 ft, to find the probability that the bee is less than 4 feet from the ceiling, realize that he is within the upper 4/10 foot section of the room. $\frac{4}{10}$ reduces to $\mathbf{\frac{2}{5}}$.

8. There are 18 marbles all together. The first marble has an $\frac{8}{18}$ or $\frac{4}{9}$ chance of being yellow. The second marble has a $\frac{7}{17}$ chance of being yellow, since there are now only 7 yellow left out of 17. Finally, the third marble has a $\frac{6}{16}$ or $\frac{3}{8}$ chance of being yellow. Multiplying all three of these gives us the probability that all three marbles are yellow: $\frac{4}{9}\times\frac{7}{17}\times\frac{3}{8}$. As a percent, this is about **6.9%**.

9. **(a)** Add all of the outcomes that contain two tails and a head: $65 + 63 + 51 = 179$. Thus, 179 out of 500 tosses resulted in two tails and a head; this is equivalent to **.358**.

   (b) The theoretical probability of tossing two tails and a head is $\frac{1}{2}\times\frac{1}{2}\times\frac{1}{2}=\frac{1}{8}\times 3$, since there are three arrangements that are two tails and a head (TTH, HTT, THT). Thus, the theoretical probability is $\frac{3}{8}$, which is **.375** as a decimal.

   (c) As you can see, the experimental probability is close to the theoretical, but with a slight difference, since it is an actual experiment and results will vary.

10. **C** When rolling two dice, the sum of 7 is most likely to occur.

11. **D** If the first marble is red, that leaves 47 marbles: 11 red, 12 blue, 12 white, and 12 green.
Thus, the probability that the second marble is not green is **35 out of 47**.

Practice Set, p. 130

1. **(a)** Positive

   (b) Yes, the person who lost 30 pounds in about 4 weeks seems to be an exception.

   (c) About 35 pounds

   (d) That it would eventually stop increasing since most people would reach their limit as to how much they can lose.

2. Positive correlation; the chirps would decrease as it gets colder.

Practice Set, p. 132

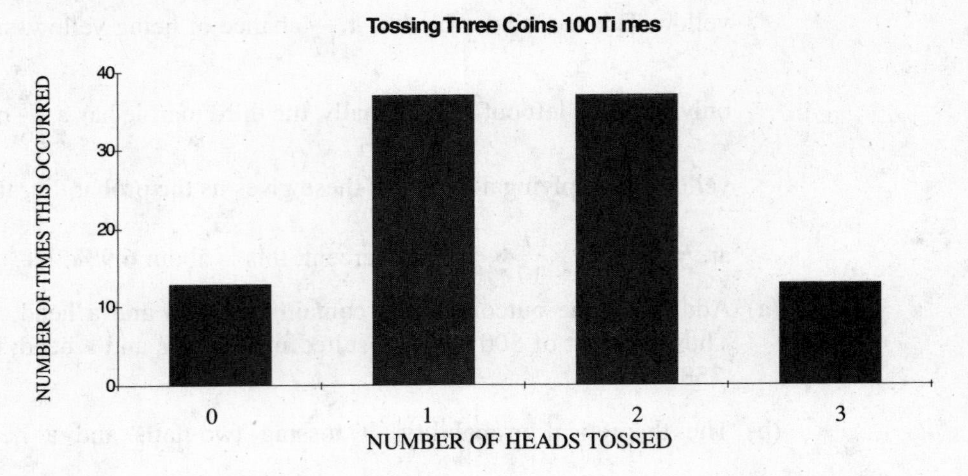

Practice Set, p. 133

1. Since a and b vary indirectly, the product of a and b remains constant. Since $a = 5$ when $b = 8$, we know the constant product is 5×8 or 40, so when $a = 4$, you have to multiply by 10 to get 40. $\boldsymbol{b = 10}$.

2. Since m and n vary directly, set up the proportion: $\frac{m}{n} = \frac{3}{4} = \frac{m}{20}$. Since $4 \times 5 = 20$, $3 \times 5 = m = \mathbf{15}$.

3. Since x and y vary directly, set up the proportion: $\frac{x}{y} = \frac{2}{5} = \frac{9}{y}$. Cross multiplying gives $2y = 45$, so $y = \mathbf{22.5}$.

4. Since r and s vary indirectly, the product of r and s is constant. When $r = 4$, $s = 9$, so the constant product is $4 \times 9 = 36$. Therefore, $s = 2$ means $r = 18$ since 2×18 is also 36.

Mixed Practice, Cluster III, Macro B, pp. 133–134

1. One would expect that as the level of education increases, so would the starting salary earned, so that would appear to indicate **positive correlation**.

2. **B** Negative correlation means one quantity goes up while the other goes down, so the most likely choice to have negative correlation would be choice B since the higher a computer's age, the lower the resale price.

3. **A** This is not an example of direct variation, since most likely, the value of a car depreciates at different rates over time.

4. If it takes 2.4 hours going at 55 mph, that means the distance traveled is $2.4 \times 55 = 132$ miles.
So, to cover that same distance going 40 mph, it will take $132 \div 40$, or **3.3 hours**.

5. **A** By definition of normal distribution, 68% of the data should fall within one standard deviation of the mean. So in this case, about 68% of the data should be between $14 - 3 = \mathbf{11}$ and $14 + 3 = \mathbf{17}$.
B By definition of normal distribution, about 95% of the data will be within two standard deviations of the mean (since our standard deviation here is 3, this

means 95% of the data should fall within 6 of the mean or between $14 - 6 =$ **8** and $14 + 6 =$ **20**).

6. Since we are within half of a standard deviation and 68% of the data falls within 1 standard deviation, 34% of it falls within half of a standard deviation. So, 17% of this amount falls *above* the mean by half a standard deviation.

7. **D** This is the only survey that is unbiased, since you are asking men over 50 and that is the group you are reporting about.

8. **A** Since 95% of the data falls within two standard deviations of the mean and our standard deviation is 3 inches, 95% of the heights should be within 6″ of the average height. So **95%** of the heights should be between **5′4″** and **6′4″**.

   B Since 76 inches is equal to 6 ft 4 in, we are asked to find the probability that the man is not within the 95% that are between 5′4″ and 6′4″. So, half of the remaining 5% are shorter than 5′4″, while the other half of the 5% are over 6′4″. The probability that he is over 6′4″ is **.025**.

Practice Set, p. 136

1. The numbers in order are: 21, 23, 25, 25, 31, 34, 37.

To get the mean, add the numbers and divide by 7: $196 \div 7 =$ **28**.
To get the range, subtract the lowest from the highest: $37 - 21 =$ **16**.
To get the mode, look for the number that occurs more frequently: **25** (since there are two).
To get the median, identify the middle number (in this case, the 4th number): **25**.

2. The numbers in order are: 91, 97, 99, 103, 105, 108.

To get the mean, add the numbers and divide by 6: $603 \div 6 =$ **100.5**.
To get the range, subtract the lowest from the highest: $108 - 91 =$ **17**.
To get the mode, identify the number that occurs most often: **NONE** (they all occur once).
To get the median, take the average of the two middle numbers (99 and 103): **101**.

Practice Set, p. 137

1. Since 102 is 36 less than 138, this means that the total of the twelve numbers is 36 less. The mean is $36 \div 12$ or **3 less**.

2. Since the median is the middle number (or average) of the two middle numbers and it has to be 24, and the average of 20 and 28 is 24, the missing number is 28. This makes the mean **23**.

3. Since the mode is the most common and the mode is supposed to be 35, that means we need to add a second 35 to the list so that the 35 is more common than the other numbers in the list. Therefore, the list in order is: 23, 25, 26, 28, 30, 32, 35, 35. The median is the average of the two middle numbers, 29, and the mean is all of them added divided by 8. This comes to $234 \div 8 =$ **29.25**.

4. By changing the 23 to an 18, we decrease the total by 5 and the average would decrease by 5 ÷ 7 or about .7.

The **18 is now the second mode** since there are now two 25's and two 18's.
The **median is not changed**, since the 23 is still lower than the median 24.
The **range is not changed** since the lowest is still 18 and the highest is still 30.

Practice Set, p. 141

Set 1

1. About an 82%. **2.** Between weeks 2 and 3.

3. About an 82%, since his average did not really change.

4. Week 1, week 2.5, week 4.5.

Set 2

1. 12 students. **2.** 15%. **3.** 2 students. **4.** 35 students.

Set 3

1. About $6,000. **2.** Approximately 64%. **3.** Between 20,000 and 40,000.

4. Between 80,000 and 100,000 or 100,000 to 120,000. **5.** About 20%.

Set 4

1. 12.5%. **2.** About .75 pounds. **3.** About 7.5 pounds.

Mixed Practice, Cluster III, Macro C, pp. 144–145

1. **C** 2. **D** 3. **C**
4. **(a)** Mean: About $200,714 Median: $170,000
   (b) The median is a better representation of the data, since the 350,000 increases the mean, making it higher than all of the numbers.
   (c) A realtor would probably use the **mean** in this case, because it makes the homes seem more valuable.
5. **(a)** 80, 80. **(b)** 90. **(c)** Any # less than or equal to 82. **(d)** 79. **(e)** 70.
6. **(a)** Reebok; New Balance.
   (b) **No,** because when you look at the scale and the numbers, about 45 like New Balance and about 23 like Reebok. 23 is more than 1/3 of 45.
   (c) **New Balance** might use this graph to make people think that brand is significantly more popular than the other brands.

Practice Set, p. 150

1. $5 \times 4 \times 3 \times 2 = \mathbf{120}$.

2.

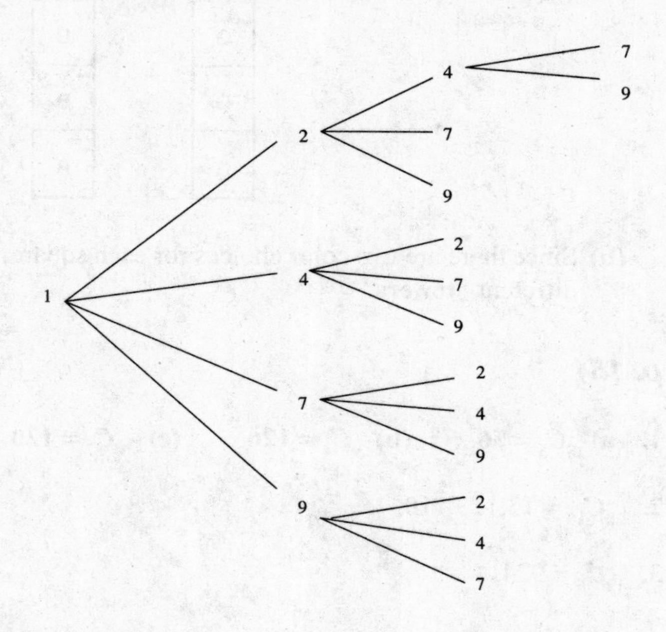

3. **12 meals**.

4. **A.** 720. **B.** 6. **C.** 40,320. **D.** 20. **E.** 210. **F.** 720.

5. $26 \times 25 \times 10 \times 9 \times 8 = \mathbf{468{,}000}$.

6. $3 \times 6 \times 5 = \mathbf{90}$.

7. $4 \times 6 \times 5 = \mathbf{120}$.

8. **C**.

9. $5! = 5 \times 4 \times 3 \times 2 \times 1 = \mathbf{120}$.

10. $3 \times 10 \times 2 = \mathbf{60}$.

11. $8! = 8 \times 7 \times 6 \times 5 \times 4 \times 3 \times 2 \times 1 = \mathbf{40{,}320}$.

12. ${}_8P_3 = 8 \times 7 \times 6 = \mathbf{336}$, $8! = 40{,}320$ and $5! = \mathbf{120}$ and $\mathbf{{}_8P_3 = 8! \div 5!}$.

13. Answers may vary, but two possible new "towers" are shown below:

(a)

B	B
G	B
G	B
G	B

(b) Since there are two color choices for each square, there are $2 \times 2 \times 2 \times 2 = 2^4 = 16$ different "**towers**."

Practice Set, p. 151

1. **(a)** ${}_8C_5 = \mathbf{56}$. **(b)** ${}_9C_4 = \mathbf{126}$. **(c)** ${}_{10}C_7 = \mathbf{120}$. **(d)** ${}_{12}C_4 = \mathbf{495}$.

2. ${}_{28}C_{10} = 13{,}123{,}110$.

3. ${}_{12}C_6 = 924$.

4. ${}_6C_2 = 15$.

5. ${}_{152}C_{12} = 203{,}360{,}683{,}733{,}321{,}660$.

Practice Set, p. 152

1. $4!/2! = 12$.

2. $12! \div (2! \cdot 3! \cdot 4!) = 1{,}663{,}200$.

3. $13! \div (5! \cdot 3! \cdot 2!) = 4{,}324{,}320$.

Practice Set, p. 153

1. **(a)** Combination. **(b)** Permutation. **(c)** Permutation. **(d)** Combination.

2. ${}_5C_1 + {}_5C_2 + {}_5C_3 + {}_5C_4 + {}_5C_5 = 5 + 10 + 10 + 5 + 1 = \mathbf{31}$.

3. $2 \times ({}_3C_0 + {}_3C_1 + {}_3C_2 + {}_3C_3) = 2 \times (1 + 3 + 3 + 1) = \mathbf{16}$.

4. ${}_{20}C_3 + {}_{35}C_2 = 1140 + 595 = \mathbf{1{,}735}$.

5. ${}_{13}C_2 + {}_{13}C_3 = 78 + 286 = \mathbf{364}$.

6. $2^5 \times 4^3 = 32 \times 64 = \mathbf{2048}$.

7. The fifth row of Pascal's Triangle corresponds to the number of sandwiches with 0, 1, 2, 3, or 4 out of 4 extras.

If we extend Pascal's Triangle another row, the following results:

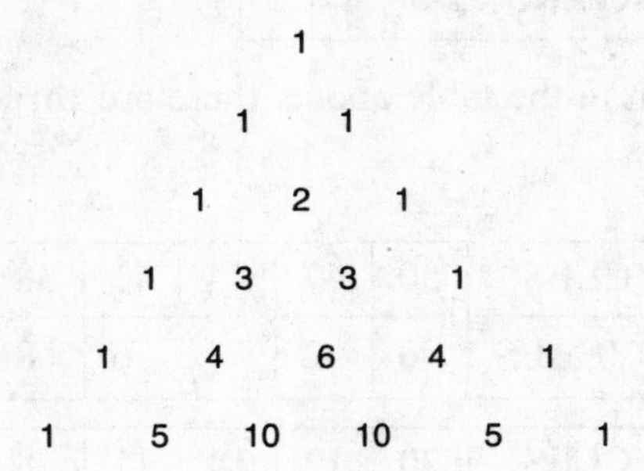

The sixth row represents the number of sandwiches with 0, 1, 2, 3, 4, and 5 out of five choices for extras. Thus, there are 10 sandwiches with 3 out of 5 extras.

Practice Set, p. 155

1.

S_1

C_1 C_1

H_1 H_2 H_1

O_1 O_3 O_3 O_1

O_4 O_6 O_4

L_{10} L_{10}

S_{20}

Thus, there are **twenty** ways to spell out "SCHOOLS" in the diagram.

2.

A_1

L_1 L_1

G_1 G_2 G_1

E_1 E_3 E_3 E_1

B_1 B_4 B_6 B_4 B_1

R_1 R_5 R_{10} R_{10} R_5 R_1

A_1 A_6 A_{15} A_{20} A_{15} A_6 A_1

The diagram with the labels at each letter is attached. There are 1 + 6 + 15 + 20 + 15 + 6 + 1, or **64 ways** to spell out the word "ALGEBRA" in the given diagram.

3. There are **19 "paths"** from *A* to *B* if you are only allowed to move South or East.

Practice Set, p. 157

1.

NICKELS	10	5	0
QUARTERS	0	1	2

As seen in the table above, there are **three ways** to make 50¢ using only nickels and quarters.

2.

BICYCLES	50	47	44	41	38	35	32	29	26	23
TRICYCLES	0	2	4	6	8	10	12	14	16	18

BICYCLES	20	17	14	11	8	5	2
TRICYCLES	20	22	24	26	28	30	32

As seen in the table above, there are **17 different combinations** of bicycles and tricycles that can be made using 100 wheels.

3.

DOGS	0	1	2	3	4	5	6	7
BIRDS	15	13	11	9	7	5	3	1

As seen from the table above, there are **8 combinations of dogs and birds** that can be made.

The pattern in the chart is that if the birds go up by 2, then the dogs go down by 1. This is because every one dog uses four legs while every two birds use four legs.

Based on the information about profits, if 2 dogs and 11 birds are made, the profit is:

$$2 \times \$10 + 11 \times \$7 = \$97$$

If five of each are made, the profit is:

$$5 \times \$10 + 5 \times \$7 = \mathbf{\$85}.$$

Practice Set, p. 158

1. You should pick one from the pile with two stones, thus leaving 1, 1, 1, and your opponent will be forced to pick up the last stone.

2. Add five to bring the total to 24. Then, no matter what your opponent adds, you can win on your next turn by claiming the 30th stone.

3. The fewest number of rolls of the dice necessary for a win would be **two 12's** (cover up 4 and 8, and then 1, 2, and 9). This is not very likely, however, because it would involve rolling double 6's twice in a row.

Practice Set, p. 161

1. All of the networks are traversable
2. The network you create should not have 0 or 2 odd vertices.
3. Two.
4. Four.
5. **B**.

Practice Set, p. 162

1. 23 mod 2 = **1**.
2. 45 mod 3 = **0**.
3. 73 mod 5 = **3**.
4. 85 mod 8 = **5**.
5. 129 mod 9 = **3**.

Practice Set, p. 163

1. Following the "formula" for ISBN Codes:

$$10 \times 1 + 9 \times 5 + 8 \times 6 + 7 \times 7 + 6 \times 6 + 5 \times 5 + 4 \times 5 + 3 \times 4 + 2 \times 4 + 0$$
$$10 + 45 + 48 + 49 + 36 + 25 + 20 + 12 + 8 + 0$$
$$253$$
$$253 \div 11 = 23$$

Thus, this ISBN code is **valid**.

2. $10 \times 0 + 9 \times 9 + 8 \times 6 + 7 \times 5 + 6 \times 3 + 5 \times 5 + 4 \times 2 + 3 \times 9 + 2 \times 5 + ? = 252 + ?$

   Since $11 \times 24 = 242$, the missing check digit has to be a **1**.

3. Verify that the check digit equals 9.

$$d_{12} = 10 - \{[3 \times (0+8+0+0+1+6) + (2+4+0+2+8)] \bmod 10\} \bmod 10$$
$$d_{12} = 10 - \{[3 \times 15 + 16] \bmod 10\} \bmod 10$$
$$d_{12} = 10 - [(61) \bmod 10] \bmod 10$$
$$d_{12} = 10 - [1] \bmod 10$$
$$\mathbf{d_{12} = 9}$$

4. $d_{12} = 10 - \{[3 \times (0+2+0+1+0+7) + (7+0+0+5+5)] \bmod 10\} \bmod 10$

$= 10 - \{[3 \times 10 + 17] \bmod 10\} \bmod 10$

$= 10 - [(47) \bmod 10] \bmod 10$

$= 10 - 7 = \mathbf{3}.$

Mixed Practice, Cluster III, Macro D, pp. 163–164

1. **A**
2. **20**.
3. **A**
4. **(a)** A **combination**, since the three points used to create a triangle can be connected in any order.

   (b) $\frac{6}{10} = \frac{3}{5}$.
5. 3 days.

Practice Set, p. 166

1. 7″ × 4″ ➡ 10″ × 4″ ➡ 13″ × 4″ ➡ 16″ × 4″.

2. 12, 17, 22, **27**, **32**, **37**.

3. $417.38.

4. –8, –5, **–2**, **1**, **4**.

5. 3, 7, 10, 17, **27**, **44**, **71**.

Practice Set, pp. 171–172

1.

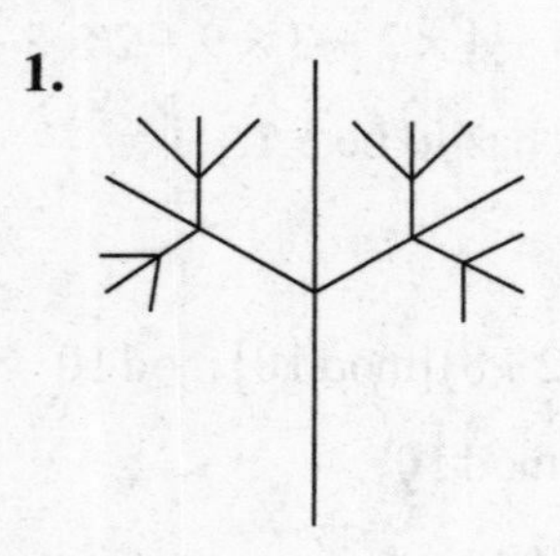

2. 64.

3. 12 units; 16 units.

Practice Set, p. 173

1. **A.** Estimate the next digit of the quotient.
   B. Multiply the guess by the divisor.
   C. If the product is too large then repeat Step A. Otherwise, go to the next step.
   D. Subtract.
   E. If the resulting difference is larger than the divisor, return to Step A otherwise go to Step F.
   F. Bring down the next digit of the dividend.
   G. Repeat Steps A–F.

2\.

3\.

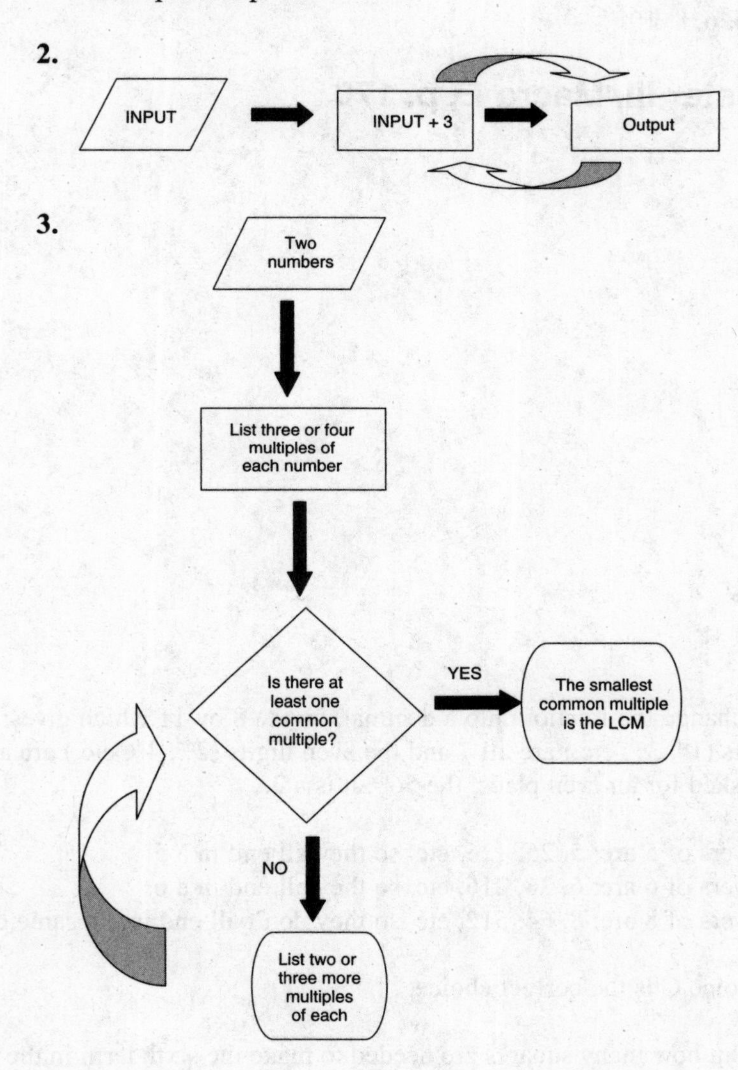

4. 30; 15; 38; 19; 50.

5. 1, 4, 3, 8, 6, 10, 5.

1, 3, 4, 8, 6, 10, 5.

1, 3, 4, 6, 8, 10, 5.

1, 3, 4, 6, 8, 5, 10.

1, 3, 4, 6, 5, 8, 10.

1, 3, 4, 5, 6, 8, 10.

Mixed Practice, Cluster III, Macro E, p. 170

1. D.

2.

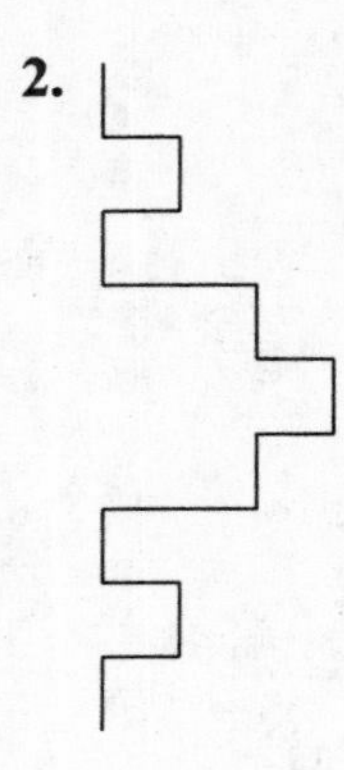

Practice Set, p. 175

1. B. To change the fraction into a decimal, divide 8 by 11 which gives: $.\overline{72}$. Thus, the odd digits (1st, 3rd, etc.) are all 7 and the even digits (2nd, 4th, etc.) are all 2. So, since we are asked for an even place, the 56th, it is a 2.

2. C. Powers of 5 are: 5, 25, 125, etc. so they all end in a 5.
Powers of 6 are: 6, 36, 216, etc. so they all end in a 6.
Powers of 8 are: 8, 64, 512, etc. so they don't all end in the same digit.

Thus, choice C is the correct choice.

3. To find out how many squares are needed to make the sixth term in the pattern, begin by looking for a pattern to determine the number of squares in each term. There is obviously one square in the first term, five squares in the second and nine squares in the third. It appears that there are four more squares in each term than in the previous term. Thus, there will be **13 in the fourth term, 17 in the fifth term, and 21 in the sixth term**.

4. Again, begin by finding a pattern to determine the number of each type of square in the tenth figure. There is 1 shaded and 5 unshaded in figure #1. There are 2 shaded and 7 unshaded in figure #2. There are 3 shaded and 9 unshaded in figure #4. Thus, there is always one more shaded and two more unshaded. So, in the fifth figure, there will be 4 shaded and 11 unshaded, in the sixth figure there will be 5 shaded and 13 unshaded, in the seventh figure there will be 6 shaded and 15 unshaded, and so on. Finally, in the tenth figure, there will be 10 shaded and 23 unshaded. Thus, the percent that will be unshaded is 23/33 or about **70%**.

5. The next sides measure: $2.4 \times 1.2 = 2.88$, $2.88 \times 1.2 = 3.456$, $3.456 \times 1.2 = 4.1472$, so the perimeter of the fifth triangle is 4.1472×3 which is about **12.4 cm**.

Practice Set, p. 178

1. $d = 5$; Next three terms: **44, 49, 54**.
2. **8, 14, 20, 26, 32**.
3. **135, 123, 111, 99, 87**.
4. $t_{50} = 4 + 3(49) = 4 + 147 = \mathbf{151}$.
5. $t_{95} = 6 + 5(94) = 6 + 470 = \mathbf{476}$.
6. $9 + 34d = 213$ so $34d = 204$ and $d = \mathbf{6}$.
7. $14 + 27d = 176$ since to get from the first to the third term, we use 2 increments and from the 3rd to the 30th we would use $29 - 2 = 27$ increments.

   Thus, $27d = 162$ and $d = 6$. So, the first term is **2** and the common difference is **6**.

Practice Set, p. 180

1. $r = 2$.
2. $r = \frac{132}{176} = \mathbf{\frac{3}{4}}$.
3. $t_{12} = 8{,}388{,}608$.
4. $1200 \cdot r^4 = 75 \text{ so } r^4 = \frac{75}{1200} = \frac{1}{16} \text{ so } r = \sqrt[4]{\frac{1}{16}} = \mathbf{\frac{1}{2}}$.
5. Approximately $2,521.05.

Practice Set, p. 181

1. 6, 8, 14, 22, 36.
2. −2, 5, 3, 8, 11.
3. **D**.
4. **272**.
5. EXAMPLE: 3, 8, 11, 19, 30.

Practice Set, p. 184

1. $S=\dfrac{45}{1-\frac{2}{3}}=\dfrac{45}{\frac{1}{3}}=\mathbf{135}.$

2. $S_{12}=12\left(\dfrac{-5+83}{2}\right)=12\cdot 39=\mathbf{468}.$

3. $S_8=\dfrac{3(4^8-1)}{4-1}=\mathbf{65{,}535}.$

4. $6+8+14+22+36+58+94+152+246+398=\mathbf{1034}.$

Practice Set, p. 185

1. $y=\mathbf{17}.$

2.

LENGTH OF CALL	1 min	2 min	3 min	4 min	5 min	6 min
COST OF CALL	.48	.65	.82	.99	1.16	1.33

EQUATION FOR COST OF A CALL LASTING "*m*" minutes: **.48 + .17(*m* – 1)**.

Mixed Practice, Cluster IV, Macro A, pp. 186–187

1. **B** The shaded section is the same in the first, third, and fourth figures.

2. **D** The common difference is $1\frac{1}{4}$, so $8\frac{3}{4}+1\frac{1}{4}$ equals 10.

3. **B** There is a repetition of four letters, so dividing 154 by 4 gives 38 (with a remainder of 2). So, the second letter, S, is in the 154th position.

4. $S=\dfrac{\frac{7}{10}}{1-\frac{1}{10}}=\mathbf{\dfrac{7}{9}}.$

5. **A** There are 0, 2, 6, 12, . . . shaded squares in the diagrams. First we add 2, then 4, then 6. So, if the pattern is continued, there will be 380 in the 20th figure.
6. **B** By definition, a geometric sequence has a common ratio.
7. **B** To find the 25th term in the arithmetic sequence, first identify the common difference, which is 6. Then, follow the formula to get $t_{25}=5+6(24)=5+144=149$.
8. **C** The numbers in the sequence are multiples of 3 (3, 6, 12, . . .). 200 can't be a term, since it is not a multiple of 3.
9. $d=7$ and $t_9=41$.
10. The common ratio is 2, thus the 20th term is $5\,(2)^{19}=2{,}621{,}440$.
11. **B** The powers of 3 end in 3, 9, 7, 1.
12. There are 1, 3, 9, 27, . . . hearts in each term, so this is a geometric sequence with common ratio of 3. The question becomes finding the sum of the first six terms, which is 1 + 3 + 9 + 27 + 81 + 243 = **364**.
13. **B** 8 + 12 + 16 + 20 + 24 + 28 + 32 = 140 so there are **7 terms**.

Practice Set, p. 190

1. Yes, since each x value is unique.

2. No, since −1 is paired with two different y-values.

Practice Set, p. 194

1. **D.** $f(-2)=4^{-2}-2^{-2}=\frac{1}{16}-\frac{1}{4}=\frac{1-4}{16}=\mathbf{\frac{-3}{16}}$.

2. **B.** Figure out what function rule works by guess and check.

3. **D.** By definition of π, π is the constant ratio of a circle's circumference to its diameter. All of the other choices will fluctuate.

4. **B.** The temperature fluctuates gradually rather than increasing or decreasing in intervals. All of the other situations represent cases where there is an interval at a steady cost and then the cost increases over the next interval.

5. JoAnn is correct because a domain element (town) has more than one range element (zip code).

Practice Set, p. 196

1. **(a)** $m=\frac{5-2}{8-4}=\mathbf{\frac{3}{4}}$. **(b)** $m=\frac{-7-(-2)}{4-2}=\mathbf{\frac{-5}{2}}$. **(c)** $m=\frac{3-(-9)}{2-(-1)}=\frac{12}{3}=\mathbf{4}$.

2. $m=\frac{-2-7}{x-5}=\frac{3}{5}=\frac{-9}{x-5}$, so $3x-15=-45$ and $3x=-30$. $x=\mathbf{-10}$.

Practice Set, p. 198

1. Slope = 3.

2. y-Intercept: (0, −7).

3.

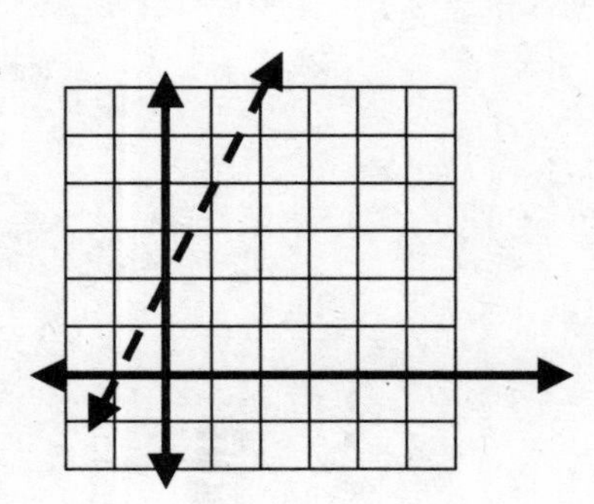

4.

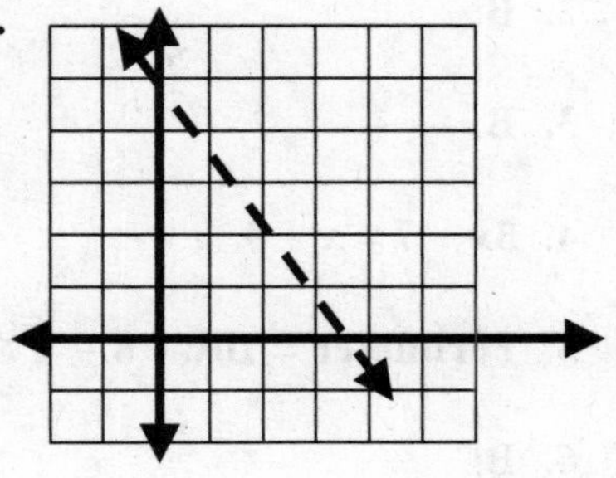

Practice Set, p. 199

1. $y=\frac{5}{2}x-6$.

2. $y=\frac{2}{5}x-7$.

3. $y=-\frac{3}{4}x+5$.

4. A.

5. $C(h) = 55h + 42$.

Yes, this is a linear equation because every time the time increases by an hour, the cost increases by $55.

Mixed Practice, Cluster IV, Macro B, p. 201

1. **C** Since a function can't match a domain element with more than one range element, II does not suggest a function, but I and III do.
2. **B** Since 0 cannot be used in the function given in (b) (because we can't divide by zero) it does NOT have the entire set of real numbers as its domain.
3. **C** If $f(a) = f(b)$, then there should be at least one point where a horizontal line crosses the function in two places.
4. Since the line has to be parallel to the line through (–6, 0) and (8, 7), it has the same slope as this line. So, find its slope: $\frac{7-0}{8-(-6)}=\frac{7}{14}=\frac{1}{2}$. Then, since the line is supposed to contain (6, –2), find the equation as follows: $-2=\frac{1}{2}(6)+b$, which means $-2 = 3 + b$ and $b = -5$. So, the equation is $\mathbf{y=\frac{1}{2}x-5}$.
5. Since $-f(x)$ represents a reflection through the x-axis, the y-coordinate becomes opposite so the point (2, 5) would become (2, –5).

Practice Set, p. 204

1. $-2x + 6$.

2. B.

3. B.

4. $3x - 7 = x + 9$.

5. Perimeter $= 10x + 8$.

6. B.

7. $3x + 5 = 5x - 7$.

Practice Set, p. 206

1. To solve $2x - 11 = 17$, add 11 to both sides, which gives you $2x = 28$. Then, divide both sides by 2 to get $\boldsymbol{x = 14}$.

2. To solve $3(4x - 9) + 10 = 7$, begin by distributing and combining like terms. This gives $12x - 27 + 10 = 7$, which then becomes $12x - 17 = 7$. Then, add 17 to both sides to get $12x = 24$. Finish by dividing both sides by 12 which means $x = \mathbf{2}$.

3. To solve $5x - 9 = 2x + 15$, begin by subtracting $2x$ from both sides. This gives you $3x - 9 = 15$. Then, add 9 to both sides to get $3x = 24$. Finally, divide by 3 to get $x = \mathbf{8}$.

4. To solve $3x + 11 < 5$, subtract 11 from both sides, which gives you $3x < -6$. Then, divide by 3 to get $\boldsymbol{x < -2}$.

5. To solve $8 - 5x \leq -7$, subtract 8 from both sides to get $-5x \leq -15$. Then, since you have to divide by a negative number, the inequality reverses so $\boldsymbol{x \geq 3}$.

Mixed Practice, Cluster IV, Macro C, pp. 206–207

1. **A** To solve $3x + 2 \geq -9$, subtract 2 from each side and then divide by 3 so $3x \geq -11$ and $x \geq -3\frac{2}{3}$. Choice A contains the correct set of integers.

2. **C** In I, they forgot that y means $1y$.
In II, everything is combined correctly.
In III, everything is combined correctly.

3. **D** $\frac{20}{-5} + -5 = -4 + (-5) = \mathbf{-9}$.

4. To solve $-5x < 30$, divide by -5. Since you are dividing by a negative number, the inequality must reverse, so $\boldsymbol{x > -6}$.

5. **B** Begin solving by distribution, which gives you $5x - 10 - 3 > x - 10$. Then, combine like terms to get $5x - 13 > x - 10$. Next, subtract x from each side to get $4x - 13 > -10$. Finally, add 13 and divide by 4 so $4x > 3$ and $\boldsymbol{x > \frac{3}{4}}$.

6. **D**

7. $x + (x + 2) + (x + 4) + (x + 6) < 200$, so $4x + 12 < 200$, and $x < 47$. Thus, the largest value for x is 45, and the numbers would be 45, 47, 49, and 51. The largest of the four is **51**.

8. $6x + 11 = 23$, so $6x = 12$ and $\boldsymbol{x = 2}$.

9. $2x\left[\frac{5x+9x}{2}\right] = \mathbf{14x^2}$.

10. **B**

PRACTICE HSPA
EXAM 1

Cut along dotted line.

ANSWER SHEET: PRACTICE TEST 1

Part One

1. Ⓐ Ⓑ Ⓒ Ⓓ
2. Ⓐ Ⓑ Ⓒ Ⓓ
3. Ⓐ Ⓑ Ⓒ Ⓓ
4. Ⓐ Ⓑ Ⓒ Ⓓ
5. Ⓐ Ⓑ Ⓒ Ⓓ
6. Ⓐ Ⓑ Ⓒ Ⓓ
7. Ⓐ Ⓑ Ⓒ Ⓓ
8. Ⓐ Ⓑ Ⓒ Ⓓ
9. Ⓐ Ⓑ Ⓒ Ⓓ
10. Ⓐ Ⓑ Ⓒ Ⓓ
11. Ⓐ Ⓑ Ⓒ Ⓓ
12. Ⓐ Ⓑ Ⓒ Ⓓ
13. Ⓐ Ⓑ Ⓒ Ⓓ

14. Use the blank page that follows to show your work.
15. Use the blank page that follows to show your work.

Part Two

16. Ⓐ Ⓑ Ⓒ Ⓓ
17. Ⓐ Ⓑ Ⓒ Ⓓ
18. Ⓐ Ⓑ Ⓒ Ⓓ
19. Ⓐ Ⓑ Ⓒ Ⓓ
20. Ⓐ Ⓑ Ⓒ Ⓓ
21. Ⓐ Ⓑ Ⓒ Ⓓ
22. Ⓐ Ⓑ Ⓒ Ⓓ
23. Ⓐ Ⓑ Ⓒ Ⓓ
24. Ⓐ Ⓑ Ⓒ Ⓓ
25. Ⓐ Ⓑ Ⓒ Ⓓ
26. Ⓐ Ⓑ Ⓒ Ⓓ
27. Ⓐ Ⓑ Ⓒ Ⓓ

28. Use the blank page that follows to show your work.
29. Use the blank page that follows to show your work.

Part Three

Use the blank page to show your work.

PART I (30 MINUTES)

1. Find the x-intercept of the graph of $2x - 5y = -10$.

A. $(-5, 0)$ **B.** $(0, 2)$ **C.** $(0, -2)$ **D.** $(5, 0)$

2. If the rectangle shown below is revolved 360° about side $\overline{RS}$ what type of solid results?

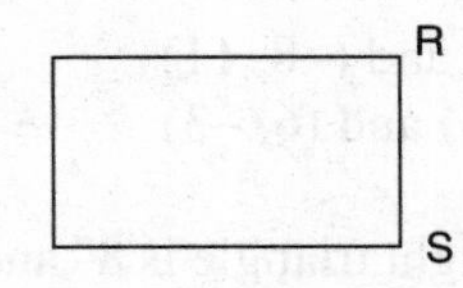

A. sphere **B.** cone **C.** cylinder **D.** pyramid

3. Which of these sets of points are possible vertices of a quadrilateral that has two right angles but no parallel sides?

A. $(-3, 0)$ $(0, 3)$ $(3, 0)$ $(0, -3)$
B. $(0, 0)$ $(0, 4)$ $(7, 7)$ $(7, 0)$
C. $(0, 0)$ $(0, 4)$ $(6, 6)$ $(8, 0)$
D. $(-3, 0)$ $(0, 3)$ $(2, 0)$ $(0, -3)$

4. Which graph below best represents what is happening to a car's resale value over time?

A.

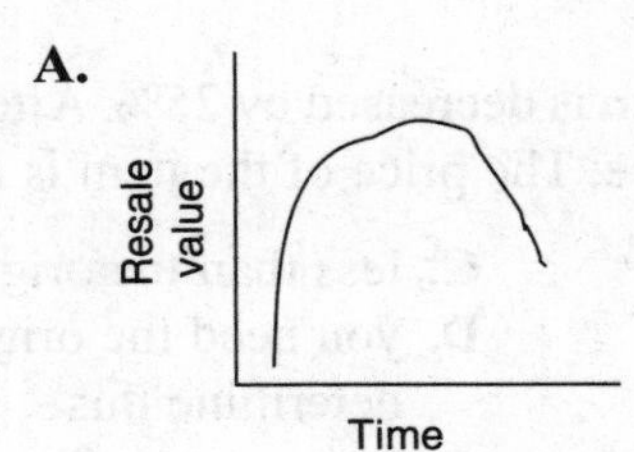

C.

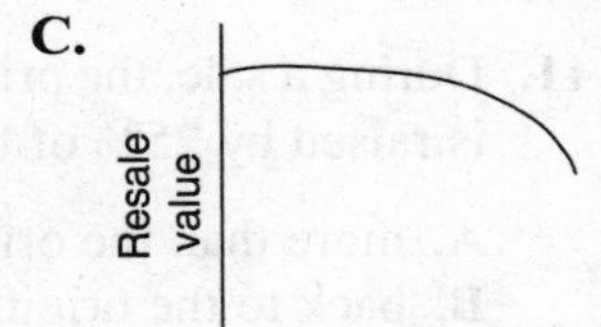

B.

Resale value
Time

D.

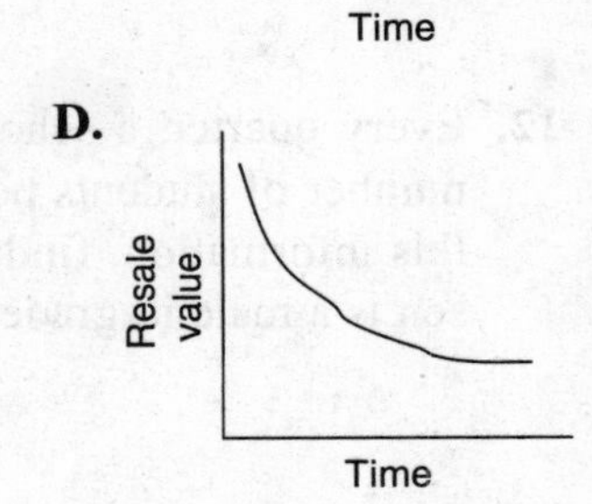

5. Which of the following polygons is always similar to a polygon of the same kind?

A. rectangle **B.** square **C.** triangle **D.** parallelogram

6. The number of parents who went to "Back-to-School Night" this year was 655. There were 115% as many parents there this year compared to last year. This means:

A. 15 more parents attended "Back-to-School Night" this year
B. There were more parents at this year's "Back-to-School" Night
C. The number of parents attending dropped from last year
D. School enrollment went up from last year

7. Using the numbers 1 through 30, remove all prime numbers, all perfect squares, all factors of 42, all multiples of 6 and all numbers in the following Fibonacci-like sequence: 2, 3, 5, 8, 13, . . .

How many numbers are left?

A. 5 **B.** 6 **C.** 7 **D.** 8

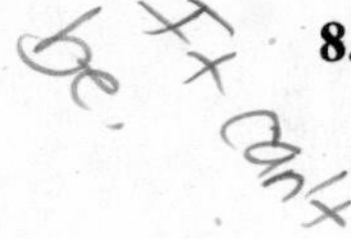

8. Which of the following are points that satisfy the equation $y = -\frac{4}{3}x + 5$?

A. (−4, 8) (4, 2) and (−8, 11) **C.** (4, 8) (4, 2) and (8, −11)
B. (3, 1) (−6, 13) and (6, −3) **D.** (−3, −1) (6, −3) and (6, −13)

9. If one side of a right triangle is 8″ and the hypotenuse is 17″, find the sine of the larger acute angle of the triangle.

A. $\frac{8}{17}$ **B.** $\frac{17}{15}$ **C.** $\frac{15}{17}$ **D.** $\frac{17}{8}$

10. There are 1,535 students who voted in the recent Student Council election. After about one fourth of the votes were counted, the leading candidate had 140 votes. Assuming that he or she obtained the same proportion of the total number of votes, the total number of students who voted for this candidate would be between:

A. 300 and 350 **B.** 400 and 450 **C.** 550 and 600 **D.** 650 and 700

11. During a sale, the price of an item is decreased by 25%. After the sale is over, the price is raised by 25% of the sale price. The price of the item is now:

A. more than the original price **C.** less than the original price
B. back to the original price **D.** you need the original price in order to determine this

12. Every quarter, a school randomly chooses a student to possibly earn a reward. The number of students per grade and broken down by gender is shown below. Based on this information, find the probability (to the nearest hundredth) that the student chosen is a male in grade 8.

Number of Students

GRADE 6		GRADE 7		GRADE 8	
Male	Female	Male	Female	Male	Female
35	47	53	55	57	53

A. .17 **B.** .19 **C.** .28 **D.** .57

13. Based on the graph below showing time versus distance, which segment is impossible?

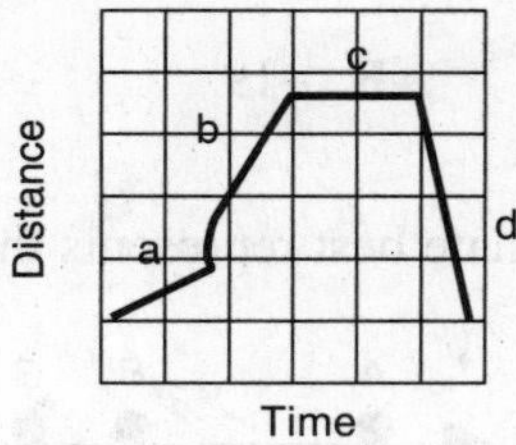

A. a **B.** b **C.** c **D.** d

14. Sketch the result if the figure shown below is reflected through the x-axis and then the y-axis.

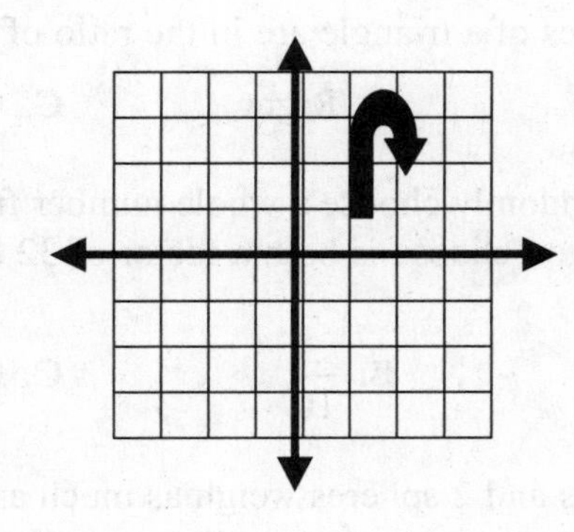

15. Solve over the set of prime numbers: $3x - 1 < 26$.

End of Part I

PART II (30 MINUTES)

16. Which of the following points is NOT on the graph of $2x - y = 10$?

A. (−2, −14) **B.** (3, −4) **C.** (4, 2) **D.** (5, 0)

17. Based on the table below, what is the equation describing the relationship between r and s?

r	0	1	2	6	10
s	−1	1	3	11	19

A. $s = r - 1$ **B.** $s = 2r - 1$ **C.** $s = r^2$ **D.** $s = 3r - 7$

18. Which of the sets of measures below cannot be used to form a triangle?

A. 3, 5, 9 **B.** 5, 5, 8 **C.** 2, 6, 6 **D.** 6, 7, 10

19. Brand A costs $5.50 for 12, brand B costs $6.95 for 20, and brand C costs $8.75 for 36. List the brands in order from the best buy to the worst value.

A. A, B, C **B.** C, B, A **C.** C, A, B **D.** A, C, B

20. Given $g(x) = 3^x + 4^x + 12^x$, find the value of $g(-1)$.

A. 0 **B.** −19 **C.** $\frac{2}{3}$ **D.** $-\frac{2}{3}$

21. Which coordinate best represents the result when *C* and *D* are multiplied?

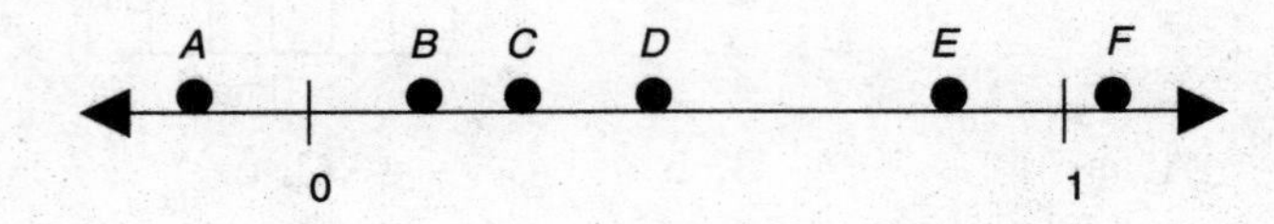

A. *A* **B.** *B* **C.** *E* **D.** *F*

22. The angles of a triangle are in the ratio of 2:3:4; what type of triangle is it?

A. Acute **B.** Right **C.** Obtuse **D.** Isosceles

23. If you randomly choose a whole number from 1 to 40, what is the probability that the number you choose is both a factor of 32 and a multiple of 6?

A. $\frac{3}{20}$ **B.** $\frac{3}{10}$ **C.** 0 **D.** $\frac{1}{40}$

24. If 5 cones and 2 spheres weigh as much as 4 cones and 6 spheres, and each sphere weighs $1\frac{1}{2}$ pounds, how much does each cone weigh?

A. $\frac{1}{2}$ pound **B.** 2 pounds **C.** 4 pounds **D.** 6 pounds

25. What percent of any circle is contained inside a square inscribed in the circle?

A. 32% **B.** 64% **C.** 75% **D.** 82%

26. On certain days, an Algebra class works in groups of four while on other days they work in groups of six or eight. If all the students are in class on any given day, there is always one extra student no matter what the size of the groups that day. Which of the following could be the number of students in this class?

A. 25 **B.** 29 **C.** 33 **D.** 37

27. The lengths of two sides of a right triangle are 4 inches and 6 inches. Which of the following could be the length of the third side of this right triangle?

I. 2 inches II. $2\sqrt{5}$ inches III. $2\sqrt{13}$ inches

A. I and II only **B.** I and III only **C.** I only **D.** II and III only

28. Rotate the figure below 90° counterclockwise:

29. If a card is chosen at random from a standard 52-card deck and then a second card is chosen without replacing the first one, what is the probability that both cards are black 9's?

End of Part II

PART III (30 MINUTES)

30. Given two independent events C and D with $P(C) = .4$ and $P(D) = .6$:

A. Find the probability that both events C and D occur.
B. Explain why your answer in Part (A) should be less than each of the individual probabilities.
C. Suppose that two events E and F are independent and not equally as likely. If you know that the probability of both events occurring is .40, what could be the probability of each?

31. A lecture hall contains twelve rows of seats and there are 15 seats in Row #1 and 48 seats in Row #12. If the number of seats in each row forms an arithmetic sequence, answer the questions which follow:

A. How many more seats are in a given row than in the row in front? Explain your work.
B. How many total seats are there in the lecture hall? Show your calculation.

32. If a pizza parlor has fixed expenses of $500 per week and their cost to make a pizza is $3, **A.** write an equation to represent their expenses per week based on selling "p" pizzas.

B. The pizza parlor sells each pizza for $8. What equation can represent revenue for selling "p" pizzas?
C. How can you use your two equations to find out the minimum number of pizzas that the pizza parlor must sell so they don't lose money? Find this number of pizzas.

33. The second baseman steps back 2 feet from the base to catch the ball. About how far does he have to throw the ball to reach the catcher who is 1 foot behind homeplate? Please round your answer to the nearest hundredth.

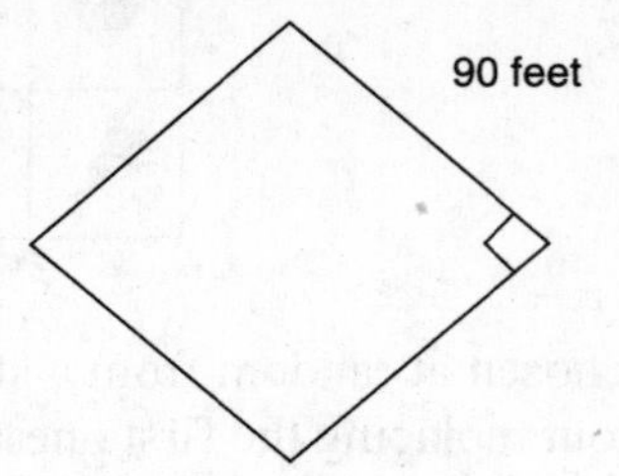

34. Suppose trapezoids are lined up side by side as shown below:

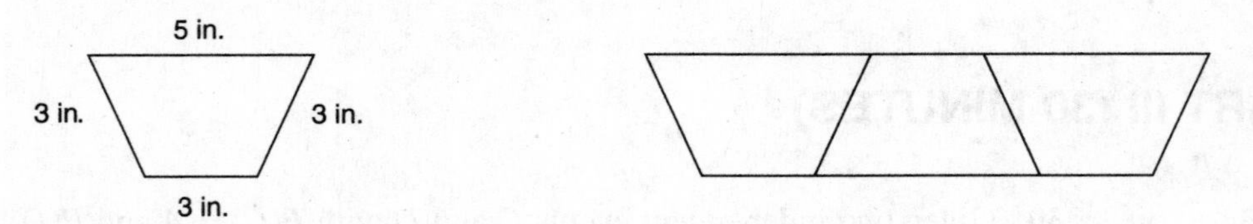

What is the perimeter of the figure if we use six trapezoids?

35. $2^2 = 4^1$, $2^4 = 4^2$, $2^6 = 4^3$, $2^8 = 4^4$, etc . . .

Describe the pattern you notice and determine what powers on 2 and 4 would produce an equal result between 1,000 and 2,000.

36. Explain the difference(s) between the two Venn Diagrams shown below:

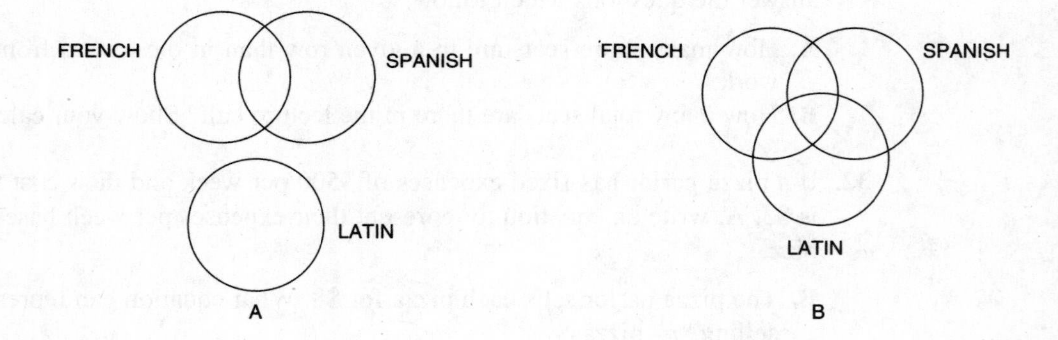

End of Part III

PRACTICE EXAM #1 ANSWERS

1. **A** To find the x-intercept of the graph of $2x - 5y = -10$, make $y = 0$. So, $2x = -10$ and $x = -5$.
2. **C** Cylinder.
3. **C**
4. **D** A car's resale value is likely to decline quickly at first, then less quickly.
5. **B** All squares are similar since they all have four right angles and four equal sides.

6. **B** An increase of 15% does not mean 15 more unless the original amount was 100 and the new amount 115. Therefore, choice A is incorrect. Choice B is correct because there was definitely an increase from last year.

7. **C** Primes from 1 to 30 to be removed are: 2, 3, 5, 7, 11, 13, 17, 19, 23, 29.

   Perfect squares to be removed from 1 to 30 are: 1, 4, 9, 16, 25.
   Factors of 42 to be removed from 1 to 30 are: 6, 14, 21.
   All multiples of 6 from 1 to 30 to be removed are: 12, 18, 24, 30.
   Numbers in the Fibonacci-like sequence given to be removed are: 8.

   Thus, the numbers between 1 and 30 which are left are: 10, 15, 20, 22, 26, 27, 28. So, there are seven numbers left.

8. **B** By trial and error, plug the first coordinate in for x and the second coordinate in for y to find out that all of the points given in choice B satisfy the equation

$$y = -\frac{4}{3}x + 5.$$

9. **C** The third side of the triangle is 15″. Thus, the larger acute angle is opposite 15″ and its sine is $\frac{15}{17}$.

10. **C** The candidate received 140 out of about one fourth of 1,535 students (about 384) and this is about 36%, so 36% of 1,535 is about **560**.

11. **C** Consider as an example an item costing $100. If it is decreased by 25%, it now costs $75. Then, increasing it 25% brings it up to only $93.75 so the price is less than the original price.

12. **B** Since there are 57 males in Grade 8 and a total of 35 + 47 + 53 + 55 + 57 + 53 or 300 students, the probability that the student chosen is a male in Grade 8 is 57 ÷ 300 or **.19**.

13. **D** This is impossible since distance can't decrease as time goes by

14.

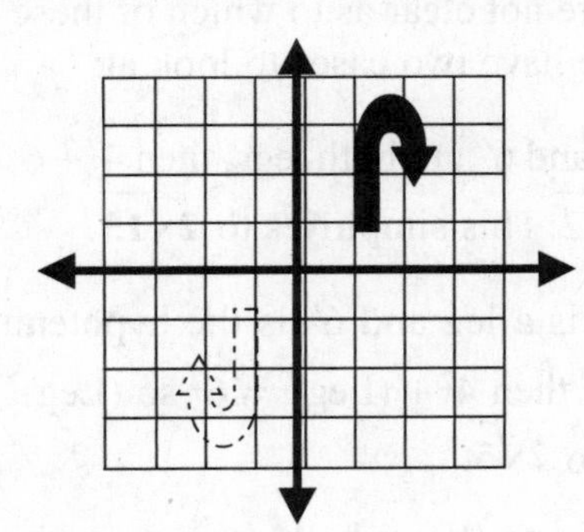

15. To solve $3x - 1 < 26$, begin by adding 1 to both sides. This gives $3x < 27$ and then dividing both sides by 3 gives $x < 9$. Since the question said "solve over prime numbers," we only want the prime numbers less than 9. Thus, the solutions are **2, 3, 5,** and **7**.

16. **C** By plugging in the first coordinate for x and the second coordinate for y, we find that (4, 2) does not work in the equation $2x - y = 10$ so the correct answer is choice C.

17. **B** By observation and trial and error, the equation is $s = 2r - 1$.

18. **A** Since 3 + 5 is not more than 9, the triangle cannot be formed using 3, 5, and 9.

19. **B** Brand A costs $5.50 ÷ 12 or about **.46** for one.
   Brand B costs $6.95 ÷ 20 or about **.35** for one.
   Brand C costs $8.75 ÷ 36 or about **.24** for one.
   So in order from the best buy to the worst buy: C, B, A.

20. **C** $g(-1) = 3^{-1} + 4^{-1} + 12^{-1} = \frac{1}{3} + \frac{1}{4} + \frac{1}{12} = \frac{4+3+1}{12} = \frac{8}{12} = \mathbf{\frac{2}{3}}$.

21. **B** Since C and D are both positive fractions between 0 and $\frac{1}{2}$, their product will be a smaller fraction so the only choice that makes sense is choice B.

22. **A** Adding $2x + 3x + 4x = 9x$, and since the angles of a triangle total 180°, $x = 20$. So, the angles are $2 \times 20 = 40°$, $3 \times 20 = 60°$, and $4 \times 20 = 80°$. Since all are acute angles, the triangle is acute.

23. **C** If you randomly choose a whole number between 1 and 40, there are no numbers that are both factors of 32 and multiples of 6. The probability of choosing such a number is 0.

24. **D** 5 cones + 2 spheres (3 pounds) = 4 cones + 6 spheres (9 pounds), so subtracting 3 pounds from each side gives us 5 cones = 4 cones + 6 pounds. Then, subtracting 4 cones from each side gives us 1 cone that equals **6 pounds**.

25. **B** The diagonal of the inscribed square is equal to $2r$ (two radii). Thus, by the Pythagorean Theorem, $(\text{Side})^2 + (\text{Side})^2 = (2r)^2$ so that $2(\text{Side})^2 = 4r^2$ and $(\text{Side})^2$, which represents the square's area is $2r^2$ and the circle's area is πr^2. To get the percent of the circle contained in the square, divide $2r^2$ by πr^2. The r^2 terms cancel, leaving $2 \div \pi \approx .64$ or **64%**.

26. **A** By trial and error, 25 can be split into 6 groups of 4 or 4 groups of 6 with one left over. 29 can be split into 7 groups of 4 or 2 groups of 6, 2 groups of 8 with one left over. 33 can be split into 8 groups of 4 or 4 groups of 8 with one left over. 37 can't be split into groups of 6 or 8 with one left over.

27. **D** We are not clear as to which of these sides is the hypotenuse or if both are the legs so we have two cases to look at:

If 4″ and 6″ are both legs, then $4^2 + 6^2 = \text{Hypotenuse}^2$, which means the Hypotenuse is $\sqrt{52}$. This simplifies to $\mathbf{2\sqrt{13}}$.

If 4″ is a leg and 6″ is the hypotenuse (remember, the hypotenuse is the longest side), then $4^2 + (\text{Leg})^2 = 6^2$ so $(\text{Leg})^2 = 36 - 16 = 20$ and Leg $= \sqrt{20}$. This simplifies to $\mathbf{2\sqrt{5}}$.

So, II and III are both correct.

28.

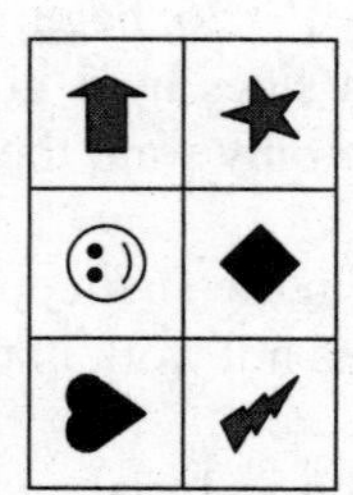

29. The first card can be any of the 52, so the chance of it being a black 9 is 2 out of 52 (it could be a 9 of clubs or spades). 2 out of 52 reduces to 1 out of 26. Once the first card is chosen, there are only 51 cards left, and a 1 out of 51 chance that it is the remaining black 9. Thus, to get the probability of choosing two black 9's, multiply: $\frac{1}{26} \times \frac{1}{51} = \mathbf{\frac{1}{1326}}$.

30. **A** To find this, multiply .4 times .6 which gives **.24**.
B Since neither event is that likely to occur, the likelihood of both occurring is even less.
C Similar to how we calculated the probability in choice A, the two probabilities should multiply to .40, so, for example, $P(E) = .5$ and $P(F) = .8$.

31. **A** $48 - 15 = 33$, so the increase is 33 and this occurs in eleven increments (to get from Row 1 to Row 12). Therefore, each increase is $33 \div 11$ or 3, meaning there are 3 more seats in every row than in the row in front.
B The number of seats is an arithmetic sequence: 15, 18, 21, . . . , 45, 48. So, by the-formula, the sum of the first twelve terms is: $(15 + 48) \times (6) = 63 \times 6 =$ **378**.

32. **A** Expenses $(E) = 500 + 3p$.
B Revenue $(R) = 8p$.
C By setting the Expenses equal to the Revenue, you can find the "break even" point (that is, the point where expenses are equal to revenue).

$8p = 3p + 500$ so $5p = 100$ and $p = 100$.

So, if they sell over **100 pizzas**, they will make a profit.

33. The hypotenuse of the right triangle formed by connecting second base to home plate is found by using the Pythagorean Theorem: $(90)^2 + (90)^2 = \text{Hypotenuse}^2$, so the Hypotenuse is approximately 127.28. Therefore, the second baseman must throw $127.28 + 2 + 1 =$ **130.28 feet**.

34. If six trapezoids are lined up, we would use six lengths of 5 inches and six lengths of 3 inches for the perimeter, plus two 3 inches for the "sides" to make up the perimeter. Thus, the entire perimeter is **54″**.

35. The exponent on 2 is double the exponent on 4. By the guess and check method, the exponent on 4 that gives a result between 1,000 and 2,000 is 5, so it would be $2^{10} = 4^5 =$ **1,024**.

36. In the first Venn Diagram, since the third circle does not overlap the first two, there are no common members in the third set. However, in the second Venn Diagram, it is possible that all three sets share members.

PRACTICE HSPA

EXAM 2

ANSWER SHEET: PRACTICE TEST 2

Part One

1. Ⓐ Ⓑ Ⓒ Ⓓ
2. Ⓐ Ⓑ Ⓒ Ⓓ
3. Ⓐ Ⓑ Ⓒ Ⓓ
4. Ⓐ Ⓑ Ⓒ Ⓓ
5. Ⓐ Ⓑ Ⓒ Ⓓ
6. Ⓐ Ⓑ Ⓒ Ⓓ
7. Ⓐ Ⓑ Ⓒ Ⓓ
8. Ⓐ Ⓑ Ⓒ Ⓓ
9. Ⓐ Ⓑ Ⓒ Ⓓ
10. Ⓐ Ⓑ Ⓒ Ⓓ
11. Ⓐ Ⓑ Ⓒ Ⓓ
12. Ⓐ Ⓑ Ⓒ Ⓓ
13. Ⓐ Ⓑ Ⓒ Ⓓ

14. Use the blank page that follows to show your work.
15. Use the blank page that follows to show your work.

Part Two

16. Ⓐ Ⓑ Ⓒ Ⓓ
17. Ⓐ Ⓑ Ⓒ Ⓓ
18. Ⓐ Ⓑ Ⓒ Ⓓ
19. Ⓐ Ⓑ Ⓒ Ⓓ
20. Ⓐ Ⓑ Ⓒ Ⓓ
21. Ⓐ Ⓑ Ⓒ Ⓓ
22. Ⓐ Ⓑ Ⓒ Ⓓ
23. Ⓐ Ⓑ Ⓒ Ⓓ
24. Ⓐ Ⓑ Ⓒ Ⓓ
25. Ⓐ Ⓑ Ⓒ Ⓓ
26. Ⓐ Ⓑ Ⓒ Ⓓ
27. Ⓐ Ⓑ Ⓒ Ⓓ

28. Use the blank page that follows to show your work.
29. Use the blank page that follows to show your work.

Part Three

Use the blank page to show your work.

Cut along dotted line.

PART I (30 MINUTES)

1. Which number below does NOT have the same value as the others?

A. 62.5×10^{-2} **B.** 625% **C.** $\frac{5}{8}$ **D.** .625

2. Given that $x - 3 = y + 3$, which of these statements MUST also be true?

A. $x = y$ **B.** $x < y$ **C.** $x > y$ **D.** $x > 3$

3. $(6.8 \times 10^8) \div (1.7 \times 10^2) = ?$

A. 4×10^4 **B.** 4×10^{10} **C.** 4×10^6 **D.** 4×10^{-6}

4. Plot and connect points A, B, and C to form one triangle and points D, E, and F to form a second triangle. Through what line would either triangle have to be reflected to coincide with the other triangle?

A (0, 2) B (0, 4) C (5, 4) D (0, –4) E (0, –6) F (5, –6)

A. $x = -1$ **B.** $y = 3$ **C.** $x = 3$ **D.** $y = -1$

5. In the diagrams below, the diameter of each larger disk is double that of each smaller disk. If the weight of the stack on the left made up of the smaller disks is 12 oz, find the weight of the stack on the right. *Note*: The disks are all the same thickness.

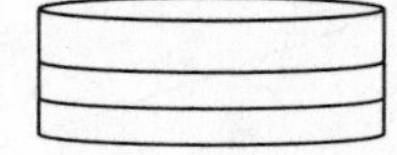

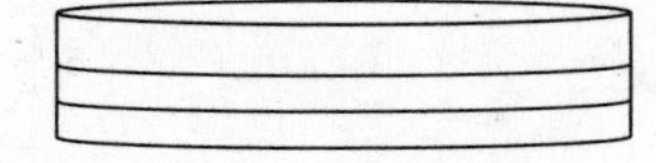

A. 48 oz **B.** 36 oz **C.** 24 oz **D.** 18 oz

6. Given that a rectangle's width is 5 cm and that its length doubles causing its perimeter to increase by 16 cm, what is the area of the original rectangle?

A. 80 cm^2 **B.** 40 cm^2 **C.** 16 cm^2 **D.** 11 cm^2

7. At a certain company, there is a security code in place with regard to employee identification numbers that are used to punch the time clock. In each valid number, the product of the sum of the first four digits and the sum of the last three digits must equal 384. Which of the following is NOT a valid employee ID number under these guidelines?

A. 9348466 **B.** 2536969 **C.** 5238659 **D.** 7296439

8. A spinner contains equal sections marked "1," "2," "3," and "4." If the spinner is spun twice, what is the probability that the sum is even?

A. $\frac{1}{6}$ **B.** $\frac{1}{2}$ **C.** $\frac{2}{9}$ **D.** $\frac{5}{9}$

9. To earn a promotion at work, Dana must handle an average of 50 customer service calls per day. If she handled 48 calls Monday, 53 calls Tuesday, 49 calls Wednesday, and 46 calls Thursday, how many must she handle on Friday to earn her raise?

A. 50 **B.** 52 **C.** 54 **D.** 56

10. Based on the pattern shown below, how many rectangles would be formed by 2 vertical lines and 25 horizontal lines?

Vertical	2	2	2	2	2	2	2
Horizontal	2	3	4	5	6	7	8
Rectangles	1	2	3	4	5	6	7

A. 22 **B.** 24 **C.** 26 **D.** 28

11. Which problem below works out to be a negative number?

A. $2 + 20 \div 2 \times 5 - 12$
B. $6 - 2^3 \times 3 + 20$
C. $8 \times (12 \div 4) - 24$
D. $12 \div 6 - 4$

12. In the triangle below, the ratio $\frac{12}{5}$ is:

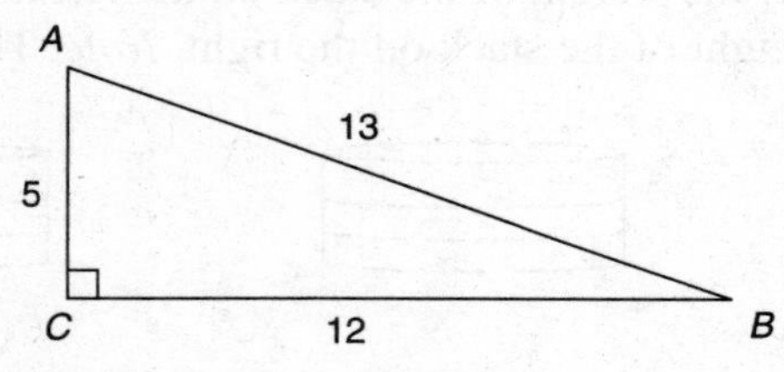

A. $\sin A$ **B.** $\cos B$ **C.** $\cot B$ **D.** $\sec A$

13. Using the numbers 1, 2, 5, 7, 8, 9, how many more three-digit numbers are possible if repetition of digits is allowed than if repetition of digits is not allowed?

A. 96 **B.** 104 **C.** 112 **D.** 120

14. At Anytown High School, $\frac{4}{7}$ of the freshmen class are boys. If the 9th grade class has 280 students, how many of these students are females?

15. Sketch the top view of the figure shown below:

End of Part I

PART II (30 MINUTES)

16. The price of a first-class postage stamp just went up from 34 cents to 37 cents. This change is what percent increase?

A. 7% **B.** 8% **C.** 9% **D.** 10%

17. Which of the following represents all of the real numbers at least 5 units away from 2 on the number line?

A. $|x| \geq 5$ **B.** $|x + 2| \geq 5$ **C.** $|x - 2| \geq 5$ **D.** $|x + 5| \geq 2$

18. Which of the following does not show a 10% decrease?

A. $100 \rightarrow 90$ **B.** $60 \rightarrow 50$ **C.** $10 \rightarrow 9$ **D.** $\frac{11}{100} \rightarrow \frac{1}{100}$

19. A florist has the following flowers available: roses, carnations, violets, pansies, daisies. If Tom wants to create a bouquet with at least three different types of flowers in it, how many possible combinations are there?

A. 60 **B.** 10 **C.** 15 **D.** 16

20. If the product of 9 integers is positive, at most how many of the integers could be negative?

A. 2 **B.** 4 **C.** 8 **D.** 9

21. If a company charges a \$15 set-up fee and then \$7 per T-shirt, how many T-shirts were ordered if the total is \$190.

A. 20 **B.** 25 **C.** 30 **D.** 35

22. If a card is drawn at random from a standard deck of 52 cards and then replaced so that a second card can be drawn, what is the probability that both cards' suits are hearts?

A. $\frac{3}{13}$ **B.** $\frac{1}{2}$ **C.** $\frac{1}{4}$ **D.** $\frac{1}{16}$

23. Suppose you begin with \$150.75 in the bank, and each day you deposit \$1.75 more than you did the day before. That is, on Day 1 you put in \$1.75 and on Day 2, you deposit 1.75 + 1.75 or \$3.50. On Day 3, your deposit is 3.50 + 1.75 or \$5.25. How much money will you have in the bank on the 10th day?

A. \$168.25 **B.** \$229.50 **C.** \$247 **D.** \$387.75

24. Which of the following would not change the mean for these five scores?

25, 30, 35, 40, 45
I. Add two new scores: 20 and 50
II. Add three more scores of 35
III. Add 7 points to each score

A. I only **B.** II only **C.** I and II **D.** I, II, and III

25. In the figure below, all eight squares shown are congruent. Which of the following squares could be removed without changing the perimeter of the entire figure?

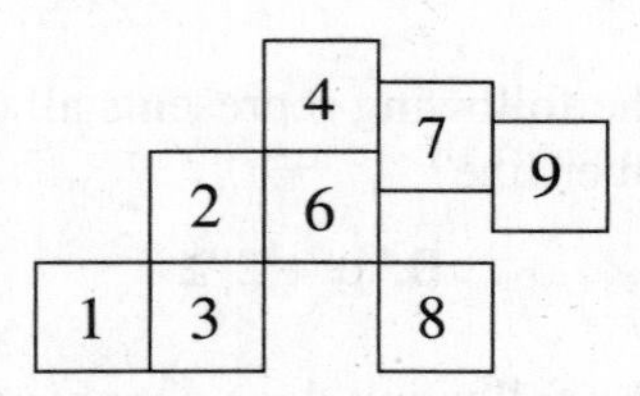

A. Square #4 **B.** Square #2 **C.** Square #6 **D.** Square #8

26. If the product of five integers is positive, then AT MOST how many of the integers can be negative?

A. 2 **B.** 3 **C.** 4 **D.** 5

27. If the measures of the angles of a triangle are in the ratio of 1 : 3 : 4, what type of triangle is it?

A. Acute **B.** Right **C.** Obtuse **D.** Equilateral

28. Which pair of points determines a line parallel to $y = \frac{3}{4}x - 2$?

A. (0, 0) (1, –2) **B.** (0, 0) (3, 4) **C.** (1, 2) (–3, 5) **D.** (3, 6) (–5, 0)

29. Sketch a vector diagram to model the following problem: $-3 + 5 + (-6)$.

30. What is the length of each edge of a cube that has the same surface area as a rectangular prism with dimensions $3'' \times 6'' \times 10''$?

End of Part II

PART III (30 MINUTES)

31. In probability, we call events complementary if their probabilities add to 1. As an example, when flipping a coin, the events "flipping a heads" and "flipping a tails" are complementary. Which of the following events would NOT be complementary events? Explain!

A. Picking a red card from a standard deck and picking a black card from a standard deck
B. Rolling a sum on two dice that is less than 8 and rolling a sum on two dice that is more than 8
C. Spinning an odd number on a spinner and spinning an even number on a spinner (*Note*: Spinner sections are 1, 2, 3, 4, 5, 6)
D. Picking a number that is divisible by 4 from 1 to 30 and picking a number that is not divisible by 4 from 1 to 30

32. Given that *ABCD* is an isosceles trapezoid with *A* (0, 5), *B* (12, 5), *C* (10, 0), answer each of the following:

A. Find point *D*.
B. Find the area of the trapezoid and show your steps.
C. Write an equation in slope-intercept form for the line containing side *AD* of the trapezoid.

33. If two sheets of 8.5″ by 11″ paper are rolled to form two cylinders as shown:

A. Find the volume of each cylinder. Show your work.
B. Find the surface area of each cylinder. Show your work.

34. If a boat starts at point A and travels 16 miles north and then turns and travels 30 miles west to point B:

A. Create a vector diagram to show the boat's movement and also include a vector that would represent the boat's direct path from A to B.
B. About how far would the boat have traveled directly from A to B? How do you know?
C. About how many degrees from north would the direct path be? Explain how you arrive at this solution.

35. A. A cylindrical can with height 10″ and radius 2″ is filled to 75% of its capacity with water. This water is then poured into a second cylindrical can with height 8″ and radius 3″. What percent of the second can is filled? Show your work.
B. A third container filled with the same amount of water is only 30% full using this same quantity of water. Give one possible set of measurements for this can.

36. Two baseball players, Jones and Adams, currently have batting averages of .325. Jones has earned this batting average after 100 at bats while Adams has earned this batting average over 200 at bats.

A. If each player goes into a "slump" and goes for 15 consecutive at bats without a hit, show the impact this will have on each player.
B. Suppose Adams still has a .325 batting average at the end of the entire season after 550 at bats. If he goes into the same "slump" of 0 for 15 at the end of the season, does it have as much of an impact on his batting average as it did in choice A? Why or why not? Explain.

PRACTICE EXAM #2 ANSWERS

1. **B** For A, remember that an exponent of -2 on 10 means to move the decimal point two places to the left, so that becomes .625, which is the same as D. Dividing 5 by 8 gives .625 for C as well. However, in B, to change a percent to a decimal, you move the decimal point two places to the left, which only makes $625\% = 6.25$, so that one is different.
2. **C** Given that $x - 3 = y + 3$, we can solve for x by adding 3 to both sides and thus, $x = y + 6$. So, since x is six more than y, we know that x and y are not equal, which eliminates A. There is also no guarantee that $x > 3$, so D is eliminated. And, since x is 6 more than y, $x > y$, so C is the correct answer.
3. **C** To compute $(6.8 \times 10^8) \div (1.7 \times 10^2)$, divide 6.8 by 1.7, which is 4. Then, subtract the exponents, which gives you $8 - 2$, an exponent of 6.
4. **D** When you plot and connect the points, the distance between the two triangles is closest at the points A and D, and the distance between these two points is 6. Thus, half of this distance, or 3, has to be measured from A or D, which puts us at -1 and we need a horizontal line, which has the form y = constant.
5. **A** Since the diameter of the disks in the stack on the right is twice the diameter of the disks on the left, we need to examine the effect this has on the volume of the cylindrical stacks. Volume is calculated by the formula $V = \pi r^2 h$. Since π is a constant and both stacks are the same height, we can ignore π and h in the formula. So, we are really comparing r^2 for each stack. If the diameter doubles, so does the radius. So, for the stack on the left, r^2 and for the stack on the right, $(2r)^2$ or $4r^2$, which means the volume quadruples or multiplies by four. So, the volume of the stack on the right is 12 oz $\times$ 4 or 48 oz.

6. **B** If a rectangle's width is 5 cm and its length doubles causing its perimeter to increase by 16 cm, we can figure out that the length goes up by 8 (so that the perimeter goes up by 16). If the length goes up by 8 and was doubled, this means the length was 8 so the original dimensions are 5 by 8, which means the area is **40 cm²**.
7. **C** Solve by guess and check.
8. **B** The best way to organize your results here is with a simple table:

	1	2	3	4
1	2	3	4	5
2	3	4	5	6
3	4	5	6	7
4	5	6	7	8

Looking at the sums, there are 16 possible combinations of spins and out of those, eight are even so the probability is 8 out of 16 or $\frac{1}{2}$.

9. **C** Adding her calls for the first four days gives: $48 + 53 + 49 + 46 = 196$. To have an average of 50 customer service calls per day over five days, she must handle $50 \cdot 5$ or 250 total calls in 5 days. Subtract 196 from 250 to figure out that she must handle **54** calls on Friday.
10. **B** Looking at the pattern, it appears that the number of rectangles is one less than the number of horizontal lines. Thus, if there are 25 horizontal lines, there would be $25 - 1$ or **24 rectangles**.
11. **D** **A.** $2 + 20 \div 2 \times 5 - 12 = 2 + 10 \times 5 - 12 = 2 + 50 - 12 = 52 - 12 = \mathbf{40}$.
B. $6 - 2^3 \times 3 + 20 = 6 - 8 \times 3 + 20 = 6 - 24 + 20 = -18 + 20 = \mathbf{2}$.
C. $8 \times (12 \div 4) - 24 = 8 \times (3) - 24 = 24 - 24 = \mathbf{0}$.
D. $12 \div 6 - 4 = 2 - 4 = \mathbf{-2}$.

As you can see by the above calculations, D is the one that has a negative answer.

12. **C** Since 12 is a leg and so is 5, this trig function has to be one that does not involve the triangle's hypotenuse. So, $\frac{12}{5}$ is not sin, cos, sec, or csc.
13. **A** Using the numbers 1, 2, 5, 7, 8, 9, there are $6 \times 6 \times 6$ (or 216) three-digit numbers possible (if repetition is allowed) versus $6 \times 5 \times 4$ or 120 (if repetition is not allowed). Thus, to find out how many more three-digit numbers are possible if repetition of digits is allowed than if repetition of digits is not allowed, subtract $216 - 120$ to get **96**.
14. To find the number of females, we need to realize that if $\frac{4}{7}$ of the freshmen class are boys, the remaining $\frac{3}{7}$ are females. Thus, find $\frac{3}{7}$ of 280 by multiplying $\frac{3}{7} \cdot 280$, which equals $3 \cdot 40$ or **120**.

15. 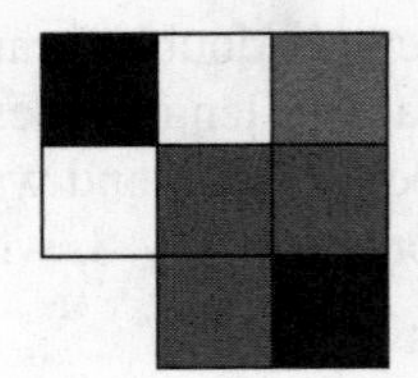

16. **C** The price increased by 3 cents out of 34 cents and dividing 3 by 34 gives about **9%**.

17. **C** Since the distance is supposed to be at least five and absolute value means "distance," we can immediately narrow our choices down to A, B, or C. The distance away from 2 is measured by subtracting the unknown number, x, and 2.

18. **B** To get the percent of change, divide the amount of change by the original. So, in A, the change is 10/100 or 10%. In B, the change is 10 also, but this time the original was 60 so the percent of change is 10/60, which is NOT 10%.

19. **D**

20. **C** If the product of 9 integers is positive, all 9 can't be negative, since the product of an odd number of negatives is always negative. At most, **8** could be negative since then their product would be positive.

21. **B** An equation can be used: $15 + 7t$ (where t is the number of T-shirts) = $190. To solve this, begin by subtracting $15 from each side to get 7t$ = $175 and then divide by $7 to get t = **25**.

22. **D** The probability of drawing a heart is 13 out of 52, or $\frac{1}{4}$. Since we are replacing the card, the probability of getting two hearts is $\frac{1}{4} \times \frac{1}{4}$ or $\mathbf{\frac{1}{16}}$.

23. **C** To find the amount on the 10th day, begin with $150.75 and add $1.75 (Day #1), $1.75 + $1.75 or $3.50 (Day #2), $3.50 + $1.75 or $5.25 (Day #3), $5.25 + $1.75 or $7 (Day #4), etc. Thus, on Day #10, you have **$247**.

24. **C** The mean for 25, 30, 35, 40, 45 = (25 + 30 + 35 + 40 + 45) ÷ 5 = 175 ÷ 5 = 35.

    I. Adding two new scores of 20 and 50 adds 70 to the total; thus the average is (175 + 70) ÷ 7, which equals 245 ÷ 7 = 35.

    II. Adding three more scores of 35 adds 3 × 35 or 105 to the total; thus, the average is (175 + 105) ÷ 8 = 280 ÷ 8 = 35.

    III. Adding 7 points to each score adds 5 × 7 or 35 to the total; thus, the average is (175 + 35) ÷ 5 = 210 ÷ 5 = 42.

    Thus, the average remains the same in I and II.

25. **B** To remove a square without changing the perimeter of the entire figure, the square we remove must have the same number of sides contributing to the perimeter as it exposes to contribute to the computer when it is removed. The only square that does this out of the choices is **square #2**.

26. **C** Please refer to the explanation for #20.

27. **B** If the measures of the angles of a triangle are in the ratio of 1:3:4, the angles can be represented as x, $3x$, and $4x$, which add to $8x = 180$ (since the angle sum of a triangle is 180) and solving gives $x = 22.5$. Thus, the smallest angle is x (or 22.5), the next angle is $3x = 3(22.5) = 67.5$ and the third angle is $4x = 4(22.5) = 90$. Since the triangle contains a right angle, it is a **right** triangle.

28. **D** If a line is parallel to $y = \frac{3}{4}x - 2$, it must have the same slope $\left(\frac{3}{4}\right)$. So, find the slope between each set of points looking for the one that equals $\frac{3}{4}$, which is (3,6) (–5,0).

29.

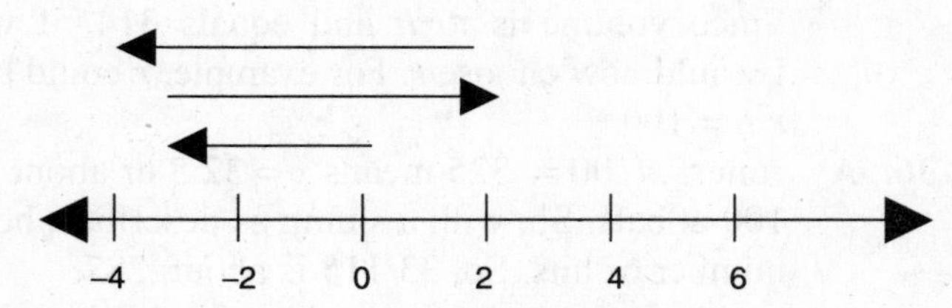

30. Surface area: $10 \times 6 \times 2 + 10 \times 3 \times 2 + 3 \times 6 \times 2 = 216$.
$216 \div 6 = 36$ for each surface of the cube so each edge is **6″**.

31. **B** The events in choice B are not complementary since neither event accounts for the number being exactly 8.

32. **A** Point D has coordinates **(2, 0)**.

B The area of the trapezoid is: $5 \times \left(\frac{12+8}{2}\right) =$ **50 square units.**

C The slope of segment AD is: $\frac{0-5}{2-0} = -\frac{5}{2}$ and the y-intercept is point A.

Thus, the equation is $\mathbf{y = -\frac{5}{2}x + 5}$.

33. **A** To get volume, we need the radius of each cylinder. Since $C = \pi d$, we can get the diameter and consequently the radius of each cylinder as follows:

$8.5 = \pi d$ and $11 = \pi d$ so $d \approx 2.7$ and $d \approx 3.5$ and $r \approx 1.35$ and $r \approx 1.75$.

Thus, the volumes are:

$V = \pi(1.35)^2 \times 11 \approx 20.05$ and $V = \pi(1.75)^2 \times 8.5 \approx$ **26.03**.

B To get surface area, we need to simply follow the formulas:

S.A. $= 2\pi(1.35)^2 + 11(\pi)(2.7) \approx$ **104.70**.
S.A. $= 2\pi(1.75)^2 + 8.5(\pi)(3.5) \approx$ **112.65**.

34. **A** The vector diagram:

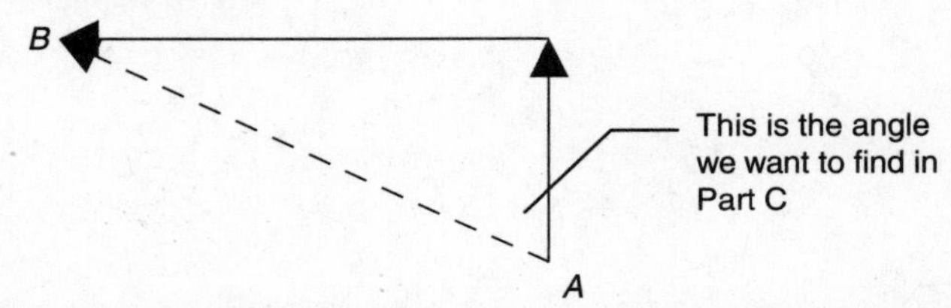

B By the Pythagorean Theorem, $(AB)^2 = (30)^2 + (16)^2 = 900 + 256 = 1156$, so $AB = \sqrt{1156} =$ **34**.

C For this, you would set up the following trig ratio:
$\sin A = 30/34$, and then $\angle A = \sin^{-1}(30/34)$.

Practice Exam 2

35. **A** Volume = $\pi r^2 h = 10(2)^2\pi \approx 125.6$.
Since it is filled 75%, find $.75 \times 125.6$ which is 94.2.
Volume = $\pi r^2 h = 8(3)^2\pi \approx 226.08$.
Then, find out what percent of 226.08 the 94.2 is by dividing 94.2 by 226.08 to get .41666666 . . . which is about **42%**.

B Solve the equation: $94.2 = .30x$ to get 314, which is the volume of the entire cylinder.

Since volume is $\pi r^2 h$ and equals 314, if we divide by π, we get $r^2 h = 100$. I would now choose r. For example, r could be 5. Then, $r^2 = 25$ and $h = 4$ to make $r^2 h = 100$.

36. **A** Jones: $x/100 = .325$ means $x = 32.5$ or about 33. This means he had 33 hits out of 100 at bats. So, with a slump as described, he now goes 115 at bats with the same number of hits. So, 33/115 is about .287.

Adams: $x/200 = .325$ means $x = 65$, which means that he had 65 hits out of 200 at bats. So, with a slump as described, he now goes 215 at bats with the same number of hits. So, $65/215 = .302$.

As we can see, it has less of an effect on Adams since he already had more at-bats.

B If Adams still has a .325 after 550 at bats, this means he has about **179 hits**.

If he goes into a 0 for 15 slump now, he still has 179 hits out of 565 at bats and his average is 179/565 or about **.317**.

Again, since he already had so many at-bats, dividing 15 by a larger and larger number means less of a change in batting average.